Solar Power

Contents

Illustrations

FIGURES

Acknowledgments

This manuscript would not have been possible without the support of friends, family, colleagues, research participants, and the great team assembled by University of California Press. My mother, Celeste, encouraged me to go to college in the first place, then graduate school, and beyond. Her own hard work and dedication inspired me to work toward my own goals. She made it clear to me through my entire life that anything is possible. My wife, Catherine, made the writing possible, tolerating late nights and early mornings. She is a remarkable woman, whose kindness sparks my enthusiasm every day. I promise next time to fit book-writing into my everyday work schedule and routine. I also received a lot of love and support from the rest of my immediate family, my brothers Sean, Darren, and Ryan, and sisters Holly and Mikaela. Thanks to my stepmom Mary, stepdad David, and my late father Denis for also being there. Uncles Timmy, Andy, Mark, Patrick, Jimmy, and Matt, all of you played such important roles in my life. Aunt Debbie in particular was a key part of my early life. My in-laws Dr. Deborah Ott and Dr. Steve Wade contributed more than just moral support, frequently talking about their energy experiences as economists working for the U.S. Federal Energy Regulatory Commission and Department of Energy, respectively. Finally, my attention often drifts to my pets, past and present: my dogs, Griffin and Wilder, and my cats, Chipko and César.

Professionally, this work benefited from a cast of mentors, starting with my dissertation advisor in environmental studies at the University of California, Santa Cruz (UCSC), Professor David Goodman, who helped me write and think about the political ecology of emerging technologies. Professor Melanie Dupuis (now at Pace University) mentored me at UCSC not only through my dissertation but also through a post-doctoral teaching fellowship supported by the National Science Foundation. During my time at UCSC I benefited from seminars by Donna Haraway, Julie Guthman, Alan Richards, and our political ecology working group and agrofood studies group. Professor Daniel Press hired me to develop and teach an energy politics and policy course after I completed my dissertation. Daniel helped me realize the value of my chemical engineering background and experience, and encouraged me to utilize it in my next ventures.

In 2007, as I wrapped up my dissertation, my next research topic sort of fell into my lap. It happened that there was a new dawn rising in the semiconductor industry, this time promising to reinvent the economy with thin-film photovoltaics—a novel technology that layered photovoltaic semiconductors on glass. It immediately became clear that the topic was ripe for environmental justice research. The green jobs discourse was prevalent, and the Intergovernmental Panel on Climate Change had just won the Nobel Peace Prize for its work on climate science and policy, but no one was asking about the impacts of these new technologies. I was soon put in touch with the executive director of the Silicon Valley Toxics Coalition, Sheila Davis, who hired me to write the report *Toward a Just and Sustainable Solar Industry* and eventually develop the annual Solar Scorecard with the support of a Switzer Leadership Grant. I am very grateful to Sheila for this opportunity and collaboration, and to Erin and Lissa and the foundation for supporting this work. Colleagues in the sustainable investment community also supported this work, and I'd like to specifically thank Stephen Heim and Seb Beloe for being clear early on about how important sustainability is to the future of this industry.

Professor Alastair Iles, in the Environmental Science, Policy, and Management Department at the University of California, Berkeley, worked with me to develop what I learned through the work with SVTC into a competitive National Science Foundation postdoctoral fellowship in science, technology, and society. The proposal was successful, and I had an office in Giannini Hall from the start of 2010 through the end of 2011, when I started at San José State University. This grant supported my fieldwork at utility-scale solar power plant sites, trips to manufacturing facilities, and other research and professional development activities, such as conferences.

I thank my environmental studies colleagues at San José State University, who have mentored and worked with me to become a better teacher and to create space to continue this research. Thank you to Professors Gary Klee, Rachel O'Malley, Lynne Trulio, Katherine Cushing, and Will Russell for your kindness and support as I joined the faculty, and later Professors Kate Davis and Jason Douglas. We also have amazing lecturers and adjunct faculty Terry Trumbull (who incidentally drafted the legislation forming the EPA during the Nixon administration), Beniot Delaveau, who teaches a number of our energy courses, and not least Bruce Olszewski, whose knowledge of waste flows helped me think about end-of-life photovoltaics management. Thanks also to Professor Bruce Cain, Dr. Iris Hui, and Geoff McGhee for wonderful conversations at the Bill Lane Center for the American West at Stanford University, where I reworked this manuscript during a sabbatical for the 2016–17 academic year.

There were many folks who helped me learn about these systems as I moved across the commodity chain. Jim Andre, Laura Cunningham, Kevin Emmerich, Chris Clarke, Sean Gonzales, Tasha Ladoux, and many other informants helped me learn about the issues in siting solar across the American West. Jan Clynke, Ricky Sinha, Andreas Wade, Karen Drozniak, Ben Santariss, Karsten Wambach, and numerous other players in the industry were extremely helpful over the years. Thanks to Professor Vasilis Fthenakis of Columbia University, who left a tremendous paper trail of research that helped inform the background of this work. Professors Paul Robbins, Rebecca Hernandez, Davis Hess, and countless others have been influential.

Colleagues and friends Professor Jill Harrison, Mike and Kimba Sammet, Professor Brian Uribe-Dowd, Professors Brian Gareau and Tara Pisani-Gareau, Dr. Ethan Mora (drums), Ken Morgan (guitar), Professor Nick Bader (bass), Dr. Mark Buckley (raft), Dr. Michael Dorsey, Dr. Constanza Rampini, Professor Max Boykoff, Professor Mike Goodman, and so many more. To all those I forgot to mention, I'm sorry; I will buy you a beer.

Lastly, thanks to four peer reviewers, one who looked at two versions of the manuscript. This feedback informed the final organization of this book. Thanks to Blake Edgar, who approached me at the Ecological Society of America conference after a talk, and asked if I'd consider a manuscript. I told him I had something almost done. At the time, I didn't realize it was barely done. Thanks to Merrick Bush, who shepherded the first round of reviews. I am most indebted to Kate Marshall,

Bradley Depew, Cindy Fulton, and others from the publishing team, who were immensely patient with me as I did some heavy reorganization after a second set of reviews, and steered toward the finish line. Lastly, I received tremendous and knowledgeable support from my copyeditor, Roy Sablosky. Without Roy's help this manuscript would be much less readable, and I owe him a great deal of gratitude. This was not an easy book to prepare, and I appreciate all the help needed to make it come together. Go team! We made it.

Introduction

We are at the start of a solar revolution. The solar energy industries have taken off in the past decade, growing forty-fold globally from 2008 to 2018 with few signs of slowing down.[1] More than 99.9% of all photo-voltaic modules and concentrated solar power plants ever built were installed after 2008. That year the photovoltaic industry produced 170 megawatts' worth of modules, enough to power a quarter of a million homes on a sunny afternoon.[2] At the end of 2018, cumulative photo-voltaic installations surpassed 500 gigawatts (GW), over a thousand times the annual production a decade earlier, and enough to power over two hundred million homes.[3] Solar power is no longer alternative energy. New records for solar generation are broken every month in California, Germany, and China, and headlines regularly announce the opening of the next largest solar power plant in the world.[4] Solar electric technologies—photovoltaic and concentrated solar power technologies such as parabolic troughs and solar power towers—are making meaningful contributions to electricity supplies in some places, albeit geographically unevenly across the globe. On sunny afternoons in 2017, solar provided over 50% of peak electricity in Germany and California.[5] These records will be broken repeatedly as more solar power is installed.

When solar power is measured in terawatts—a hundred times the annual production today—the industry will be making inroads as a major source of electricity. Today solar remains a small portion of the overall energy supply. In June 2015, solar surpassed 1% of total energy supply globally.[6] Solar triumphalists contend that solar power's biggest growth period lies in the future. Based on growth rates since 2009, that prospect is taking shape today. However, not all energy experts share this vision. Concerns about "value deflation"—the idea that increasing penetration of solar power becomes less valuable to the grid and to investors over time—could lock photovoltaics into low contributions of electricity.[7]

From the perspective of environmental studies, the ways that solar power development unfolds will transform society and the environment in different and important ways.[8] The net social and environmental benefits of solar power are well documented and generally uncontested—more jobs, higher quality of life, and much less air pollution and greenhouse gas emissions, compared to an equivalent energy supply from fossil fuels. Yet, all forms of energy development have impacts or pose new or different risks to specific communities, ecosystems, and landscapes. The transition to solar power will be no different, requiring greater land-use changes for solar energy landscapes, production of silica and various metals from mines, processing in smelters, blast furnaces, glass factories, chemical plants (with their effluents), and manufacturing facilities ("fabs") to fabricate components and assemble devices. Exploring the environmental challenges of scaling up photovoltaic manufacturing to the terawatt level can help society plan for the coming environmental impacts of the solar energy transition. Analysts with the SunShot Initiative, an effort led by the U.S. Department of Energy, estimate that to get photovoltaic module manufacturing levels to 20 GW per year, production of supply chain materials would need to increase 6% for glass, 520% for polysilicon, 38% for tellurium, 160% for indium, and 30% for silver, from current levels.[9] To bring solar power to terawatt levels implies a hundred-fold increase in these numbers. Environmental and energy justice outcomes are more likely if impacts can be identified and proactively governed.[10]

This book aims to identify the challenges facing a sustainable and just transition to solar power, and to inspire us to solve these challenges before they become problems. Much of the analysis will focus on the production of *photovoltaic modules* (or *photovoltaics*), which are devices that covert the energy of sunlight into electricity. Electric power generation currently contributes one-third of global greenhouse gas emissions,[11] but photovoltaics generate electricity with no emissions at all (putting aside the manufacturing process), making them key technologies for decarbonizing electricity.

Photovoltaics are manufactured with chemical feedstocks and processes similar to those used in electronics and semiconductor production.

This raises some concern, as the legacy of electronics and semiconductor industries in places like Silicon Valley left workers and communities around manufacturers like Fairchild Semiconductor and IBM exposed to toxic vapors and solvents and other wastes in groundwater.[12] Public health and occupational questions followed these semiconductor and electronics factories as they offshored to East Asia, Mexico, and other places in the global economy that attract chemical and manufacturing industries.[13] Scaling up photovoltaic manufacturing will raise environmental, health, and safety challenges that may require planning, management, and regulatory interventions, especially where some workers and communities might bear disproportionate burdens or risks.

Integrating solar into electricity grids will require the deployment of new landscapes. Energy from the sun is diffuse compared to the concentrated forms of fossil and nuclear fuels, so solar power requires a lot of space.[14] New requirements for electric utilities and falling costs for utility-scale projects could provoke conflicts over land use in regions with rich solar energy resources.[15] And while they provide low-carbon electricity, utility-scale solar energy projects could negatively impact certain communities, workers, and ecosystems. Land-use changes for utility-scale solar energy development could permanently alter ecological communities, disturb cultural artifacts, and transform landscapes.[16] Conflicts over what the Nature Conservancy terms "energy sprawl" are already occurring, most prominently at a few sites in the California deserts.[17] The insolation that blankets southern Spain, China's Gobi Desert, the Atacama Desert of Chile and Peru, and Africa's Sahara, Kalahari, and Sahel deserts are starting to attract proposals and investments for utility-scale solar power plants.[18] Communities and land managers in these places will face tough decisions about how to resolve land-use conflicts over renewable energy development. But conflicts are not inevitable: future patterns of solar deployment can be guided to work toward climate adaptation and biodiversity conservation goals rather against them.[19] Photovoltaics can be readily integrated into the built environment, on abandoned agricultural or disturbed land, parking lots, landfills, and other landscapes where conflicts may be minimal. "Floatovoltaics" can be placed on reservoirs and other open water bodies.[20] Most fittingly, few electricity sources are so well suited for people to live under.

Energy justice aims for "a global energy system that fairly distributes both the benefits and burdens of energy services, and one that contributes to more representative and inclusive energy decision-making."[21] The aim of this research is to raise questions about solar power transitions: Who bears the burdens? Where might collateral effects manifest? How can these aspects be integrated into energy policy, planning, and practice?[22] Energy justice can inform energy policy by providing a framework capable of incorporating distributional, recognitional, and procedural tenets—these include accepting that impacts are unevenly distributed, representing silent or marginalized voices, and ensuring that processes are fair and democratic.[23] Attaining energy justice in solar energy transitions requires overcoming power asymmetries at several scales, from inequitably arranged political interests shaping regulation and legislation, to decisions embedded in our everyday routines.

How can this solar revolution be scaled rapidly and at the same time be kept *sustainable* and *just*? Identifying and resolving issues with solar power supply chains, construction activities, operation, decommissioning, and end-of-life management can ensure more sustainable and equitable outcomes. This requires building effective institutions to coordinate decision-making processes and planning efforts, and social movement engagement, as well as sustainability leadership from industry. Geographers and energy policy experts on the low-carbon energy transition note that energy justice needs sustained theoretical and empirical attention.[24]

Solar power is a term colloquially used to describe electricity generated from a photovoltaic module, by steam boiled by the sun, or even the solar heating of water. In electronics, power is the voltage times the electric current; it is the ability to do work with electrical energy. But power is also a social concept. Social scientists think of power operating in society in several ways, depending on the entry point of the respective research community. Power can describe control over others. People, communities, social movements, and nations are said to "have power over" or to "influence" outcomes. For other social scientists, no one holds power because power is a relational effect, produced through interaction.[25] Power is diffuse or discursive and a factor that conditions our everyday behavior and ways of thinking, or subjectivities. In this book, I point out how power structures shaped certain environmental outcomes, but also how discourses and subjectivities configured particular consequences in specific ways.

Anticipating future environmental justice issues is an emerging research theme in energy transitions research.[26] Political ecology is an area of inquiry in environmental studies and human geography that focuses on the multi-scalar and interconnected aspects of what humans make and use. Its roots are in disaster studies, cultural ecology, and development studies, and much of this research connects disparate places of production to sites of consumption, showing how decisions made in one part of the world might be connected to environmental problems somewhere else. Researchers in political ecology consider questions related to how landscapes and communities become vulnerable to environmental degradation from commodity production and how uneven power relations sustain these effects. Drawing from sociology, political ecologists have borrowed the "commodity chain" as a conceptual apparatus to connect raw material extraction, through supply chains, to sites of manufacturing, use, and eventual disposal. This allows the researcher to follow a commodity through the all the stages of production and use. The narratives taken from these commodity stories often focus on how political-economic factors and power structures shape nature–society relations in natural resource struggles or environmental change. This political ecology lens will help highlight potential frictions associated with scaling up solar technologies, as it will be attentive to the challenges associated with the political economy of solar energy development in an uneven world.

This book does not argue that solar power is in any way a poor technical choice or worse than conventional energy sources. The evidence emphatically points to the benefits of solar power.[27] The argument in the book does emphasize opportunities to make solar energy commodity chains more just and sustainable. Photovoltaic production has a green halo compared to other electronic and chemical industries, which means it sometimes escapes the scrutiny deserved by all systems of commodity production if the goals are sustainability and environmental justice. If few question the environmental bona fides of photovoltaics, opportunities to green design, production, and deployment along the life cycle will be missed. The following chapters take a closer look at solar power commodity systems and their implications for energy transitions. Solar power remains the most attractive and sustainable option to supply society with low-carbon energy, but it will require careful planning, assessment, and practice to ensure that socio-environmental impacts are minimized and equitably distributed.

OVERVIEW OF THE CHAPTERS

Chapter 1 introduces the synergies and tensions between solar power innovations, green jobs, and environmental justice. In the United States, a "green jobs" discourse emerged starting around 2005 and manifested in government investments in economic stimulus through the American

Recovery and Reinvestment Act (ARRA) several years later. Tens of billions of dollars in ARRA investments went toward renewable energy technologies and projects, and this created thousands of jobs during a time of recovery from a global economic calamity. Rust Belt communities embraced the idea of reinventing the economy around renewable energy, as it was an opportunity to retrain skilled workers in industries experiencing automation or offshoring. Urban communities saw opportunities to employ people in traditionally underserved communities. Silicon Valley was flush with a new round of semiconductor industries, as solar equipment and thin-film photovoltaic manufacturers numbered in the dozens in the late 2000s. Federal investments in a green jobs workforce led an early wave of the solar energy transition in the U.S.

Several environmental justice organizations began to focus on green jobs training in the installation of photovoltaic modules. However, very few were asking if the green jobs being created would be linked to other jobs that exhibit patterns of environmental inequality. Other electronics and semiconductor manufacturers are frequently in the news for chemical contamination or worker health and safety issues.[28] Would the growth in green jobs be linked to environmental inequality elsewhere in the commodity chain? Several organizations, including the Silicon Valley Toxics Coalition, led an effort to investigate the risks new semiconductor manufacturers posed to communities. Critical approaches like these are important to energy justice research related to the solar energy transition because the lessons learned can shape policy and practice. Chapter 1 introduces the primary environmental, health, and safety issues in solar power commodity chains against the backdrop of green jobs.

Chapter 2 describes in more detail the investments in solar innovations made through ARRA—the policy architecture that helped invest $90 billion in renewable energy. A set of institutional forces set into motion by the U.S. Departments of Interior, Energy, and Treasury would provide inertia to projects that transformed landscapes and commodity chains, with implications for socio-ecological systems. Starting with activities initiated for energy development on public lands by the Department of the Interior in 2001 on the recommendations of vice president Richard Cheney's Energy Task Force in president George W. Bush's administration,[29] the three agencies all implemented incentives that would favor developers of large-scale solar power projects, including power plants and manufacturing facilities. This chapter describes the important changes in governance that aided solar deployment. One key set of policies were state-level renewable energy portfolio standards, requiring investor-owned utilities to acquire greater amounts of electricity from renewable sources every year until they hit some predefined target percentage (20%, 33%, 50%, etc.). This effectively created guaranteed markets for renewables, making long-term investments less risky, as the most economically viable projects were offered power purchase agreements by investor-owned utilities. The Energy Policy Act of 2005 asked the Bureau of Land Management (BLM) to develop 10 GW of renewables on public lands (expanded to 20 GW in 2015). This opened over 33,000 square miles (about 22.5 million acres, or 87,336 km²) of public land to solar energy development across the American Southwest.[30] These federal decisions about public lands enabled the expeditious processing of ARRA expenditures within the short window that they were available, but also led to a series of what some might call rash or hasty decisions. Finally, a set of programs to subsidize the risks of solar energy finance and investment made it possible to leverage more capital from firms that otherwise would not fund such endeavors. These policies helped reduce costs for solar and allowed several large-scale projects to access capital that would otherwise not be available, particularly at the height of a global financial crisis. One of these projects would be the largest photovoltaic installation in the world for a few years.

Chapter 3 explores some of the environmental, health, and safety impacts of innovations in the life cycle of thin-film photovoltaics. As innovations in solar technologies make it cost-competitive with conventional energy sources, they introduce new manufacturing risks that deserve consideration. The ARRA investments in thin-film photovoltaics, for example, relied on semiconductors containing cadmium compounds. Sometimes these new risks take the form of unknown impacts of novel materials, such as carbon nanospears or cadmium quantum dots. Other new risks emerge from innovations in social organization. For example, contract manufacturing could be seen as an innovation in production, but also as a new environmental health risk, as accountability shifts and social distance is increased between sites of production and consumption.[31] Life-cycle-assessment experts at the Photovoltaic Environment Research Assistance Center at Brookhaven National Laboratory and the National Renewable Energy Laboratory have developed a comprehensive literature on the environmental, health, and safety hazards of photovoltaics.[32] The chapter describes how these risks are articulated through environmental performance metrics produced through life cycle assessment. While the framework offers much in telling stories about systems of production, the construction of performance metrics can often obscure or silence other ways of understanding environmental impacts, particularly ones that are unevenly distributed.

Chapter 4 tackles the important questions around disposal and end-of-life management accompanying the widespread adoption of photovoltaics. The rare and precious materials in photovoltaics are compounded with or embedded in more toxic ones, the same recipe that fuels the global trade in e-waste that poses public health problems in West Africa and Southeast Asia. Separating the cadmium from tellurium and the lead from silver in end-of-life photovoltaic modules might be done by artisan e-waste collectors with crude tools, raising concerns about occupational exposures and public health.[33] The top manufacturer of thin-film photovoltaics from 2005 through 2017 has a "filter cake" recycling system able to recover 95% of the tellurium from processed modules. The narrative here explains the background that led to this arrangement and identifies best practices to increase the recyclability and recycled content of photovoltaics. Other environmental benefits of recycling photovoltaics include avoided mining and obtaining a secure supply for rare substances that could be subject to price volatility or material scarcity.[34] There are already viable recycling schemes, based on extended producer responsibility, throughout Europe.

Chapter 5 describes controversies involving public lands, managed by the BLM across six states in the American Southwest, that were offered to solar energy developers in the name of climate protection. Several utility-scale projects in California that received loan guarantees were sited on lands that many viewed as having important conservation value and biological and genetic significance, and as habitat for threatened and endangered species. A species facing severe habitat loss, Agassiz's desert tortoise (*Gopherus agassizii*), was at the center of many of these controversies. With 80% of desert tortoise habitat on public lands in the U.S., the BLM plays a special role as steward for this species. Opening large swaths of habitat to leasing for solar development put ecological considerations into direct conflict with climate change mitigation strategies, putting the BLM in the familiar position it is in elsewhere across western public lands, where it balances domestic oil, gas, and coal production against ecosystem conservation. Solar energy policies were intended to create opportunities for solar development, but also created several intractable conflicts and deep rifts among environmental groups over land use across the Southwest.

Chapter 6 describes new planning institutions, policies, and practices put in place in reaction to early ecological and cultural resource controversies. After an overview of the socio-ecological impacts of utility-scale solar power plants and proposed mitigations, public policymaking processes are described that sought a framework to resolve some of the land-use conflicts between renewables and ecosystems. The first is the Western Solar Plan, a public process initiated by the BLM via Secretarial Order 3285A1, which set the policy goal of identifying and prioritizing land for solar energy development. With the help of several agencies and national energy labs, the BLM proposed Solar Energy Zones, where development would be incentivized because they were deemed to have fewer ecological and cultural resource conflicts. The process through which Solar Energy Zones were proposed and reshaped is an example of how public participation can have meaningful impacts on energy landscapes. From the time the assessment was initiated in 2008 until 2014, tens of thousands of public comments led to the elimination of several proposed Solar Energy Zones. The Desert Renewable Energy Conservation Plan (DRECP) is a framework to guide solar development to focused areas on public lands in California with minimal land-use conflicts and will extend the analysis to private lands. The public participation and extensive agency coordination required to work through the DRECP make it one of the most comprehensive planning analyses for solar energy transitions ever prepared. But the impact of this planning effort is now up in the air because in early 2018 interior secretary Ryan Zinke announced plans to dismantle the DRECP.

Chapter 7 explores the challenges to public policies designed to foster innovation through case studies of loan guarantees to venture capitalists in the solar space. By establishing the industry context for the investments made through the Department of Energy loan guarantee program to solar power startups like BrightSource, Solyndra, Abound Solar, and SoloPower, the chapter explains why particular projects were believed to represent "breakthrough technologies"—the term that framed the Department of Energy loan program's mission. Many of these investments became controversial, because the innovation process for venture capital has certain tendencies and logics that make for politically vulnerable public policy bets. The public has a different perspective on risk from a venture capitalist or a hedge fund manager. After a description of the trends in the venture capital sector around clean technology, the issues related to the bankruptcies of thin-film photovoltaic manufacturers Solyndra and Abound are detailed. Overproduction of photovoltaics in China and crony capitalism were common explanations for the projects' failure. And tensions between thin-film manufacturing and justice were highlighted when cadmium compounds were left behind after the bankruptcies. The chapter documents the rapid ascendance of the Chinese photovoltaic industry, which sparked several ongoing trade disputes. Starting in 2010, the U.S. and Europe engaged in retaliatory trade measures with China over solar industry subsidies. The accusation made by the U.S. Department of Commerce was that China was illegally subsidizing its solar energy industry, allowing Chinese manufacturers to sell photovoltaic modules below cost in foreign markets. U.S. policymakers claimed that these subsidies had led to the downfall of several ARRA investments in thin-film technologies.[35] The eventual outcome would be a significant tariff on modules imported into the U.S. from China, and soon after that, from Taiwan; and eventually the tariffs would be proposed for all photovoltaic module and cell imports to the U.S.

Trade disputes in the solar industry are not limited to China. Japan sued Canada over Ontario's domestic sourcing requirement, which required a specified portion of the module to be made or assembled within the Canadian province to take advantage of a feed-in-tariff—a valuable consumer incentive that usually rewards a solar-producing customer with a high rate for electricity delivered to the grid. The U.S. is calling on India to repeal its domestic sourcing requirement for similar reasons. India argues that the requirement is critical to developing its own photovoltaic manufacturing capacity. These trade conflicts add cost and shape where solar manufacturing capacity will take root and expand. The solar trade war remains active on all these fronts. In February 2018, the U.S. commerce secretary under President Donald Trump declared that all imports of crystalline silicon photovoltaics (with a handful of exceptions) would have a 30% tariff levied on them. Anticipation of the tariff led to a massive increase in imports prior to the ruling; Bloomberg New Energy Finance reported an eleven-fold increase in photovoltaic imports to the U.S., which will be warehoused until needed, defying the administration's ruling.[36]

Chapter 8 outlines a vision for a solar energy transition that simultaneously promotes decarbonization, environmental justice, sustainability, and community resilience. There are opportunities to improve community livelihoods, reduce chemical exposures in the workplace, and eliminate solar waste by pursuing innovations in green chemistry, worker health and safety, industrial ecology, and design for recycling. This chapter draws from experiences with efforts to establish a sustainability leadership standard for photovoltaics through a stakeholder-led process. Developing any system of industrial production around principles of sustainability and environmental justice will be challenging, especially as efforts continue to prioritize measures that drive down the cost of solar electricity. All energy sources have impacts, and there is ample evidence that the social and environmental costs of fossil fuels are significantly greater than the impacts of solar power. For example, the fossil fuel industries have significantly higher rates of occupational injury and death than renewable energy industries.[37] Keeping with the themes of the book, the goal of this chapter is to offer a vision for a more just and sustainable solar power in policy and praxis. Taking stock of the full picture of all of the environmental and energy justice impacts we have learned about and can foresee now, hopefully brings us closer to that objective.

SOLAR POWER TECHNOLOGIES

The Earth is bathed in electromagnetic radiation—that is, light from the sun, often called *insolation* or *solar radiation*—which can also be thought of as a stream of packets of energy called photons. Going back to the start of sedentary civilization, humans have developed numerous contraptions to harness power from the sun. Many point to the "solar death ray" designed by Greek inventor Archimedes as an early instance of a technological device specifically designed to harness solar energy for human use. The device may be mythical rather than historical, but according to some Greek historians, it used mirrors to concentrate sunlight—enough to burn the masts and sails of incoming warships at the battle of Syracuse in 212.[38] Other early uses of the sun include drying crops, which was probably done even before sedentary agriculture, and which is critical for storing food.[39] Efforts to harness solar energy in human civilization are nothing new.

The solar energy technologies addressed in this book are those that generate electricity, with photovoltaics being the most widely featured in the case studies. Photovoltaics use semiconductors to directly generate electric current in response to photons collected from sunlight. Recall from chemistry and physics that the electrons that surround atoms represent quantities of energy. In a solar cell, the photons carrying some portions of the spectrum of solar radiation deliver enough energy to raise an electron's energy level from the valence band to the conduction band. Electrons in the conduction band are free to move within the material, which means that a current can flow. The solar cell architecture allows electrons to flow from a layer with extra electrons toward a layer that loses electrons when exposed to light. There are many variations on these basic principles, with different devices constituted by different semiconductor materials (sometimes in combination with an electrolyte).

French physicist Edmond Becquerel built the first photovoltaic cell in 1839, an electrolytic solution that generated electric current proportional to light exposure. Becquerel learned that electric current could be generated across plates of platinum submerged in a solution of silver chloride in acid when exposed to sunlight. Hence, the photovoltaic effect is sometimes referred to as the "Becquerel effect." In 1873, while evaluating materials for the trans-Atlantic underwater telegraph wire, English electrical engineer Willoughby Smith discovered that selenium was photoconductive.[40] The physical properties of selenium were deemed excellent in the lab, but in the field, where they were exposed to variations in sunlight, the selenium equipment did not perform as expected. Smith would later realize that the resistance of selenium changed with incident light. In 1883, inventor Charles Fritts made the first solid-state photovoltaic device, a selenium-based solar cell with gold conductors. At the turn of the century, Wilhelm Hallwachs started making solar cells of copper and cuprous oxide, and what he would learn would evolve into the foundational principles for making CIGS (copper indium gallium diselenide) thin-films. Scientists had very little understanding of the mechanism behind the photovoltaic effect (in which photon energy is absorbed and an energized electron is drawn into a circuit, creating voltage) until 1905, when Albert Einstein described the the first part of the effect (where a photon's energy is transferred to an electron, but no circuit is present). Robert Millikan, the University of Chicago physicist whose famed oil-drop experiment verified the charge of the electron, would provide an experimental apparatus for the photoelectric effect in 1916. This experimental proof resulted in Einstein winning the Nobel Prize in 1921 (the prize was not given, as many people assume, for his special theory of relativity, also published in 1905).

Electricity is also generated by concentrated solar power (CSP) plants, which rely on steam-powered electric generators. While photoconductivity was still being explored and tested in the early twentieth century, other engineers and inventors began to investigate harnessing the sun to drive steam engines. In the late 1800s, French inventor Augustin Mouchot and his assistant Abel Pifre used the sun to make steam and drive a motor for a printing press. One of the first CSP parabolic troughs was in built in Egypt, where it was used to make steam that powered irrigation systems.[41] Today, solar thermal energy is concentrated as solar flux, using mirrors or heliostats, primarily to make electricity. The solar flux heats a fluid, which then boils water to make steam, which turns a generator to make electricity. There are several types of CSP technologies, but each makes steam. The oldest, called solar parabolic troughs, uses large curved mirrors that track the sun, directing heat at a fluid carried in a pipe that runs down the center of the trough. Solar "power towers" like the Ivanpah Solar Electric Generating System use fields of heliostats, large mirrors that track the sun and direct solar flux toward a boiler atop a tall tower. Stirling engines are external combustion engines, where a dish of mirrors focuses solar power on a heating element that warms to create the temperature difference needed to drive a piston. More obscure is the solar chimney concept, where solar energy is directed toward heating air, which turns a turbine as it rises up the chimney.

Solar devices like solar hot water heaters, which collect thermal energy from the sun to warm water, are not much discussed in this book. Passive technologies like solar hot water heaters and passive solar design for interior living spaces were not the focus of ARRA support, despite their widespread use in some parts of the world, including China and Israel. The ARRA investments focused more on technologies that make electricity and infrastructure that would be integrated into the electricity grid. This should not be read as an indictment of the maturity of solar hot water heaters, as they have tremendous potential to displace natural gas and electricity used to heat water. But public policymakers did not see an opportunity to generate game-changing technologies out of solar hot water heater investments because the technology remains basically similar to the kinds that have been commercially available for over a century. Anaheim, California, for example, saw widespread solar hot water heating in the 1890s, and the U.S. saw growth in solar hot water heaters again in the late 1970s. Yet solar hot water heaters inexplicably remain a fringe technology for hot water in the U.S. today; they are far less common than photovoltaics atop residential rooftops.[42]

PHOTOVOLTAICS

Photovoltaic modules are colloquially called solar panels. A single photovoltaic device is referred to as a module and is made up of numerous solar cells. When several photovoltaic modules are interconnected, it becomes a photovoltaic or solar array. Some refer to it as a solar system, though this is obviously a confusing term given its more widespread astronomical use. Most photovoltaic modules are flat plates of silicon cells or thin films; flat plate means that the entire surface collects light. Concentrated photovoltaics use a glass or plastic lens to concentrate light on a much smaller, but expensive, semiconductor surface. Instead of silicon, these are typically several stacked p–n junctions made of gallium arsenide, indium phosphide, or similar compounds. Concentrated photovoltaics are mostly limited to specialized satellite, telecom, and military applications because they are expensive, although there have been several commercially available concentrated photovoltaics for utility-scale solar power plants.

The most common photovoltaics use crystalline silicon semiconductors. Pure crystals of silicon are sliced into thin wafers and made into solar cells by "doping" the crystal with small amounts of impurities on each side, one with extra electrons, the other devoid of electrons, to facilitate electron flow. The solar cells are most commonly sandwiched between a sheet of glass and a back cover, and encapsulated in a polymer to protect the module from the weather.

While a community of scientists sought to better understand the photovoltaic effect and the behavior of electrons, in 1916 a Polish materials scientist named Jan Czochralski developed an important understanding of how to make very pure silicon. Czochralski's key innovation took advantage of different ways to cool and crystallize silicon. The basic approach is still widely used to make the silicon chips in transistors, which made possible the twentieth-century revolution in electronics. Even into the twenty-first century, crystalline silicon remains the backbone of electronic devices. Photovoltaic manufacturing depends on experience, techniques, and knowledge developed to improve the transistors that underlie the computer revolution.

All crystalline silicon photovoltaics are indebted to work at the iconic Bell Laboratories in Berkeley Heights, New Jersey, where important inventions such as the transistor were born. In 1954, three scientists— Daryl Chapin, Calvin Fuller, and Gerald L. Pearson—demonstrated a 6% conversion efficiency with a crystalline silicon photovoltaic device.[43] A year prior, Pearson had made solar cells of silicon better than those of selenium, the starting material, which the research team had found to be limited to 0.5% efficiency. Chapin and Fuller refined Pearson's work and made cells powerful enough to power small electrical equipment. Meanwhile, the university-military-industrial complex anchored by Stanford University and the University of California, Berkeley, was transforming the Santa Clara Valley into Silicon Valley. Among the major players in crystalline silicon, Hoffman Electronics produced solar cells at 8% efficiency in 1957, 10% by 1959, and 14% by 1960, mainly for NASA space and communications applications.[44] Today, premium-brand crystalline silicon photovoltaic modules routinely have conversion efficiencies better than 20%. With the help of several internal advocates of photovoltaic technologies, the U.S. military adopted solar power for satellites, culminating in the flight of the Vanguard 1 in 1958, a solar-powered satellite launched in reaction to the unexpected launch of the battery-powered and short-lived Sputnik satellite by the USSR in 1957. Federal government investments were instrumental in the development of crystalline silicon solar cells.

There are three kinds of crystalline silicon photovoltaics, which are produced by several hundred manufacturers and have over 95% of the photovoltaic market share.[45] *Monocrystalline* silicon is named for the ingot, made of a single crystal, that forms when using a process of heating and cooling polysilicon—the Czochralski method. The process requires placing a rod of pure silicon in the core of a reactor holding molten silicon. As the silicon is cooled, the crystalline silicon grows around the rod. In 2015, the Czochralski method was used to make roughly half of the volume of the silicon-based photovoltaics sold. Other processes that turn polysilicon into crystalline silicon include the fluidized bed process, which uses silane as the primary input. Silane is used in small quantities for doping crystalline silicon photovoltaic modules. But companies that use silane as their sole silicon source use very large volumes of the gas, which is responsible for more worker deaths than any other chemical in this industry, with twelve documented deaths (context matters here—this is orders of magnitude lower than worker death rates in other energy industries).[46]

Prior to casting in crucibles, small amounts of impurities like boron are added to the molten silicon, doping it to be intrinsically a positive layer, or p-layer, able to accept an incoming electron. The pure ingots of monocrystalline silicon, round when they are drawn out, are cut into rectangular bricks using diamond-bladed wire saws. They are then sliced into wafers. These silicon wafers are next cleaned, textured, and doped with a second impurity to form the negative layer (n-layer) that gives this part of cell its negative charge and makes it an electron donor. With the n-layer and p-layer now integrated into the solar cell, an antireflective coating is applied to maximize light absorption. Finally, the contact grid lines and busbar are added to the solar cell surface. These contacts harvest the electrons freely moving in the conduction band, drawing them into the circuit.

There are several other processes used to turn molten silicon into *multicrystalline* or *ribbon-crystalline* silicon photovoltaics. Whereas monocrystalline silicon is cooled into one single crystal, multicrystalline silicon (sometimes called polycrystalline silicon) is cast into crucibles and when it cools forms an ingot composed of many crystals. Ribbon-crystalline silicon is made by pulling wafers directly out of the molten silicon, as opposed to slicing wafers from ingots. In the 2000s, there were several companies exploring ribbon silicon, most notably a Massachusetts firm, Evergreen Solar, but it is no longer commercially produced. Evergreen Solar made headlines in 2008 when it was named by the *Boston Herald* as one of the major hazardous waste generators in the state and caused controversy when it moved its manufacturing to China shortly after taking several large local government grants to expand a factory in the U.S.[47]

Thin-film photovoltaics use semiconductor layers on the order of hundreds of nanometers in thickness that are applied to a substrate as it moves along the production line. The substrate is often glass, but these layers can also be applied to flexible materials like plastics and metal foils. Thin-films use less semiconductor materials and lower energy inputs, making them in principle less expensive to manufacture. They also can be made more rapidly in a continuous manufacturing process, which further reduces costs. The time from when a piece of glass starts down the production line until when the product is ready for inspection is on the order of hours, instead of weeks or months with silicon-based technologies. Common types of thin-film cells now in commercial production include cadmium telluride (CdTe), copper indium gallium diselenide (CIGS), and amorphous silicon (a-Si). Table 1 shows the major technologies used in the photovoltaics industry.

Many scientists, investors, and technologists see thin-films as having the greatest potential for long-term, widespread adoption as they can be significantly cheaper to make, requiring less material and energy. But excitement about thin-films in the investment community seems to have waned since the 2006-to-2010 window. Some interest remains, but much of the enthusiasm has faded, because many of the cost reductions either did not come to fruition or were outpaced by the falling price of crystalline silicon (allegedly because Chinese manufacturers were dumping crystalline silicon photovoltaics onto the U.S. market below cost, according to the U.S. Department of Commerce). Since the time that thin-films spurred widespread investments in the mid-2000s, many of the thin-film companies that garnered support have folded, or were sold to larger firms; some have even switched to silicon-based technologies. The major exception to the decline in thin-film manufacturing is a major actor in many of the cases presented in this book, First Solar of Tempe, Arizona. It is one of the world's largest and most innovative players in photovoltaics. First Solar was the number-one photovoltaics manufacturer in 2010, and a top-five producer for several subsequent years.

TABLE 1

Crystalline silicon	Thin films	Crystalline gallium arsenide
Monocrystalline	Cadmium telluride	Monocrystalline
Multicrystalline	Copper indium gallium diselenide	Concentrator
Thick silicon film	Amorphous silicon	Thin-film crystal
Thin-film crystal	Nano-silicon	
Silicon heterostructures (HIT)		
Multi-junction cells	**Emerging**	
Three junction (concentrator)	Dye-sensitized solar cells	
Three junction (non-concentrator)	Organic cells	
Two junction (concentrator)	Organic tandem cells	
Two junction (non-concentrator)	Inorganic cells	
Four junction (non-concentrator)	Quantum dots	
Multijunction silicon	Perovskites	

Founded by inventor Harold McMaster and seeded by the National Renewable Energy Laboratory and True North Partners—an investment arm of the Walton family, owners of Walmart—First Solar receives many accolades for its environmental and sustainability policies around chemical handling and product stewardship, including an extended producer responsibility program.[48] But at times it has been embroiled in land-use controversy, including lawsuits from major environmental NGOs such as the Center for Biological Diversity, Sierra Club, and Defenders of Wildlife against projects proposed on public lands.[49]

The evolution of thin-film photovoltaic manufacturing is rooted in a separate line of semiconductor innovations. Scientists began to experiment with cadmium-based thin films in the 1930s. The most commonly explored materials for thin-film solar cells at the time included cadmium sulfide and cadmium selenide. Cadmium compounds are very good semiconductors for photovoltaics because the energy required to elevate their electrons to the conduction band—what physicists call the band gap—closely matches the energy found in the most powerful part of the solar spectrum. The availability of materials that produce the photovoltaic effect is quite limited by the laws of physics and the electromagnetic spectrum of the sun. By the 1950s, cadmium telluride specifically was garnering attention from solar researchers. Companies exploring cadmium-based photovoltaics as far back as that time include many household and Fortune 500 names, including General Electric, Kodak, Matsushita, and British Petroleum.

By the 1960s, the first commercial photovoltaics were coming to market, alongside other solid-state electronics powered by semiconductors. Space satellites and telecommunications equipment were among the first devices powered by photovoltaics. In the 1970s, a researcher named Elliot Berman, with support from major oil producer Exxon, made important innovations that led to a much less costly crystalline silicon photovoltaic module. Amorphous silicon thin-film photovoltaic technologies were also coming into commercial applications, most notably in consumer electronics. Companies were researching and investing in crystalline silicon, thin-films, crystalline gallium arsenide, and multijunction solar cells by the end of the decade. The key technologies undergirding the solar revolution were already being commercialized.

By the 1980s, photovoltaics were powering many devices too far from the electricity grid to warrant running copper wires. Photovoltaics were increasingly competitive and sometimes cheaper than running copper wire to remote places. Among the companies investing in photovoltaics were oil and gas companies, who were using photovoltaics to power small devices at remote or offshore drilling platforms. Oil companies would later embark on a buying spree of photovoltaic companies, sparking speculation about conspiracies to undermine the success of photovoltaics. Many began to question the motives of the oil and gas companies, noting that developing renewable energy was not in the interest of industries that derive their profits from fossil fuels. Some went so far as to suggest that the oil industry was buying up patents to prevent others from developing renewables. More likely is that the oil and gas industry viewed photovoltaics a way to diversify their energy product portfolio and control their supply chain, as photovoltaics were increasingly being installed in remote operations such as drilling platforms and pipeline compressor stations.

Today, photovoltaic and CSP technologies are competing with mainstream electricity technologies. Transitioning toward increasing amounts of solar power can have minimal impacts if planned well and deployed with environmental best practices. All energy technologies come with social and environmental externalities that can produce, maintain, or reproduce environmental inequality. Fossil fuels are widely documented to cause the most environmental harm. The following pages present some lessons learned from the early years of solar power deployment in the hopes that the benefits of these rapidly emerging technologies will foster positive social and environmental change.

Solar Power

When you think about the emerging green economy, don't
think of George Jetson with a jet pack. Think of Joe Sixpack
with a hard-hat and lunch bucket, sleeves rolled up, going off
to fix America. Think of Rosie the Riveter, manufacturing
parts for hybrid buses or wind turbines. Those images will
represent the true face of a green-collar America.

—Van Jones, Founder of Green For All[1]

THE RISE OF GREEN JOBS DISCOURSE

In October 2004, a white paper, later widely discussed, was presented at
the annual meeting of the Environmental Grantmakers Association. The
authors, Ted Nordhaus and Michael Shellenberger, who had recently
co-founded the Breakthrough Institute, argued that the "death of environmentalism" was near.[2] The environmental movement had changed
things for the better, but its effectiveness was waning. They based this
assessment on marketing data about American consumer advertising
and voting trends, which showed that the traditional tools of environmental protection, such as legal intervention and state regulation, were
viewed less favorably among the public. They directed this message at
the prominent environmental NGOs—the Sierra Club, Natural Resource
Defense Council, and Environmental Defense Fund—and underscored
the importance of shifting strategies. Their data suggested that environmental messaging carried less resonance with the public and failed to
enroll broad support for the kinds of coalitions needed to respond to
urgent environmental problems, particularly issues related to climate
change. Polling and market data showed American public opinion moving away from the messages long held by the environmental community,
such as the need for the state to protect communities through regulation.

Other solutions promoted by environmental organizations, such as reducing consumption in a "limits to growth" framing, also failed to appeal to the same public. Messaging for environmental change needed to be organized around opportunity, they argued. People were more concerned about job security than any threat posed by climate change. Nordhaus and Shellenberger were arguing to reframe environmentalism around innovation and job creation, appealing to working-class people.[3]

Green jobs—manual jobs in enterprises whose products and services improve environmental quality—became currency for political claims about the mutually reinforcing benefits of pursuing economy, health, and environment simultaneously. The authors later extended this thesis to critique the strategies of environmental justice movements. They argued that environmental justice movements needed to focus on alternative economic paradigms and advocate for green jobs, not only against the actions of polluting industries. Claims about environmentalism, or more specifically environmental justice, were failing to enroll public support for climate and clean energy policy. Coalitions between labor unions and environmentalists would be needed for this to be effective. John Kerry's 2004 presidential campaign was the first to emphasize the climate policies that his colleagues were crafting in the House and Senate. The Center on Race, Poverty, & the Environment released a report speculating on the potential for green jobs in urban communities.[4] Environmental justice and human rights leaders such as Green For All founder Van Jones echoed the sentiments calling for green jobs explicitly in a justice frame—to ensure that the incoming "green tide lifts all boats."[5]

Before long, the "green jobs" moniker entered the lexicon of American politics, and by 2008 it was a mainstream storyline of U.S. presidential hopefuls Barack Obama, John McCain, and Hillary Clinton.[6] McCain, the Republican nominee, argued that "green jobs and green technology will be vital to our economic future. There is no reason that the U.S. should not be a leader in developing and deploying these new technologies."[7] The call for green jobs had bipartisan support.

Numerous journalists, political analysts, activists, economists, and politicians argued in commentaries and op-eds that public investments in green jobs could stimulate job growth in the same Rust Belt and urban areas most severely impacted by the simultaneous global financial crisis and automobile industry collapse of the late 2000s.[8] Green jobs entered national debates about how to foster economic recovery and job creation. Around the same time, forecasts of jobs in thin-film photovoltaic manufacturing were hinting at an employment boom across U.S. urban and Rust Belt communities, in Silicon Valley, and even in the high-tech corridors of East Asia. The California-based non-profit Solar Richmond invoked the specific symbolism of Rosie the Riveter, a cultural icon of the female workforce that was critical to the U.S. effort in World War II.[9]

The push for a green jobs economy manifested as public investments in clean energy infrastructure and innovations instigated by metaphors of the 1930s New Deal and the 1960s Apollo program to get a human to the moon. Clean energy investments were viewed not just as public works programs but also as investments in innovations and technologies. With clear messaging around green investments from major political institutions, environmental and social justice organizations—including the Ella Baker Center for Human Rights, Tradeswomen, Inc., Green Communities Online, Solar Richmond, the Solar Living Institute, and Groundwork USA—began advocating for and offering green jobs training. Green jobs training focused on good, well-paying jobs in solar sales and installation, weatherization, and energy retrofitting. These initiatives forged new coalitions, such as the partnership between the U.S. Steelworkers, the Sierra Club, and numerous other environmental and employment-focused NGOs known as the Blue–Green Alliance. This particular coalition capitalized on aims to link "blue-collar" sensibilities with an emphasis on green workforce development, understanding that appeals to the working class would be critical to broaden support for climate and renewable energy policy. The discursive groundwork was done to garner public support for investments in clean energy projects.

PUBLIC INVESTMENTS IN CLEAN ENERGY AND ECONOMIC RECOVERY

Job creation figures centrally in many justifications for public policies that use taxpayer dollars or lead to forgone taxes, and solar policies are no exception. The investment tax credit is a critical tool in the U.S. that shapes the retail economics of homeowner photovoltaic adoption and the cost of electricity from utility-scale projects. It is a tax equity financial incentive that allows the photovoltaic system owner to deduct some portion of the cost of the system from their taxes. Through 2018, the policy allows 30% of the total system cost to be deducted, meaning that the owner of a $10,000 system could deduct $3,000 from their tax liability over a five-year period. The credit will step down and phase out from 2019 through 2022. Public investments in the solar industry through policies like the investment tax credit are paying dividends. According to the Solar Energy Industries Association, "the solar industry . . . pumps $15 billion a year into [the] economy" and created "one out of every 78 new jobs" in the U.S. in 2015.[10] So the first thing to know about public expenditures for renewable energy is that many, like the investment tax credit, pay for themselves through other economic activities that generate tax revenues.

On the presidential campaign trail in 2008, candidate Barack Obama asserted that his policies would generate five million "green-collar jobs" in a decade through an investment of $150 billion. The American Recovery and Reinvestment Act (ARRA) offered the opportunity to fund this ambition. After taking office, President Obama created a leadership post focused on green workforce development—the "green jobs czar," formally known as the Special Advisor for Green Jobs, Enterprise and Innovation. Van Jones would have a fleeting appointment in this role. Soon after, the U.S. Bureau of Labor Statistics codified green jobs as an official statistical measure.[11] These actions formally linked the presidential campaign discourse to scores of billions of dollars the federal government would invest in "innovative" clean energy companies.

From 2009 through 2016, the U.S. invested over $90 billion in green jobs and clean energy through ARRA, the oft-maligned $780 million economic "stimulus" passed by the U.S. Congress and signed into law by President Obama.[12] A significant portion (65%) of the $90 billion went toward solar technologies, innovations, and utility-scale power plant deployment.[13] ARRA investments aimed to create green jobs and at the same time kindle climate efforts.[14] The emphasis on green jobs and clean technology led some observers to call the ARRA stimulus investments a "Green New Deal."[15] As the global financial crisis of the late 2000s rippled through national economies, government investments in solar energy infrastructure and research were seen as good investments for job growth and breakthrough innovations.

Two programs would dispense the majority of these investments, the Department of the Treasury's Section 1603 program and the Department of Energy's loan guarantee program. Some 13.5 GW of renewable energy projects were aided by Treasury 1603 wind and solar energy support, which comprised about 94% of the portfolio.[16] By 2017 the 1603 program had been granted nearly $25 billion from ARRA.[17] The second program allowed startup renewable energy companies to obtain loans that were cosigned by the U.S. federal government. About $12 billion was spent on investments in solar power.

TABLE 2

Program	Jobs created	Public investment
Treasury 1603 Program	55,000–72,000	$25.0 billion
1703 & 1705 loan guarantee programs	2,298	$26.3 billion
Advanced Vehicle Manufacturing Program	35,000	$8.4 billion

From 2012 to 2014, the Bureau of Labor Statistics tracked official data on the number and types of green jobs and workers' wages. The bureau defined green jobs as "jobs in businesses that produce goods or provide services that benefit the environment or conserve natural resources" and "jobs in which workers' duties involve making their establishment's production processes more environmentally friendly or use fewer natural resources."[18] The U.S. solar industry employed over 200,000 workers in "green jobs" by 2016, rising four times since ARRA.[19] Investments in renewable energy industries generate three times as many jobs as a similar investment in the fossil fuels industries.[20] According to the International Renewable Energy Agency, 3.7 million people were employed in the photovoltaics sector worldwide in 2017.[21] The Renewable and Appropriate Energy Laboratory at the University of California, Berkeley, reports that a 1 MW solar farm produces more jobs per unit energy than a wind, biomass, coal, or natural gas plant of the same power.[22] A 2017 Department of Energy report found there are 79 times more jobs per million MWh of solar energy than the same amount of energy from coal.[23]

Of course the investments in solar were about more than just jobs. From 1970s back-page appearances in the *Whole Earth Catalog,* to the front cover of *Alternative Energy Magazine* or *Green Biz* in the 2000s, solar power has long been at the vanguard of ethical consumption. Solar went from a fringe alternative energy to a mainstream solution to household electricity greenhouse gas emissions and decarbonizing the electricity grid. While there are many motivations for consumers to invest in photovoltaics, most studies suggest that cost is still the primary consideration.[24] Cultural and political beliefs such as utility or grid independence and concern about the power of monopolies, and psychological reasons such as personal expectations and "neighborhood" or "peer" effects, also factor in.[25] Some photovoltaic consumers are motivated by environmental considerations and specifically climate action, but some researchers have found that environmental values and concern are not enough.[26] Utilities also make solar investments because of state laws requiring renewables purchases, which are supported by rate-payers and voters. The evidence is quite clear that solar power delivers both jobs and other benefits (environmental, public health, and climate) overall.[27] The challenge to making solar more sustainable is ensuring that worker health and safety and environmental benefits are maximized, by minimizing the negative impacts of solar power production.

ARE GREEN JOBS JUST JOBS?

Even as overall public health benefits increase from solar power use, there may be green jobs that are actually dangerous or hazardous for workers and fenceline communities in and near high-tech manufacturing or deep in the supply chain.[28] Particular places, communities, and people could bear more burdens than others under the green economy, just as energy justice is disproportionately allocated today. Explorations of energy justice are an emerging research theme in energy transition studies.[29] Calls for green jobs grew louder as initiatives began to frame the benefits of green jobs for disadvantaged communities. Most major economic and technological transitions leave behind the most vulnerable and economically disadvantaged. These calls highlighted this issue of environmental equity. Slogans like "jobs not jails" offered an optimistic and hopeful narrative of economic opportunity, much as Nordhaus and Shellenberger suggested.[30] Environmental justice organizations embraced green energy and green jobs. But a small number of organizations wanted more assurance that solar industries would not burden communities and workers with air or water pollution, or other public health or occupational hazards.

Preparing for the environmental risks of scaling up photovoltaics deserves consideration because the crystalline silicon technologies rely on materials and manufacturing processes similar to those used to make silicon transistors and other electronic devices. The most advanced commercial solar energy innovations, such as thin-film photovoltaics, borrow techniques first used in flat-panel display and hard drive manufacturing. These processes use materials that may pose occupational and public health risks, such as heavy metal compounds of cadmium, gallium, and indium, some of which can be toxic both in compounds and in elemental form. While semiconductor manufacturing poses a number of environmental, health, and safety risks to workers, most can be managed with best practices. However, this notion of best practices comes with a serious caveat owing to the uneven geography of labor laws and environmental regulation, in addition to the occasional unscrupulous actor. Working conditions and the environmental impacts of industrial production can be notably worse in some developing economies, where regulation can be weak, lacking enforcement, or absent altogether; although the notion that some countries have become industrial pollution havens is somewhat controversial.[31]

Several waste disposal incidents in the solar industry, described in the following paragraphs, reinforce the point that poor waste management practices are the primary environmental issue in photovoltaics manufacturing. Energy and water use also have major impacts, according to research discussed in later chapters, but the direct impacts on workers and nearby communities that manufacturers control are from accidents, exposure to hazards, fugitive and permitted emissions, or sloppy chemical stewardship. Anticipating where similar accidents or poor practices might be repeated in photovoltaic commodity chains is important as they develop and are scaled up. Since creating green jobs in clean energy industries is often proposed as an antidote to urban poverty, underdevelopment, and persistent unemployment, there are arguably moral obligations to ensure that occupational, environmental, and energy injustices are not reproduced in low-carbon futures. Importantly, it is not the intrinsic issue of scale itself, as in the size and magnitude of the industry, but the speed of the scaling that makes mistakes more likely, invites policy design that fosters unintended outcomes, or increases the likelihood of mishaps a mature industry would not allow.

While the photovoltaics sector is like the electronics and semiconductor industries in some ways, even overlapping in many instances, in other ways they are very different. Photovoltaics are made to outlast 20-to-30-year warranties. This is very different from telecom and computing devices, which have shorter life spans and are already contributing to electronic waste (e-waste) flows. Electronics manufacturing generally uses a wider variety of hazardous materials than photovoltaic manufacturing. The pervasive use of plastics in electronics, particularly circuit board materials where metals are embedded in plastics, also makes them more toxic than photovoltaics. Yet, there are enough overlaps in production, techniques and modes of organization, and ideas about environmental management to warrant drawing lessons from the more established electronics industry to understand the risks to workers employed in a rapidly scaling solar industry.

Photovoltaics manufacturing relies heavily on the chemical industry for inputs and materials. Historically, environmental organizations have been critical of waste and improper or illegal chemical handling or disposal by chemical industries and manufacturers. But because photovoltaic modules generate renewable energy, they are not imagined as part of the same industrial complex encountered elsewhere in environmental studies, such as Bhopal, Love Canal, Times Beach, or Silicon Valley. The lack of any significant disaster so far and the environmental benefits of photovoltaics warrant a positive framing. But as the photovoltaics industry grows, so will the scale of chemical use and the need for chemical stewardship. Early in the industry there were a handful of very hazardous chemicals in use, including arsine, diborane, hydrogen selenide, and phosphine.[32] But these chemicals were phased out over time and replaced with chemical inputs that pose fewer hazards.

Crystalline silicon photovoltaic manufacturing in East Asia grew dramatically in 2008 through 2018 due to business model innovations and other competitive advantages that enabled lower production costs. Figures 1 and 2 show the 2016 market share for photovoltaic modules and cells and tell the story of Chinese and Taiwanese dominance of the industry over the past decade. National and local policy efforts in China, Malaysia, Vietnam, the Philippines, and Taiwan attracted photovoltaic manufacturers with enticements including land, discounted electrical or water service, subsidized energy and water, and other typical local grants and tax breaks, which allowed manufacturers to sell their modules at very low prices. These overseas crystalline silicon manufacturers were able to outpace the cost reductions promised for thin-films in the U.S., with the exception of the very successful First Solar. And even First Solar's competitive advantage comes in part from offshoring its thin-film production. This growth shifted the industry's workforce from places with strong worker protections like the U.S., Germany, and Japan, where most photovoltaics were manufactured until 2008, to Asia, where labor practices in electronics and semiconductor manufacturing have been questioned.[33] The reality is that photovoltaic modules contain parts from all over the world. So even a module made in China may have polysilicon made in Tennessee or South Korea. Nonetheless, given the larger trend in electronics and chemical manufacturing, we should understand where these supply chains reach and how they impact local labor and communities so that policymakers can prepare for just and sustainable solar energy transitions.

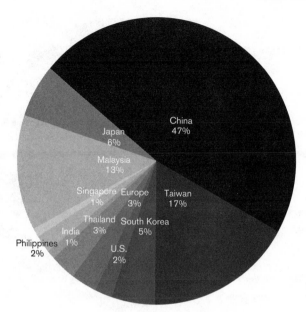

FIGURE 1. Crystalline silicon and thin-film photovoltaic cell manufacturing capacity, 2016. Photovoltaic cells are later assembled into photovoltaic modules.

A number of idiosyncratic stories about the solar industry have emerged that led to several media accounts of environmental pollution from manufacturing. In 2008, the *Washington Post* published a dystopian narrative of the production of polysilicon, the key feedstock used to make crystalline silicon. In China's Sichuan Province, trucks were dumping silicon tetrachloride waste in nearby farmers' fields, before returning to the gated compound of a polysilicon manufacturing facility.

> About nine months ago, residents of Li's village, which begins about 50 yards from the plant, noticed that their crops were wilting under a dusting of white powder. Sometimes, there was a hazy cloud up to three feet high near the dumping site; one person tending crops there fainted, several villagers said. Small rocks began to accumulate in kettles used for boiling faucet water. Each night, villagers said, the factory's chimneys released a loud whoosh of acrid air that stung their eyes and made it hard to breath [*sic*]. "It's poison air. Sometimes it gets so bad you can't sit outside. You have to close all the doors and windows," said Qiao Shi Peng, 28, a truck driver who said he worries about his 1-year-old son's health. The villagers said most obvious evidence of the pollution is the dumping, up to 10 times a day, of the liquid waste into what was formerly a grassy field. Eventually, the whole area turned white, like snow.[34]

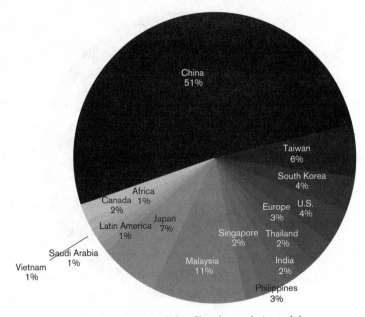

FIGURE 2. Crystalline silicon and thin-film photovoltaic module manufacturing capacity, 2016.

The growth in polysilicon production in China came during a time when polysilicon prices were rising as demand for the key feedstock outstripped supply after a surge in demand for photovoltaics in Germany, Spain, and Italy. When the photovoltaic industry produced fewer modules, crystalline silicon cell manufacturers had acquired silicon from microchip-makers, which rejected wafers that did not meet the computer industry's purity requirements. Microchips require silicon pure to 99.99999999%, or "8N" silicon, and less pure silicon ignot discards would suffice for converting photons into electricity. But by the mid-2000s, the boom in photovoltaics outstripped this supply. This prompted the construction of many new polysilicon refineries in China and elsewhere, and many of these facilities did not take common-sense precautions. The *Washington Post* investigative piece framed the silicon tetrachloride waste dumping incident as a story of industrial malfeasance by an irresponsible company that was taking advantage of high prices. Most mature manufacturers would invest in and install the proper processing equipment to convert silicon tetrachloride waste into trichlorosilane feedstock to produce more polysilicon. Figure 3 shows a worker loading a polysilicon chunk to be made into monocrystalline silicon.

FIGURE 3. The Czochralski method for purifying silicon is still the basis for mono-crystalline silicon photovoltaics today. (Photo: SolarWorld Industries Americas, Inc.)

At the time, few countries had stringent rules covering the storage and disposal of silicon tetrachloride waste, and China was no exception, as *Washington Post* reporters discovered in their investigation. In March 2008 they profiled a Chinese polysilicon facility owned by Luoyang Zhonggui High-Technology Company, near the Yellow River in Henan Province.[35] This facility supplied polysilicon to Suntech Power Holdings, at the time the world's largest solar cell manufacturer, and several other high-profile photovoltaics manufacturers.[36] The reporters found evidence that the company was dumping silicon tetrachloride waste on the community's agricultural fields. Silicon tetrachloride can be recycled into the chlorosilane feedstock used to make polysilicon. But instead of investing in equipment that could reprocess it, the company chose to dispose of the waste in fields near the manufacturing facility, inflaming the eyes and throats of nearby residents. Community members told the reporters that the land was now useless for growing crops. The article suggested that the company was not alone in this practice.

The story about Luoyang Zhonggui and silicon tetrachloride did not go completely unnoticed. After the publication of the *Washington Post* story, solar companies' stock prices fell. Investors began issuing research notes to investors on the negative media coverage of photovoltaics. Green attributes are the primary attraction for photovoltaic industry investors, and questions about the greenness of the solar industry could limit market penetration. Stocks in the solar industry were widely held by investors in the socially responsible investment (SRI) community. These investors use environmental or social responsibility criteria to screen investments and create SRI funds. When SRI funds divest from certain solar holdings, it can have severe financial implications. Investors feared that the revelations would undermine an industry that relies so much on its green credentials. The green attributes attract most customers and draw public support for policies that foster solar power adoption.

Greenpeace and the Chinese Renewable Energy Industries Association called attention to the problem of silicon tetrachloride pollution in a report claiming that in 2010, two-thirds of polysilicon manufacturers failed to meet national standards for environmental emissions and energy efficiency.[37] As China scaled up manufacturing of photovoltaics, the environmental costs were amplified by lax enforcement and poor regulatory oversight.[38]

To protect the industry's reputation, photovoltaic module manufacturers began to inquire about the environmental practices of their own polysilicon suppliers. In 2011, China set standards for polysilicon requiring that companies recycle at least 98.5% of their silicon tetrachloride waste back into trichlorosilane feedstock to make more polysilicon.[39] These standards are relatively easy to meet so long as factories invest in the proper equipment. Consequently, the situation is improving, and there have been no negative media stories about dumping of silicon tetrachloride waste since 2011. One industry expert suggested that the rules were put in place to drive consolidation in the country's polysilicon industry, as many smaller companies would not have the capital to invest in new equipment and might shutter their factories instead. Indeed, the rules set off a wave of polysilicon facility closures starting in 2011 as several large-scale operations replaced many smaller ones.[40]

Two major environmental and social challenges confront the manufacture of photovoltaics. The first is that many photovoltaic technologies use hazardous materials and involve labor processes similar to those in the electronics and semiconductor industries. Hence, technological innovations may disproportionately impact workers and communities the same ways they do in similar industries. As the frontiers of photovoltaic innovation rely on even more exotic chemical combinations, metal compounds, and nanoparticles, regulation will need to stay ahead of the risks of emerging technologies, without undermining innovation.

Second, questions about safe and healthy working environments are elevated by the rise of key institutional innovations in electronics: contract manufacturing and offshoring of production. A critical supply chain innovation is the growing use of contract manufacturing, in parallel with what has occurred in electronics production.[41] The model is increasingly used in the solar energy industry, with several manufacturers, such as Jetion, Flextronics, and Foxconn, leading contract manufacturers. Successful electronics companies like Apple depend as much on changes in supply chain management and overseas contract manufacturing as they do on technological and scientific innovations. Companies collaborate with original equipment manufacturers (OEMs) to manufacture their products. Contract manufacturing can allow electronics manufacturers to scale rapidly at lower cost, as they do not take on the risk of owning manufacturing capacity. Contract manufacturing drives down the cost of solar by offering manufacturers flexibility and regional locations near major project sites or customers. Flextronics, a manufacturer based in Milpitas, California, contracted to produced SunPower modules for SunPower's 250 MW California Valley Solar Ranch, a utility-scale solar farm in San Luis Obispo County. Typically, SunPower would import modules from their factories in the Philippines for a project they develop.[42] In this case, contract manufacturing brought jobs from an industrializing country to one of the most expensive labor markets in the world, California's San Francisco Bay Area. So not all stories about contract manufacturing fit the received narrative of production moving toward lowest-common-denominator environmental and labor standards.

Environmental and social justice organizations contend that contract manufacturing and offshoring to areas lacking environmental and labor regulation tends to hurt marginalized workers and communities.[43] Contract manufacturing offers greater flexibility for companies, provides greater access to outside expertise, and does not require sunk costs in manufacturing equipment and factories.[44] But these structural arrangements sometimes rely on a migrant workforce with weaker environmental, health, and safety rules, less job security, lax labor laws, or prohibitions on collective bargaining. Electronics industry watchdogs have singled out companies for poor factory conditions, demanding increased monitoring of production conditions, particularly in factories in Asia.[45] Contract manufacturing is common in demand-responsive economies like China, whose government can selectively choose industries to expand by offering financial incentives such as subsidized land, water, power, and finance. The separation of the owners and designers from the producers of products in contract manufacturing arrangements impairs accountability for working conditions and environmental impacts, most notably in textile, electronics, and food supply chains, raising concerns about social responsibility. If the manufacturers do not own the factories, it becomes more problematic to connect factories to brand names, which makes it more difficult to organize against products—a thesis made popular by Naomi Klein in *No Logo*.[46] So long as the contract-manufacturing model is a profitable way of constructing commodities and some regions of the world lack strong environmental governance and worker rights protections, the anonymity of global supply chains will pose challenges for accountability.

A few years after the *Washington Post* story on silicon tetrachloride dumping, another incident in China, with crystalline silicon photovoltaic cell and module manufacturer Jinko Solar, grabbed international headlines. Crystalline silicon photovoltaic manufacturers rely on hydrofluoric acid to clean silicon wafers, remove sawing damage, and texture cell surfaces to better collect light. When hydrofluoric acid comes into contact with unprotected skin, the highly corrosive liquid can destroy tissue and decalcify bones. Handling hydrofluoric acid requires extreme care, and it must be treated and disposed of properly. The acid is also a very simple molecule and simple to treat and neutralize.

In August 2011, a factory owned by Jinko Solar in Haining, Zhejiang Province, spilled hydrofluoric acid into the nearby Mujiaqiao River.[47] The company was in dispute with its waste hauler, and the drums it stored hydrofluoric acid in were washed into the river after a major rainstorm. The spill killed hundreds of fish, led to the death of livestock that were washed in river water, and led villagers to riot against the factory. Even though the manufacturer was a large producer overall, it had limited experience in manufacturing and chemical stewardship. The company was founded only five years prior and quickly became one of the largest crystalline silicon manufacturers in the world.

Farmers working adjacent land used the contaminated water to clean their animals, accidently killing dozens of pigs. In investigating the death of the livestock, Chinese authorities found levels of hydrofluoric acid in the river ten times the permitted limit, and they presumably took these measurements long after much of the hydrofluoric acid had washed downstream. Hundreds of local residents, upset over the incident, stormed and temporarily occupied the manufacturing facility. Again, investors retaliated: when major media outlets such as the BBC carried the news the next day, Jinko's stock price dropped by more than 40%, translating into nearly US$ 100 million in lost value.[48] These two incidents may have been isolated, but they had financial implications as well as stirring interest in greening photovoltaic manufacturing.

THE SILICON VALLEY TOXICS COALITION'S GREEN JOBS PLATFORM FOR SOLAR

A large contingent of technology pundits picked thin-film photovoltaics as the "best bet" semiconductors that would reinvent Silicon Valley and encourage clean-tech clusters of firms, allowing the U.S. to lead in the clean energy race. Because thin-films rely on proprietary processing technologies, patents and other forms of intellectual property protection can be invoked for them, unlike crystalline silicon, where such protections are less useful. Because of its experiences with the electronics and semiconductor industries, an environmental justice organization, the Silicon Valley Toxics Coalition (SVTC), took notice. Most environmental groups and even environmental justice organizations were promoting photovoltaics, not investigating their impacts, and SVTC filled this niche as the industry watchdog. The organization worried that despite the benefits of solar energy, environmental inequality could be reproduced if the industries and jobs introduced on behalf of green transitions present high levels of occupational injury or community exposure to pollution.

After several media accounts of pollution in China and as an extension of its work on e-waste, SVTC announced a campaign to green the solar industry. Based in San Jose, California, SVTC has a deep history of working on issues related to semiconductor and electronics production and pollution issues in Silicon Valley, also known as the Santa Clara Valley. SVTC is most widely known in the environmental justice community for its grass-roots activism on groundwater pollution and the community's right to know in Silicon Valley in the 1970s and 1980s. One hidden legacy of Silicon Valley's acclaimed semiconductor industry is the largest concentration of Superfund hazardous waste cleanup sites in the U.S. SVTC worked for many years on these challenges to hold companies accountable for pollution. As semiconductor and electronics companies offshored or dissolved, SVTC shifted to other emerging issues, such as household hazardous waste and e-waste. By the late

TABLE 3

Manufacturer	Founded	Technology	Location	Status
Bloo Solar	2008	CdTe	El Dorado Hills	Bankruptcy 2017
Maisolé	2001	CIGS	Santa Clara	Acquired by Hanergy 2013
Nanosolar	2002	CIGS	San Jose	Assets auctioned off 2013
NuvoSun	2008	CIGS	Milpitas	Acquired by Dow Chemical 2013
OptiSolar	2004	Amorphous silicon	Hayward	Acquired by First Solar 2008
Solextant	2006	CdTe/CIGS	San Jose	Acquired by Wakanda 2011
Solyndra	2005	CIGS	Fremont	Bankruptcy 2011
SoloPower	2005	CIGS	San Jose	Reorganization 2013; Acquired by BASF 2017
Stion	2006	CIGS	San Jose	Assets auctioned off 2013

2000s, SVTC began a campaign to identify the risks and hazards to workers and communities from the different photovoltaic industries and technologies that were coming to Silicon Valley. Acting early would allow the group to get ahead of any pollution issues before they affected workers and communities.

Around this time, thin-film photovoltaic manufacturers—SoloPower, Nanosolar, Solyndra, Miasolé, and OptiSolar, to name a few listed in Table 3—were flocking to build campuses across Silicon Valley. Much enthusiasm surrounded these new manufacturers in the local press. The blogosphere and even the scientific journals *Nature* and *Science* proclaimed a new dawn for the semiconductor industry in Silicon Valley based on thin-film photovoltaics.[49] Under the leadership of Executive Director Sheila Davis, SVTC began in earnest to identify technologies and materials used in their manufacture. Lists of major semiconductor technologies, processes used to apply thin films, material inventories, company names, and information about venture capital or private equity moving toward these companies were tracked as the industry evolved. SVTC's Green Jobs Platform for Solar was launched in January 2009 with the aim of publicizing the issues to shape the industry's trajectory by engaging in dialogue about best practices, chemicals of concern, and safe end-of-life disposal.[50]

GREEN JOBS PLATFORM FOR SOLAR: PRINCIPLES

1. The workers' activity and the products they produce must contribute to improving the quality of the environment.
2. Workers are paid a living wage. Everyone has the right to a standard of living adequate to support the health and well-being of himself/herself and of his/her family, including food, clothing, housing and medical care, child care, adequate transportation and utility costs.
3. Adequate health benefits are provided to workers and their families at an affordable cost.
4. Workers have an opportunity for job advancement, and, wherever possible, share in the wealth of their company.
5. Worker rights are protected, including whistleblower rights and the right to organize. Workers have the right to form and join unions for the protection of their interests without interference, intimidation, threats or harassment from the employer. Workers are not discriminated against based on race, religion, gender, or sexual orientation.
6. Reduce toxic exposure to workers by
   a. Phasing out chemicals currently used in products and production that are or are suspected of being hazardous to human health and the environment.
   b. Protecting workers by reducing exposure levels in accordance with the principles of the hierarchy of controls—(i) constantly striving for safer alternatives, while using (ii) engineering controls to keep exposure levels as low as possible and (iii) using personal protective equipment only as a "last resort" temporary stop-gap measure.
   c. Using the green chemistry principles and the precautionary approach to develop new products and manufacturing processes.[51]
   d. Providing workers with useful, meaningful, and material information in appropriate languages related to worker injuries and illnesses and other job hazards, and recognizing that workers have the right to know and to act when they are handling or being exposed to toxic materials.
7. The company's goal should be to produce products that do not contain chemicals that are hazardous to human health and/or the environment.

8. The company is a good corporate citizen and invests in the community in which it is located. The company hires locally, pays fair share of taxes, and contributes its earnings to support sustainable social and physical infrastructure development such as roads, schools, transportation, housing, waste and recycling systems.

9. The company abides by the strictest environmental, labor, and health and safety laws of the country in which they are located striving at all times to protect its workforce from toxics to the same extent that the company strives to protect the environment from any adverse consequences from its operations. Green jobs protect the environment and local communities with an ethical code that reflects the laws of the country with the strictest environmental rules and not the most lenient. The company will not only uphold the laws, but will also openly support sustainability and environmental health standards for the solar industry. This includes not lobbying against the Green Job Principles and self-disclosing any lobbying efforts.

10. Environmental health and safety standards are shared throughout the global product supply chain. Information on health hazards and chemicals used in the workplace, tests to measure chemical noise and radiation levels on employees, precautions employees should take and procedures to be followed if employees are involved in an incident or are exposed to hazardous chemicals or other toxic substances should be made available and implemented by the company at all stages of the product's lifecycle.[52]

11. The environmental burden created by the product, the disposal and recycling of the product, or the company that makes the product should not disproportionately impact people of color, women, poor communities, or developing nations. All communities should be ensured equal protection under the law. The product's design should minimize waste, thus minimizing environmental burdens on communities that dismantle the product.

Numerous environmental organizations signed on to the platform, including the Apollo Alliance, Asian Communities for Reproductive Justice, Basel Action Network, Bayview Hunters Point Community Advocates, Center for Environmental Health, Center on Race, Poverty & the Environment, Clean Production Action, Clean Water Action, Communi-

TABLE 4

PV type	Chemical hazards
Crystalline silicon (c-Si)	Silicon tetrachloride waste, lead in solder and metallization pastes, strong acids (HF, HCl), caustics (NaOH), solvents, dopants, pyrophoric gases (silane)
Amorphous silicon (a-Si)	Pyrophoric gases (silane), solvents, indium tin oxide
Cadmium Telluride (CdTe)	Cadmium compounds, solvents
Copper indium gallium selenide (CIGS)	Cadmium, selenium, and indium compounds
Gallium arsenide (GaAs) crystalline	Arsenic compounds, phosphine gas, trichloroethylene
Polymer/organic	Ruthenium, indium compounds, nanoparticles
Dye-sensitized	Indium compounds, nanoparticles, ruthenium

SOURCE: Selected list compiled by the author from various sources, mainly compiled by the National Photovoltaics Environmental Research Center.

ties for a Better Environment, Electronics TakeBack Coalition, Environment California, Environmental Health Coalition, Friends of the Earth, Green For All, Just Transition Alliance, Physicians for Social Responsibility, Science & Environmental Health Network, United Steelworkers Local 675, and Worksafe. Notably missing were major environmental groups such as the Sierra Club, Natural Resources Defense Council, and Environmental Defense Fund, all of which have programmatic work on e-waste and chemical stewardship issues.

Later that year SVTC released a white paper I helped prepare called *A Just and Sustainable Solar Energy Industry*.[53] The report identified numerous environmental, health, and safety risks from several different solar technologies and was widely reported on in the media. The paper led several SRI funds to reach out to SVTC, most notably Henderson Global Investors and Boston Common, two firms invested in the solar and renewables sectors. Table 4 lists the main chemicals with environmental, health, and safety issues related to selected photovoltaic semiconductor materials.

SVTC's Green Jobs Platform for Solar captures the essence of the "just sustainability paradigm," a meta-concept developed by environmental justice scholars Julian Agyeman, Robert Bullard, and Bob Evans, which emphasizes "the need to ensure a better quality of life for all, now and into the future, in a just and equitable manner, while living within the limits of supporting ecosystems."[54] With "just sustainabilities" questions of inequality remain central, but tensions between justice and sustainability do not pose an impasse to further exploration of concepts of sustainable development. One can explore just sustainabilities without remorse over whether deepening conversations about sustainability reproduce systems of inequality because they justify the status quo for some future state of production. This epistemological barrier to blending themes of environmental justice and sustainability discourses has long plagued environmental studies, and so far efforts to bring these two worlds of environmental studies together remain somewhat tangent to the mainstream climate change and clean energy movements, with the exception of groups like SVTC and other groups for justice in energy transitions.

A POLITICAL-ECOLOGY APPROACH TO UNDERSTANDING SOLAR POWER

Political ecology is a field of research into nature–society relationships and geographically of uneven forms of environmental change and governance. The research spans several fields, including geography, sociology, political science, and anthropology, but also includes some ecologists and physical scientists. The field of inquiry aims to uncover stories that connect the products that comprise everyday life to where they come from and unveil the hidden dynamics that drive social and environmental problems.[55] Commodity chain analysis is one tool political ecologists use to understand material and social conditions across the global systems that link production and consumption.[56] This approach employs case-study approaches and typically involves ethnography, participant observation, semi-structured interviews, or survey research to understand the impacts of natural resource extraction through the production of commodities, and sometimes through disposal.[57] These studies tend to focus on conflicts over natural resource access and control with marginalized communities or social movements and how these issues are connected through political-economic issues.[58] Global commodity chain dynamics can reveal how power operates at a distance, connecting places across disparate global spaces by sometimes seemingly unrelated institutions, markets, social norms, and consumer subjectivities, behaviors, and preferences.

The phrase "commodity fetish" describes how things produced through economic exchange are connected through social relations that are hidden from view.[59] Investigations to unveil the environmental and social impacts of commodity production are becoming increasingly popular, particularly with the rise of alternative systems that attempt to unmask the relations between consumers to producers. Deeper investigation into the objects and devices used and purchased by consumers every day—or the energy they take for granted but that makes their day possible—is revealing unexpected connections to environmental degradation and labor exploitation. Interest in green consumption, whereby consumers seek out alternative products with purported environmental attributes, like organic strawberries, bird-friendly coffee, fair-trade bananas, and in this case solar power, has taken hold in reaction to a desire to understand the social and environmental relationship to production systems.

Unmasking the commodity fetish requires exploring the social lives and materiality of commodities and may result in designing new ways of making these connections more transparent.[60] The human subjectivities that animate these objects with social life—such as how solar panels will assist with climate action—need to be understood against the natural resources and labor base that make the objects possible. Commodity chain analysis is widely used in political ecology, economic and human geography, and economic and rural sociology to understand systems of production and how they might lead to or sustain "world-economic spatial inequalities."[61] Commodity chain analysis provides a framework to investigate complex and multifaceted economic sectors and industries. The framework allows researchers to explore socio-ecological relationships among labor, industry organizations, science, regulation, and culture.[62] Specialization is making supply chains more complex, and many distant lands link ecologies across spaces, but these distant spaces can be less well governed, with weaker environmental health and safety regulations.[63] Some speculate that "green products" and "ethical consumption" can reproduce homogeneous categories that warrant further critical examination; ethical consumption itself can fall prey to the commodity fetish without critical consumer engagement.[64]

This book argues that photovoltaics as ethical, green products are not subject to enough critical examination; consumers, the public, and even environmentally conscious minds may reproduce a commodity fetish with photovoltaics that masks socio-environmental relations while crowning them with a "green halo." Social and environmental problems are often influenced by power asymmetries between different stakeholders, and this commodity fetish can further obscure these relations. Outcomes such as social vulnerability and environmental degradation can be produced through places interconnected by economic transactions, and these relations are shaped by how power operates across space. While communities and landscapes are connected through commodity chains, some are positioned in distinct ways from these flows. Some are more in control of the flows, while others are controlled by the flows. These power asymmetries are important expressions of power to study to better understand energy justice for communities and workers.[65]

The commodity chain as a social science concept is "a blueprint appropriate for multi-sited research."[66] Production–consumption linkages can reach across great distances as commodities flow through multiple places. As production networks globalize, environmental impacts are increasingly offshored and hidden from view.[67] The commodity chain analytic aims to reconstruct these interconnections by envisioning that the "global is collapsed into and made an integral part of parallel, related local situations."[68] As raw materials are made into components and products, they move across different regulatory and legal boundaries and jurisdictions, and between classes, cultures, and social norms. This multi-sited approach contains "de facto comparative dimensions" for examining questions of power and justice.[69] This permits the researcher to juxtapose sites of production across the commodity chain, such as portraying the differences in occupational safety between working in mines for materials in some device and jobs in sales or retail of the same product. The structure and dynamics of commodity chains help illustrate the causes of environmental change by recreating production linkages of "human activities and bio-physical processes."[70] This opens up opportunities for researchers interested in production systems to "to examine the circulation of cultural meanings, objects, and identities in diffuse space time."[71] This approach of looking for differences along the commodity chain is helpful for exploring emerging trends that are claimed to be implicated in social and environmental injustice, such as contract manufacturing and migration patterns in global factories and supply chains in demand-responsive economies (for example, countries such as China and Malaysia, which can quickly mobilize manufacturing resources).[72]

Several variants of the commodity chain approach coincide along different scholarly and disciplinary research agendas, including world-systems theory, economic geography, rural sociology, and political ecology.[73] They all have different matters of emphasis, but they share an interest in explaining linkages between consumption and production. This work of unveiling the productive forces and materials that constitute commodities can reveal sites of injustice and enliven our understanding of the politics of consumption and the possibilities for social change.[74] Commodity production touches down in local communities and impacts human bodies and ecosystems in different ways. Take for example the U.S. system of delivering fuel for transportation. Tracing this commodity chain can reveal how disproportionate harms to communities living near oil wells or pipeline infrastructure in Nigeria are connected to refineries and transfer railyards in Richmond, California, and eventually neighborhoods with wells possibly contaminated by gasoline, near stations where consumers purchase fuel. The U.S. electricity system, too, is delivered by commodity chains with differential if not contradictory experiences, extending from the coalfields of Appalachia, from eastern Kentucky to West Virginia, where mountains are carved into valleys by mountaintop removal mining, to the front ranges of the Bakken shale oil fields in North Dakota, where prostitution, crime, and methamphetamines have become widespread in so-called "man-camps."[75] These are symptoms of the boom-and-bust cycle in oil and gas production. A commodity chain approach can reveal how these seemingly interdependent socio-environmental changes are interlinked and why these activities occur or are sustained. For example, research might illustrate how community benefits from increased economic activity, employment, and other side effects of resource development are dependent on global economic conditions or events.

Sometimes commodity chain analyses reveal generalizable patterns, while other times they reveal idiosyncrasies or power asymmetries at work. An energy justice perspective asks why some people and communities are more exposed to pollution from energy production than others.[76] The approach uncovers environmental justice issues that can be missed when research involves case studies of single factories because of how it examines linkages between different segments of production. For the study of energy transitions, this is helpful because policies promoting green jobs in one part of the world could be complicit in the reproduction of environmental inequality elsewhere. Some commodity chain research shows how the spatial composition of production can mask environmental inequality.[77]

While the global commodity chain concept is a widely mobilized analytic, it is worth remarking on the term "commodity chain" itself, because the chain metaphor is too linear and implies links that are difficult to break. In an era of globalization and expanded contract manufacturing, global supply chains are not made of solid links. They can be fluid and dynamic just-in-time production networks, changing with prices, speculation in markets, industrial planning in demand-responsive economies, cultural desires for novel products, and even the weather or natural disasters. The commodity chain metaphor also implies a set of strong material linkages, when often there are other, more ephemeral actors, ideas, and concepts that weave global production networks together. Nonetheless, I will continue to use the term except when a more meaningful synonym is available.

New technologies can pose novel and unintended risks. Scholars in science and technology studies help us understand that all technologies, no matter what they are intended to accomplish, have unintended consequences, because our understanding of how they will impact society is always partial.[78] Sometimes technologies can cause damage that does not manifest until after many years of commercial production.[79] Due to the lack of certainty about outcomes, policymakers increasingly treat emerging technologies with the precautionary principle when the effects could be severe or irreversible.[80] Tools such as risk assessment, alternatives assessment, and life cycle assessment are also used to evaluate the impacts of technologies, but these too are only incomplete narratives. Sustainability concepts are increasingly incorporated into industrial practice with the mainstreaming of policy and practice around green chemistry, environmental design, and extended producer responsibility. Social justice considerations such as providing a safe work environment and fair wages, avoiding forced labor, and allowing freedom of association are also gathering momentum in many places in the world.

This research used mixed methods to collect data, drawing from approaches in political ecology and environmental studies. Political-ecology approaches seek to understand environmental change from a vantage point that privileges questions of justice and the ultimate causes of environmental degradation. These cases tend to be idiosyncratic rather than nomothetic, which works well when looking for scenarios to illustrate concerning matters: things that might go wrong, but not necessarily.

Commodity chain analysis starts with a general sketch of the production system, connecting the suppliers of inputs needed to make the product, and follows and describes the socio-ecological processes that are encountered and connected through them.[81] Next, frictions or unexpected connections in these commodity chains are highlighted or juxtaposed as an entry point into a case that warrants deeper examination or helps tell some story about a product or production practices. Fully piecing together commodity chain stories can be complicated by supply chain complexity, anonymity, or proprietary information. Studies of high-tech industry often face challenges with access to research subjects along the commodity chain. Sites vary in "quality and accessibility" to research subjects, sites, and other considerations, so often commodity chain analyses are partial stories.[82] Data access is difficult where companies must be secretive for competition, legal, or regulatory reasons.[83] Semiconductor manufacturing facilities contain highly guarded trade secrets, ranging from the types of machines and equipment to the composition of chemical inputs and energy use. Accusations of industrial espionage occur frequently in this space when companies buy machines and make their own copies, ignoring patent law or intellectual property regimes. Health records are informative for energy transition planning, but not available in this research for reasons of employee privacy. Data used as proxies for environmental health and safety issues include descriptions of best practices, information from company reports, regulatory emissions and exposure standards, case studies, and other sources such as patents, chemical inventories, manufacturing process diagrams, and overviews of toxicities and routes of worker exposure to hazardous materials.

Despite all these challenges to studying industry, many photovoltaic manufacturers invited me to visit their facilities in the U.S., Asia, and Europe. I made several trips in 2008 and 2009 as a technical advisor to SVTC and had annual correspondence with the environment, health, and safety officer or the public affairs person for ten to fifteen major photovoltaic manufacturers for SVTC's Solar Scorecard. Several other site visits occurred during my time as a postdoctoral scholar at the University of California, Berkeley, from 2009 to 2011, and later as an assistant professor at San Jose State University, starting in fall 2011. I conducted interviews and had many personal communications with numerous players in the photovoltaic industry, including First Solar, SolarWorld, Solyndra, Yingli Solar, Jinko Solar, SunPower, Suntech Power, Sharp, MiaSolé, SoloPower, REC Group, Abound Solar, BP Solar, LDK Solar, Solon, Calyxo, GE-PrimeStar, Avancis, Q-Cells (a German cell manufacturer from 1999 through 2015), and Hanwha Q-Cells (a South Korean module manufacturer that bought Q-Cells in 2015).

Primary qualitative data I collected included over 150 transcribed in-depth and semi-structured interviews from 2009 to 2014. Key informants included the NGO communities working on energy and climate issues, wildlife and wilderness advocacy, and solar and renewable energy promotion, as well as people working in socially responsible finance, the venture capital sector, investment banks, government officials and reg-ulators, ecologists and wildlife biologists, semiconductor manufacturers, project developers, life cycle analysis researchers, union representatives, and the real estate sector. These semi-structured interviews covered topics including chemical use and stewardship in facilities, solar energy project siting, and recycling management strategies. I started with an initial list of people to interview that represented the categories of informants listed above. These interviews generated new informants using snowball sam-pling. I interviewed photovoltaic companies receiving Department of Energy loan guarantees as well as other investors in the solar energy space. Not all companies were willing to be interviewed on the record, so I used participant observation at industry association meetings, trade shows, public hearings, and workshops to collect additional data. To clarify minor points and gather information, I used personal communications and email exchanges.

In addition to primary data, secondary sources included patents, gov-ernment reports and transcripts of hearings, industry and venture capi-tal newsletters, newspapers, blogs, and podcasts. Several news outlets and journalists engaged in tremendous coverage of solar issues and pro-vided some data and information that would otherwise not be available. The *New York Times, Washington Post,* and *Forbes* all dedicated sig-nificant space to coverage of solar manufacturers and utility-scale solar energy projects discussed throughout this book. Online blogs, podcasts, and other electronic media also have in-depth clean-tech journalism, serving the venture capital and clean-tech investment crowds. I unearthed some information from congressional and Government Accountability Office investigations. Electronic media were sources of detailed informa-tion on financial transactions, production projections, technologies, key suppliers, contracts between companies and suppliers, and patterns of ownership, including analyses from contributors to *Gunther Portfolio, Greentech Media, Climateer Investing, Solar Curator,* the *Energy Col-lective, CleanTechnica,* the *Interchange,* and *Seeking Alpha.* I used many of their interviews with executives of multinational and venture capital–financed photovoltaic firms, as they candidly air their motivations and frame the debate about incentives and policies. The subjects of this study—solar energy, green technology, and innovation—are relatively overrepresented on media such as blogs, listservs, and podcasts, which were invaluable sources of data. I had personal off-the-record commu-nications or anonymous interviews with people in the clean-tech invest-ment and financial analyst community, which helped direct my attention to emerging issues. I also attended various talks around the San Francisco Bay Area and Silicon Valley, including meetings held at the Palo Alto Research Center, the Northern California Solar Association, the University of California (Berkeley), Stanford University, the Electric Power Research Institute, and several annual solar trade shows, including PV America, Solar Power International, and InterSolar North America.

I collected data from technical papers, manufacturing-site visits, and patents to understand the various manufacturing processes and chemical inputs used to make different photovoltaic technologies. Academic publications, journals, and conference proceedings on the environmental, health, and safety impacts of semiconductor manufacturing I read provided context for interpreting the environmental justice impacts of solar development and manufacturing. Brookhaven National Laboratory maintains an invaluable repository of environmental, health, and safety and life cycle assessment publications on photovoltaics dating back to the late 1970s. These research resources are also forums where data to evaluate the performance and life cycle impacts of particular emerging clean technologies are verified and debated. I reviewed and catalogued over 150 life cycle assessments of solar power technologies to understand where the most significant impacts occur and their extent.

Beginning in 2008, I visited many areas for proposed utility-scale solar projects across the American West. I went to both private and public lands under development, before construction. Because of private property restrictions, I observed projects on private lands only from the site border. The first project site I visited was in the Panoche Valley in San Benito County, California, while the first proposed project I visited on public lands (in the Mojave Desert) was the Calico Solar Energy Project in San Bernardino County, California, in spring 2009. The first operating utility-scale solar energy project I went to was the 25 MW Blythe Photovoltaic Power Plant, built by First Solar. At the time it was the largest photovoltaic farm in California. In all, I travelled to over fifty proposed solar project sites. This includes over thirty solar power projects that were either selected for fast-track status or had an active right-of-way application. I went to twenty-four Bureau of Land Management fast-track and priority-project Solar Energy Zones, including several proposed and later withdrawn, from 2008 through 2014. I would make site visits sometimes with an interviewee, or with a group, and other times alone. The work I completed on-site (in addition to just experiencing the place) involved photographing landscapes, interviewing research subjects, or walking the site with developers or environmental organizations. To research utility-scale solar power plant project development, I conducted participant observation at public comment meetings, semi-structured interviews, and reviewed public comments submitted during environmental review processes. I also collected data from the spoken and written public comments from individual projects' environmental impact statements and the broader solar energy Programmatic Environmental Impact Statement process, as well as regional initiatives such as the Desert Renewable Energy Conservation Plan. Public meetings offered opportunities to ask clarifying questions of officials not able to go on record in formal interviews. For example, I was not able to conduct formal interviews (on the record) of U.S. Bureau of Land Management, Fish and Wildlife Service, or Department of the Treasury staff, or at any other government agency for that matter, because of the controversial nature of the topics covered. I used agency official statements, press releases, or comments and quotes to journalists for these viewpoints.

I collected these data from all these different sources to piece together solar power commodity chains to help us understand the social and environmental dimensions of scaling up solar power. This book does not try to weigh in on the magnitude of future contributions of solar relative to other energy sources. Whether it provides 100%, 80%, 30%, or 1% of our electricity, solar power is no longer alternative energy, and there are social and environmental considerations that deserve attention and planning. I did not choose these specific case studies of policies, companies, and projects to imply generalizations or inevitable characteristics of solar power manufacturing or project siting at scale. Rather, I chose cases to offer real-world examples of lessons learned and best practices so that communities, policymakers, and practitioners can better design institutions and incorporate ideas that mitigate social and environmental burdens in an emerging green jobs economy based on solar power.

CHAPTER 2

Green New Deal

"This is our generation's Sputnik moment. Two years ago, I said that we needed to reach a level of research and development we haven't seen since the height of the space race. In a few weeks, I will be sending a budget to Congress that helps us meet that goal. We'll invest in biomedical research, information technology, and especially clean-energy technology—an investment that will strengthen our security, protect our planet and create countless new jobs for our people."

—U.S. President Barack Obama, State of the Union address, January 25, 2011

OUR GENERATION'S SPUTNIK MOMENT

On October 4, 1957, the Soviet Union (USSR) launched the Sputnik satellite, causing great angst in U.S. politicians, military, and the general public. The apprehensive reaction was led by national defense experts, scientists, and others aiming to convince the public of the critical need to spur public investment in rocket and space technologies to catch the USSR, which they feared would soon be able to launch intercontinental missiles capped with hydrogen bombs. These "space race" headlines ultimately helped spur the public support necessary for the United States to catch the Soviets: the widely supported, taxpayer-funded investments in research, development, and demonstration of space technology, culminating in trips to the moon, a successful space shuttle cargo program, and forty years of building Cold War nuclear stockpiles.

In his 2011 State of the Union speech, U.S. president Barack Obama argued that the time had come to seize another "Sputnik moment" and invest in clean energy technology. He emphasized that the U.S. risked falling behind Europe and China in the clean energy race, pointing to the proliferation of photovoltaic manufacturing facility construction in China and Germany. He argued that large-scale strategic investments in clean tech, biomedicine, and information technologies were a geopolitical imperative that required bipartisan collaboration and public support. Sputnik was also invoked to help justify the American Recovery and Reinvestment Act (ARRA) of 2009. To win the clean energy race, policymakers would develop land-use, electricity-market, financial, and tax-equity policies to help construct the first major utility-scale solar projects and build out several thin-film manufacturing facilities.

The ARRA investments echoed Roosevelt's New Deal emphasis on a role the state can play, through public works, in investing in tomorrow, while employing people today. The *Green* New Deal bridges public works and green development. *New York Times* columnist Thomas Freidman argued that the U.S. needed such a project to create jobs, provide clean energy, and spur the breakthrough innovations that will lead to a more prosperous world: "We need a Green New Deal—one in which government's role is not funding projects, as in the original New Deal, but seeding basic research, providing loan guarantees where needed and setting standards, taxes and incentives."[1] Friedman simultaneously was writing about ARRA and the Waxman-Markey cap-and-trade legislation that was being floated early in the administration. Many viewed this initial ARRA infusion into clean tech as the start of something new, but also just a beginning. But ultimately the effort to price carbon through cap-and-trade failed to become law.

Starting in 2009, the U.S. Department of the Treasury invested $90 billion in grants and loans in numerous clean energy innovations, including photovoltaics, advanced batteries, electric vehicle manufacturing facilities, new wind farms, geothermal power stations, energy-efficiency devices, grid modernization projects, smart grid applications, integrated gasification clean-coal technology, and nuclear power plants.[2] Critics asserted that this was far too little to meet the ambitious commitments made earlier that year at the Copenhagen meeting of the Intergovernmental Panel on Climate Change, and only a fraction of the $700 billion stimulus bill. Green jobs and the clean energy race became common refrains in public speeches with backdrops of solar energy manufacturing and electricity generating stations, such as the Copper Mountain Solar Project near Las Vegas, the Solyndra factory in Fremont, California, and Fort Irwin in the Mojave Desert.[3] Despite the direct references to clean technology and climate change that headlined announcements of ARRA project milestones, at its core these programs were most committed to economic stimulus, because the global economy was in a state of a financial meltdown not seen since the Great Depression. ARRA spending was framed in the context of the great race to be the clean energy superpower, but in reality it had another aim: to keep the economy from sinking deeper into economic recession. Estimates are that ARRA raised the gross domestic product by two to three percent.[4] The timing was important because loans and grants became available just as projects seeking finance were being frozen out of capital markets after the collapse of Bear Stearns and Lehman Brothers late the prior year. These federal investments leveraged capital markets for additional investments.

The availability of ARRA finance for research, development, and deployment of clean technology coincided with the rise in popularity of thin-film photovoltaics. ARRA directed nearly $5 billion in loans and grants toward projects involving thin-film photovoltaics, with one utility-scale power plant as large as 550 megawatts (MW) and several more measuring in the hundreds of megawatts. The period from 2005 to 2009 saw unprecedented growth, media coverage, and enthusiastic interest from investors in thin-films. Investors from Wall Street in lower Manhattan to Sand Hill Road in Silicon Valley flocked to thin-films because some key attributes seemed to be ripe with opportunity: lower energy and material input requirements, relatively quick residence time at the factory, and the ability to patent or protect innovations with trade secrets. For fifty years, researchers in public and private research labs had explored many different semiconductor types and different means of depositing thin films of conductors and semiconductors onto substrates. But now investors, scientists, and energy futurists alike saw thin-film photovoltaics as an important technological innovation worthy of substantial attention and investment.[5]

The burgeoning clean-tech media, such as *Inside Renewable Energy* and *PV Tech,* published stories based on thin-film company announcements about new efficiency records or new tranches of venture capital triggered by company milestones, reinforcing this narrative of thin-film dominance over the photovoltaic industry. The most optimistic forecasts projected that it was only a matter of years until all kinds of surfaces would become thin-film photovoltaic solar cells—windows, flexible plastics, building facades and awnings. More importantly, thin-films would displace the industry workhorse, crystalline silicon photovoltaics. ARRA investments targeted the creative destruction of the incumbent silicon-based technology by insurgent thin-film photovoltaics.

Some projected that thin-films would dominate the market, with up to 90% market share by 2020. Others were more modest, at 50%. In 2017, the market share for crystalline silicon photovoltaics remained at 5%, with one manufacturer (First Solar) constituting the majority of thin-film production.[6]

Interest in thin-films ramped up when polysilicon feedstock hit record-high prices in 2007 and 2008. Proponents argued that one key attribute of thin-film photovoltaics is that they do not depend on inputs and feedstocks that fluctuate wildly in cost, like what happened when demand outstripped supply for polysilicon feedstock, needed to make crystalline silicon photovoltaics. Crystalline silicon manufacturers experienced a sharp spike in the cost of production until a new round of suppliers began to come online a year later, prompting companies like SolarWorld and Solon to explore recycling technologies. Thin-film proponents promised that the higher margins from lower-cost production would further insulate manufacturers from price shocks.

At the time ARRA passed, investors generally viewed solar energy as a risky proposition, an impression made worse by the economic recession. Investors are more comfortable investing in familiar facilities, such as natural gas power plants, because there is a well-established track record of building these projects on time and at the anticipated costs. Solar power plants did not have such a track record. Investments in fossil fuels can yield high returns from speculation, including price volatility, geopolitical conflict, or the discovery of large new reserves. Utility-scale solar energy projects require large amounts of upfront capital, and large power plants do not start selling electricity until three or four years after the projects are proposed, and they can sell only the amount of electricity contracted, so there is not the potential bonanza seen in other natural resource extraction or energy production.

This high risk and moderate (or at least predictable) reward translates to higher interest rates for solar developers and higher overall costs in power purchase agreements, which are passed on to electric utility ratepayers. In 2009, the largest utility-scale solar power plants in the world were on the order of 20 MW; a decade later the largest exceed 600 MW.[7] Renewable energy companies seeking to cross the "valley of death"—the deep debt burden that accumulates as a new company begins production but is not yet profitable, and where many startup companies go bankrupt[8]—were particularly vulnerable because of higher borrowing costs and inability to leverage much out of their own assets. Many companies fell short on capital right before their final push to become a profitable manufacturer. Policymakers saw this as undermining investment in important clean-tech innovations. A loan guarantee program would provide the bridge financing necessary to scale up from pilot to commercial production to help companies through the pre-commercial period.

UNDERWRITING CLEAN ENERGY INNOVATION AND INFRASTRUCTURE

The loan guarantee was an institutional innovation that helped overcome the problems clean-tech entrepreneurs had accessing capital. The Energy Policy Act of 2005 allows the Department of Energy (DOE) to cosign and underwrite some of the interest on loans for projects that "avoid, reduce, or sequester air pollutants or anthropogenic emissions of greenhouse gases; and employ new or significantly improved technologies as compared to commercial technologies in service in the U.S. at the time the guarantee is issued."[9] Many energy policy experts suggest this was mainly to incentivize nuclear power plant construction, because they have very large capital requirements and a history of significant cost overruns, making it impossible to attract investment in private capital markets. Before the loan program, the last nuclear power plant ordered in the U.S. was in 1978. Cost overruns and delays on prior nuclear power projects kept investors away from financing these 5-to-10-billion-dollar (or more) reactors. Congress authorized the loan guarantee program to invest in commercial technologies that the private sector eschewed.

President Obama elevated his administration's green jobs agenda by making loans, grants, and tax equity available for clean tech through ARRA. With a loan guarantee, the risk is shared between the federal government and the investors in the project. It provides a source of finance for the phase where investment risk is highest, at the time when a startup firm has taken on the most debt, but may have never sold a product. Section 1703 of Title 17 of the Energy Policy Act of 2005 authorizes the DOE to underwrite loans for innovative clean energy technologies and thus to be, in the words used on the DOE website, "the financing force behind America's clean energy economy."[10] It is intended to support technologies typically not capable of conventional private financing because of "high technology risks."[11] Only a thin-film photovoltaic manufacturing facility for Solyndra in California and the Vogtle nuclear power plant in Georgia used ARRA funding through the

Section 1703 program.[12] Most projects would receive loans through the Section 1705 program (Table 5).

On February 17, 2009, ARRA expanded the DOE Innovative Technology Loan Guarantee Program (loan guarantee program) by creating Section 1705 of the Energy Policy Act for "projects that employ innovative energy efficiency, renewable energy, and advanced transmission and distribution technologies."[13] Energy secretary (and Nobel laureate) Steven Chu appointed former venture capitalist Jonathan Silver executive director of the loan guarantee program. Silver came to Washington, D.C., after a successful career at an early-stage technology investment firm. Before that, he worked at one of the nation's largest hedge funds. He ran the DOE loan portfolio alongside the Advanced Vehicle Manufacturing Technology program, made famous for its loans to two juggernaut electric vehicle manufacturers, Tesla and now-defunct Fisker. The 1705 program aimed to assist startups working on renewable energy technologies and infrastructure: geothermal, wind, solar, batteries, flywheels, and advanced biofuels. The caveat was that loan applicants needed to be developing pre-commercial technologies, just on the cusp of being commercially ready, but starved out of capital markets that were still reeling from the financial crisis.

The legislation authorized $12 billion in loans between 2009 and 2011 for solar energy manufacturing facilities and utility-scale solar energy projects. Silver believed these investments would enhance U.S. competitiveness in science and technology. He stated in a public interview, "Deploying innovative clean energy technologies will have an enormous impact on our global economic competitiveness, energy security and the environment, as well as on our continued economic recovery. Equally as important, deploying commercial technologies will help the country regain control of its energy future in the near term, reduce oil consumption and strengthen our domestic supply chain."[14] Silver commonly referred to the loan guarantee program as a "shadow bank" that should take risks that the private sector refuses. This would be reflected in the poor grades for solar projects provided by investment ratings firms like Standard & Poor's and Fitch Ratings (Table 6).

The loans were accompanied by $26 billion in Department of the Treasury grants, which transformed an important tax equity tool, the investment tax credit, into a cash grant program. It was called the Treasury 1603 program, and startup companies that did not have tax liability (because they did not yet have profits) were eligible. The tax credit only works when there are taxes to pay. These financing arrangements helped the Obama

TABLE 5

Program	Loan guarantees	Loans defaulted	Amounts at closing ($ billions)	Remaining authority ($ billions)
Loan guarantee program, Section 1703	2	0	$6.2	$28.7
Loan guarantee program, Section 1705	31	3	$15.7	$0
Advanced Technologies for Vehicle Manufacturing loan program	5	2	$8.4	$16.6
Total	38	5	$30.3	$45.3

SOURCE: Frank Rusco, "DOE Loan Programs," Testimony by the Director of Natural Resources and Environment before the Subcommittee on Oversight and Investigations, Committee on Energy and Commerce, House of Representatives, GAO-14–645T, May 30, 2014.

administration accomplish some of its climate action and clean-tech invest-ment goals. Over 2,000 companies took advantage of this grant in lieu of the tax credit. Also, 183 companies benefited from the $2.3 billion Advanced Energy Manufacturing Tax Credit—the 48C Program—another way that ARRA spilled over into the solar industry.[15] Companies like Abound Solar received tens of millions of dollars in benefits.[16]

President Obama highlighted the green jobs investments—electric vehicle and photovoltaic manufacturing facilities, solar and wind farms—in public speaking appearances, even though these investments represented only about 2% of overall jobs created by ARRA.[17] Green jobs investments eventually resulted in a return on investment for tax-payers, outperforming other investments in the Treasury's portfolio, despite several bankruptcies that became major news headlines and the subject of several congressional investigations. By the end of 2016, the loan guarantee program would yield taxpayers $1.65 billion profit on the investments.[18] Furthermore, the forgone federal tax revenues from these projects can drive economic activities that increased tax collection elsewhere, such as with state and local sales tax from installations, elec-tricity generation sales tax, local property tax from solar developments, employee state and federal income taxes, and other taxes attributable to the profitability of private solar companies. These public investments attract accompanying private finance. A National Renewable Energy Laboratory study found that ARRA investments attracted over $30 bil-lion from the private sector.[19]

TABLE 6

Company	Project	Cost	S&P/Fitch Date	Rating***
1366 Technologies	Input manufacturing	$150 million	9/2011	B
Abengoa Solar (Mojave)	Power generation	$1.2 billion	9/2011	BB+
Abengoa Solar (Solana)	Power generation	$1.446 billion	12/2010	BB+
Abound	Thin-film manufacturing	$400 million	12/2010	B
BrightSource Energy	Power Generation	$1.6 billion	4/2011	BB+
Cogentrix of Alamosa	Power generation	$90.6 million	9/2011	B
Antelope Valley Solar Ranch*	Power generation	$646 million	9/2011	BBB–
Mesquite Solar 1*	Power generation	$337 million	9/2011	BB+
Desert Sunlight*	Power generation	$1.46 billion	9/2011	
Genesis Solar	Power generation	$852 million	8/2011	
California Valley Solar Ranch	Power generation	$1.237 billion	9/2011	
Agua Caliente*	Power generation	$967 million	8/2011	BB
Crescent Dunes (SolarReserve)	Power generation	$737 million	9/2011	CCC+
SoloPower	Thin-film manufacturing	$197 million	9/2011	BB
Solyndra**	Thin-film manufacturing	$535 million	9/2009	BB–

* Using First Solar CdTe photovoltaic modules.

** Defaulted on their loan.

*** Ratings were not available for all USSE projects.

Thin-film technologies were also viewed as important investments for the U.S. solar industry more broadly, because more aspects of the technologies could be patented or protected by keeping manufacturing processes trade secrets. Many of the basic techniques to make crystalline silicon were well known, but no company had yet perfected the recipe for thin-films. By 2009, the count of thin-film patents (4,300) had already surpassed crystalline silicon patents (3,300).[20] Winning the race to be a global clean-tech titan means not only research and development but also securing intellectual property and protecting trade secrets. The U.S. would put big bets on thin-film photovoltaic technologies via ARRA public investments to develop the solar cells of the future. Hence, the Sputnik moment was an opportunity to invest in the future of solar energy technology, securing key patents and ensuring that technological rents could be harvested from any manufacturing needing key pieces of equipment or processes that would have to be licensed from U.S.-based manufacturers. Some framed the ARRA investments as an energy security issue, raising concern that future U.S. photovoltaic manufacturers would have to license technologies developed in foreign countries. Henry Waxman, a Democratic congressman from California and ranking member of the House Energy and Commerce Committee, justified the investments, exclaiming, "*We can't lose this race!*"[21]

Energy secretary Steven Chu echoed the importance of investing in emerging photovoltaics and the clean energy race in testimony before Congress.[22] Dr. Chu pointed to research from the Energy Information Agency, forecasting that by 2030 the wind and photovoltaic markets would increase in value between $1.2 trillion and $1.5 trillion, to call attention to the economic opportunity of gaining market share in the clean energy.[23] The Chinese Development Bank offered domestic solar manufacturers over $34 billion in credit and was investing billions more in wind, smart grid technologies, transmission infrastructure, and energy storage.[24] Chu lamented:

> The United States, meanwhile, has fallen behind. The world's largest turbine manufacturing company is headquartered in Denmark. 99 percent of the batteries that power America's hybrid cars are made in Japan. We manufactured more than 40 percent of the world's solar cells as recently as the mid 1990s; today, we produce just 7 percent. When the starting gun sounded on the clean energy race, the United States stumbled. But I remain confident that we can make up the ground. When we gear up our research and production of clean energy technologies, we can still surpass any other country.[25]

By early 2011, there were murmurs that several of the ARRA investments in thin-film manufacturing facilities were weathering tough financial times.[26] The bankruptcy of Solyndra in August 2011 would direct national attention to ARRA and the DOE loan program, as would several *New York Times* and *Los Angeles Times* stories about utility-scale solar projects that were embroiled in wildlife conflicts, most notably a concentrated solar power project in the Ivanpah Valley. Within another year another thin-film manufacturer, Abound Solar, went bankrupt, and another, Solo-Power, stopped taking loan payments and mothballed a factory in Oregon before returning to pilot-scale production. The solar energy industry landscape underwent significant transformation and restructuring as China's crystalline silicon photovoltaic sector climbed from a $2 billion industry in 2006 to over $25 billion in 2010, the same year China became the first nation to surpass annual production of 10 GW.[27] By the time thin-film manufacturers were having financial trouble, manufacturers in China had built out production capacity so rapidly that they were liquidating inventories—selling modules below cost—which further imperiled the struggling companies trying to commercialize production.[28] That year, several U.S. manufacturers began to accuse China of dumping crystalline silicon photovoltaic modules in the U.S., undercutting domestic prices.[29]

Thin-films seemed like a most promising investment early in the clean energy race, but things radically shifted. For investors in thin-film manufacturers First Solar and Solar Frontier, investments have paid off in many gigawatts of installed solar capacity around the world. However, claims that thin-films would surpass crystalline silicon photovoltaics in market share never materialized, as the better bet turned out to be taking a known commodity and making it cheaper through economies of scale. Figure 4 shows the relative market share of thin-film versus crystalline silicon photovoltaics.

A smaller but also important ARRA budget item invested $400 million in basic research in 2009 through the Advanced Research Projects Agency-Energy. ARPA-E is a research agency of the DOE, modeled on DARPA, a U.S. Department of Defense advanced research unit that develops breakthrough national security technologies for the military-industrial complex—the internet, drones, stealth technology. Investments in both basic research and pre-commercial production offer opportunities to innovate but are considered risky enterprises usually limited to financing from venture capitalists and angel investors, which overall represents only a small portion of clean-tech investment.

Using ARRA to provide loans to startup or pre-commercial solar companies complemented several other state and federal energy, land-use planning, and climate innovation policies. While the New Deal analogy implies a high degree of measured state coordination, in reality clean-tech development and deployment resulted from a patchwork of solar policies at the state, federal, and local levels and the actions of private companies. These details contrast with Roosevelt's New Deal, which oversaw extensive public works programs through the development of new public institutions.

The loan guarantee program played a critical role in the development of the first utility-scale solar farms. The availability of capital and generous policy programs (the loan guarantee program, the Treasury 1603 grant program, and accelerated depreciation tax benefits) helped scale up several developers and manufacturers working on innovative solar solutions. But finance and capital alone do not explain the rapid growth and uptake of solar power. Several other policy processes at work augmented solar power deployment in the U.S. California and twenty-nine other states created new mandated markets for renewable electricity through an energy policy scheme called renewable portfolio standards—more on this in a moment.[30]

CREATIVE DESTRUCTION AND THE MAKING OF RENEWABLE ENERGY MARKETS

The transition away from fossil fuels to systems powered by clean energy and renewables like solar power will require radical transformations in the way energy is generated, transmitted, and consumed. While some of the solar power transition will be off-grid, or distributed and fed to the grid, much of this solar power will be at the utility scale, displacing conventional utility-scale power plants. Economist Joseph Schumpeter described industrial transitions as "incessantly revolutioniz[ing] the economic structure from within, incessantly destroying the old one, incessantly creating a new one."[31] He called this an output of the innovation process caused by "the gale of creative destruction," drawing on the earlier use of "creative destruction" by Karl Marx, with a negative connotation, to describe the natural outcome of wealth creation.[32] From Schumpeter's perspective, creative destruction in energy systems describes how conventional industries are replaced by renewable and low-carbon sources in a radical market transformation.[33] But such pathways are

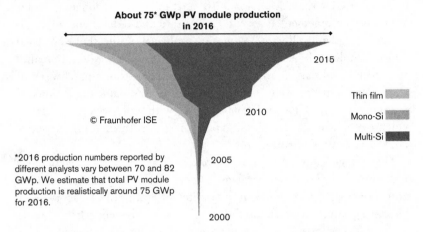

About 75* GWp PV module production in 2016

2015

© Fraunhofer ISE

2010

Thin film

Mono-Si

Multi-Si

*2016 production numbers reported by different analysts vary between 70 and 82 GWp. We estimate that total PV module production is realistically around 75 GWp for 2016.

2005

2000

FIGURE 4. Thin-film and crystalline silicon photovoltaic module production, 2000–2016 (Fraunhofer, 2018).

neither predetermined nor inevitable. Creative destruction can be stalled by path dependence caused by major infrastructure projects and policy entrenchment. Energy policy expert Gregory Unruh refers to the energy system as in a condition of "carbon lock-in" because of the commitments to fossil fuel infrastructure and the policies that continue to subsidize it.[34] Given the inertia of energy systems, creative destruction depends on orchestrating state regulation and corporate practice to move industries away from conventional stocks of energy toward flows of renewables, and often requires shifts in social norms and behavior (which can be more resistant to change).

There are ongoing debates in climate policy circles on how to most quickly achieve greenhouse gas (GHG) reductions. Economic theory says that the "polluter-pays principle" will increase costs for GHG polluters. One approach to hastening renewable energy adoption favored by many environmental organizations and policymakers is to price GHGs and treat emissions as a form of pollution, by way of either carbon taxes or cap and trade (or its many variations, such as cap and dividend). Each policy approach puts prices on carbon such that energy sources that emit little or no GHGs are priced correctly for having relatively fewer environmental externalities. Yet, despite decades of discussions about carbon taxes and offsets, there are few instances where these policies have led to rapid shifts in energy infrastructure, particularly of the magnitude necessary to achieve deep GHG emissions reductions.

What *has* led to major shifts in energy infrastructure toward decarbonization in the U.S. are policies that impose purchasing quotas on electric utilities called renewable portfolio standards (RPSs). These require that utilities purchase a specified amount of electricity from renewable sources, which ratchets up over time. This guarantees a market for renewable electricity and invites utility-scale renewable energy developers to invest in projects. Utilities began to sign power purchase agreements (PPAs) with developers to comply with RPSs. California's RPS began attracting proposals for utility-scale solar energy facilities in 2003, and today one of the world's largest economies also has some of the highest levels of solar electricity use in the world. Any renewable energy source (excluding large-scale hydroelectric) can qualify to sell electricity to California investor-owned utilities to meet its RPS quotas. Governor Arnold Schwarzenegger strengthened the California RPS as part of the comprehensive Global Warming Solutions Act he signed into law in 2006. By 2030, California investor-owned utilities must procure 36 GW of renewable power to reach a target of 50% of electricity from renewable sources. The proposal by the next California governor, Jerry Brown, to increase the RPS was passed by the legislature in 2015. As of 2018, thirty-seven states across the U.S. have adopted renewable energy portfolio standards or goals.[35] Some state RPS policies have separate quotas for specific kinds of renewables, like solar or wind.

RPSs are what energy and innovation scholars describe as "technology-pull" policies.[36] The goal of the California RPS is to create guaranteed markets for companies planning utility-scale renewable energy projects. Renewable energy developers compete only against other renewables, knowing that utilities will have to contract to buy renewable electricity. Technology-pull policies *pull* new technologies onto the market by making them more affordable to consumers. The most important example in the solar space is the tax rebate program known as the investment tax credit, described earlier, which allows photovoltaic module owners to deduct 30% of the entire system cost from their tax bill over one to five years.[37] The credit costs the Internal Revenue Service millions in forgone tax revenues annually, according to the U.S. government's budget watchdog, the Government Accountability Office (GAO), but other federal, state, and local tax revenues are generated elsewhere to make up the difference through knock-on economic activities.[38]

The effectiveness of the RPS in California is reflected by the queue of projects under development as of 2018. Over half of the solar projects proposed are outside of RPS schemes, no longer needing the RPS mechanism. Many of these projects are directly contracted to companies and institutions looking to green their operations. Apple, for example, is contracting with the 2,900-acre California Flats Solar Project in San Luis Obispo County, north of the community of Cholame, to "offset" the electricity used at its Cupertino headquarters over 100 miles away. Facebook, Google, Amazon, and other tech firms contract directly with renewables developers to supply solar electricity to the grid to compensate for the amount consumed elsewhere.

Electric utilities comply with the standard in various ways, depending on the particular commission. In California, they comply by demonstrating a good-faith effort to contract for solar electricity. In many states, electric utilities sign PPAs with solar project developers, agreeing to purchase electricity from the power plant for some long period, sometimes as long as twenty years. These long-term contracts are leveraged to secure the necessary financial resources to bring the project to fruition. Developers with signed PPAs can represent commitments to projects to their investors to attract the additional capital needed to build projects. Securing a PPA makes particular parcels of land more attractive for finance because it suggests that utilities have the available infrastructure (transmission lines, substations, etc.) to bring new electricity generation online. It also signals that the utility will clear the hurdle of the ratepayer cost tests, which estimate the impact on ratepayers from buying electricity at particular prices. PPAs signed between developers and California investor-owned utilities—Pacific Gas and Electric, San Diego Gas & Electric, and Southern California Edison—for RPS compliance were critical to attracting investment, because they signaled a commitment to customers for 20 to 25 years, depending on the contract length.

Back in the 1980s, the largest network of utility-scale solar projects built in California was the nine Luz Solar Energy Generating Stations. These nine solar power plants stretched from Kramer Junction to Daggett to Harper Lake in the Mojave Desert. These nine separate concentrated solar power projects were attractive to electric utilities because of the high oil prices related to the OPEC oil embargoes of the 1970s and into the early 1980s.[39] But Luz went bankrupt a decade later when natural gas and oil prices fell sharply and unpredictably, as supply constraints eased and after the expiration of key state and federal tax incentives.[40] Market conditions became unfavorable, and Luz was left with no offtaker (buyer) for its electricity, as utilities returned to cheaper oil- and gas-fired peaker plants for midday power demand. Eventually another company acquired the assets and the nine solar power plants, which still operate in the Mojave Desert as of 2018. The total cost of building and operating these plants since they were built is around $0.05/kWh. The RPS makes financing easier by providing some assurances to investors, who might otherwise be deterred by concerns about competing sources of electricity in the long run, absent any kind of contract. Figure 5 shows one of the nine Luz Solar Energy Generating Stations in the Mojave Desert.

PPA contracts are not as difficult for project developers to acquire as might be expected, because they are ultimately only promises from an electric utility to buy electricity at agreed prices. In 2008, a venture capital startup named Solaren proposed to place a photovoltaic array in low Earth orbit and beam microwaves to a terrestrially based receiver near Fresno, California. The clever proposal borrows the idea initially proposed by science fiction writer Isaac Asimov in a 1941 short story, "Reason."[41] Southern California Edison entered into a PPA in 2009 with Solaren for 200 MW of power for 15 years despite serious questions about the technical and economic feasibility of the project. Signed PPAs with electric utilities are not necessarily evaluated with due diligence for economic or technical feasibility, only interconnection potential with the grid and electricity-selling price.

By August 2011, the California Public Utilities Commission had approved 193 PPAs totaling 90,000,000 MWh, far more than the 68,000,000 MWh needed to fulfill the utilities' 2020 RPS obligations, and 2030 commitments were too far away.[42] Only four PPAs were denied from 2005 to 2011, all four for not being in the interest of ratepayers. The RPS fostered creative destruction in the electricity space as wind and even more solar was built and contracted to utilities. However, it perhaps attracted too much investment too quickly, as many projects that were proposed and obtained PPAs were eventually withdrawn.

These public investments are not risk-free for taxpayers, which makes the program subject to criticism. If a company tied to one of these loans becomes insolvent, taxpayers are on the hook for, in the case of failed ARRA projects, hundreds of millions to billions of dollars. The magnitude of the investments and the narrowness of the portfolio became major points of contention. The large loans to certain projects, noted some journalists and activists, meant that some projects might be "too big to fail," and warned that the public could eventually be on the hook for bailing them out.[43] To minimize risks, the DOE uses a metric-based assessment, which it calls "strong business fundamentals," to determine

FIGURE 5. A Luz solar power plant in Kramer Junction, California, part of the largest solar power plant complex in the word in the 1980s. (Photo: WikiCommons).

what should be safe and worthwhile investments.[44] This may help explain why so many of the projects obtaining loan guarantees were tied to large banks like Citibank and Morgan Stanley. The financial metrics help determine a project's risk. But true to the program's mission, DOE aimed to pursue breakthrough technologies, which assumes a greater amount of risk.

The loan program became a mainstream political controversy as President Obama's administration was accused of directing investments toward companies linked to campaign donors, sparking claims of "crony capitalism" by the opposing party's presidential candidate, Mitt Romney, in 2012.[45] And the DOE came under fire for inconsistent treatment of various applicants in the review process, according to a GAO report.[46]

The GAO report found fault in some places, such as questions about audit and verification, but noted that other charges were overblown, such as any charges of cronyism. The hyperbolic headlines and extensive media coverage that emerged in the wake of the Solyndra collapse left much of the public with an impression that the loan program was a failure. But most solar and clean tech innovation experts consider the loan program a success, a point confirmed by an independent audit by the GAO.[47] Despite several notable bankruptcies, the program helped drive down costs for future solar power plants and offered investors proof that large utility-scale projects could be built on time, at budget, and operate successfully. The GAO and DOE differ on whether the program has ultimately benefited federal taxpayers, though considering local and state tax revenues, taxpayers undeniably benefit overall.[48] GAO concluded that the DOE lacked a loan-monitoring program during the critical period between 2009 and 2013 when most of the funds were dispensed.[49] The program was politically vulnerable because of financial and political connections between the administration and some of the investors backing companies selected for fast-track status and loan guarantees.

Analysts at the DOE used life cycle assessment in the loan guarantee program to evaluate whether projects reduced GHG emissions. Title 17 of the Energy Policy Act of 2005 requires a life-cycle inventory of carbon savings to qualify for a federal loan guarantee.[50] The loan program tracks forecasted reductions in GHG emissions from projects receiving loan guarantees compared to "business as usual" energy generation. The DOE also forecasts reductions in air pollutant emissions (nitrogen oxides, sulfur oxides, and particulates) from projects receiving loan guarantees. The DOE would advertise the GHG emissions saved from the program and report metrics such as green jobs created for ARRA investments.

INSTITUTIONS FOR DECARBONIZATION

Public investments in clean energy infrastructure and innovation have sought to address crises in climate, unemployment, and technological innovation. Federal investments in renewable energy, mostly forgone tax revenues in the U.S., helped build some of the largest solar farms in the U.S. At the same time, they helped attract private capital that might not otherwise be there without the incentives and the proofs of concept. The incentives and policies that created the conditions to deploy new solar farms and manufacturing facilities would cause a split in the environmental movement between those willing to sacrifice desert ecosystem conservation for renewable energy, and those asking for alternative paradigms for solar energy deployment. Furthermore, efforts to politicize these issues by Republican members of the House of Representatives throughout 2011 and 2012 were enhanced by the "green revolving door" between clean-tech venture capital funds and the regulators and federal agencies empowered to finance risky investments.

The final key policy innovation that influenced some ARRA investments in renewable energy infrastructure dealt with the availability of large parcels of land to build utility-scale solar energy facilities. In 2005, to ensure sufficient land for the enormous utility-scale projects, the Bureau of Land Management, the largest landlord in the U.S., opened 22 million acres of public lands in the Desert Southwest to solar energy developers. This policy required unprecedented interagency collaboration between the bureau, the DOE, and the California Energy Commission to secure Department of the Treasury finance. Having these large swaths of land available arguably helped the DOE directly finance a large portion of many gigawatts of solar on public lands across the Desert Southwest. The Bureau of Land Management used its discretion to expedite environmental and cultural resource reviews using an agency rule to allow fast-tracking permits for energy projects.

In 2009 political scientist Timothy Luke asked, "Is a Green New Deal possible? Is this highly sought after 'greenness' only a superficial coating brushed across truer grey, brown or black qualities in urban industrial society that inescapably remain the same underneath the rhetoric? Why does labeling any state-led public policy intervention 'a New Deal' automatically turn that 'deal' into something 'new'?"[51] The answer can be explored by examining case studies of actual investments in technologies and infrastructure. Identifying lessons learned and best practices along the entire life cycle of these technologies and their supply chains can provide opportunities to manufacture and develop projects in the most responsible way.

The projects that were offered loan guarantees show that access to capital was much easier if you had the right combination of technology, land, and capital when the DOE agreed to cosign loans to renewable energy developers. Socializing the risk of pre-commercial technological innovations projects is riskier than typical investments in infrastructure, but that was the culture of innovation guiding the policy. A dynamic that favors high risk-to-reward ratios—speculative capital—has been shaping the culture of finance since the 1980s.[52] The influence of former Federal Reserve chairman Alan Greenspan put momentum behind this ideology as the high-rolling dot-com and housing bubbles funneled fantastic wealth to the top. Speculative capital shifts wealth in one direction, and risk in the other. This would foment hostility toward the DOE and the Obama administration across a wide range of actors, from local Tea Party activists to Occupy Wall Street.

Designing institutions for solar energy transitions that anticipate many of the lessons learned from ARRA projects and these policies generally could obviate controversies and target investments more effectively. Well-intentioned policies—a Green New Deal for solar energy—can result in controversy or negative outcomes, or be ineffective at achieving climate goals. What would have happened if the U.S. prioritized investments in crystalline silicon in ARRA? Would Bureau of Land Management projects have faced less controversy with better planning and participation, instead of advancing projects with fast-track authority? What can state and local officials do to facilitate the appropriate development of utility-scale solar generation? These will be important considerations in the design of institutions for decarbonization.

CHAPTER 3

Innovations in Photovoltaics

The simplest, yet most environmentally benign source of
electricity yet conceived.

—*Science* magazine, 1955[1]

OPENING A BLACK BOX: INNOVATIONS IN THIN-FILMS

There has never been a sun spill. The most obvious advantage of photovoltaics is that there is no direct air pollution or greenhouse gas emissions during operation. Photons from the sun are absorbed by electrons,
which are energized into the conductive band, allowing them to move
freely as electric current. Electricity produced from fossil fuel—coal,
natural gas, or petroleum—has higher environmental emissions. And
photovoltaics offer other environmental benefits, using less water than
thermal power plants, and causing fewer hazardous emissions than
coal and natural gas.[2] As much as 89% of U.S. air pollution emissions
could be avoided if solar replaced all combustion-based sources of
electricity.[3]

But photovoltaic module manufacturing does require plenty of
energy and materials—water, metals, plastics, glass, and other parts and
components—that have social and environmental impacts. Even if these
impacts are smaller so far, scaling up photovoltaics to the terawatt levels needed to significantly impact greenhouse gas emissions requires
new considerations about materials and land uses. What are the impacts
of the natural resources and materials used to make photovoltaics?
Where are the cradle-to-grave social and environmental impacts? How
far and where do photovoltaics' supply chains reach? What kinds of
impacts are associated with new innovations in photovoltaic technolo-

gies? The sooner impacts can be identified, assessed, and planned for, the more likely the project will be successful.

Following traditions in political ecology that seek to explore frictions at the interface of natural resource use, access, control, and governance, this chapter explores several regulatory controversies and public debates over the use of cadmium compounds in thin-film photovoltaics. This story of this technology reveals some of the tensions between hazardous materials management and clean technology, but also speaks to effective policy, practice, and sustainability leadership in industry. This chapter aims to show how claims about toxicity in photovoltaics were raised, debated, and ultimately resolved.

Scholars of science and technology studies use the term "black box" when social processes or technologies are taken for granted; where the social content in scientific and other knowledge-making processes is deemed not relevant to the matters at hand.[4] To open the box or shine light into a black box is a metaphor for a deeper exploration of how the device or process is made or its inner workings. Photovoltaics are often treated as black boxes in decarbonization conversations. Photons flow in, and electricity flows out, and little else matters to most onlookers. Few give much consideration to how these devices operate or the materialities and social relations that hold the technology together. They instead focus on the cost of the technology or the amount of land it may require.

The other way to think about the phrase "black box" looks to how engineers and scientists use it to describe a process or stage where the internal mechanisms are not specified. In the process flow diagrams used by engineers as blueprints for production processes, there are many steps and stages represented as only a box with arrows to denote the direction of movement, with inputs and outputs noted. The notion that technologies could have politics was not considered in early scholarly engagements with the evolution of technology.[5] Technologies were believed to be neutral. They could be put to negative uses, but that was a product of individuals using the technology, not the technology itself. Early efforts to explain the histories of science and technology relied on explanations based on personal or institutional aspects, their reward structures, ambitions, and norms.[6] Historian Lewis Mumford described how technologies can structure society, for example describing the effects of the automobile on the restructuring of social relations in New York City as major highways split up neighborhoods; but he attributed these effects to economic and political decisions.[7] How might solar power technologies evolve on the frontiers of innovation, and can we identify potential social and environmental dimensions of technological change that designers, regulatory institutions, and environmental agencies and organizations should prepare for?[8]

THE RISE OF THIN-FILM PHOTOVOLTAICS

The black box explored here is thin-film photovoltaics made from cadmium compounds. Technologies such as cadmium telluride (abbreviated CdTe, pronounced "cad-tel") and copper indium gallium diselenide (CIGS, pronounced "cigs"; but sometimes the gallium is dropped and CIS is said letter by letter) thin-film photovoltaics were heralded as the technology of the future in the late 1990s. Thin-film semiconductor technologies were believed to have cost-saving potential compared to the older crystalline silicon.[9]

Cadmium compounds are found in several thin-film semiconductors and may also be used in some chemical processes, such as where cadmium chloride is used for annealing, a process that aims to preserve the structure of semiconductor compounds where the different surfaces touch. CdTe can be used as the p-type or n-type layer. It is correspondingly paired with cadmium sulfide (CdS) and the other layer. Cadmium compounds can be used in the transparent conductive layer, as cadmium stannate is used in pilot and laboratory-scale projects, unless indium tin oxide, or something tin- or zinc-based, is used. In CIGS technology, CdS is commonly used as the n-type layer (and the p-type layer is the CIGS compound). These two thin-film technologies intrinsically contain cadmium, and there are no easy substitutes for the semiconductors without significantly altering the entire design. Proprietary formulations, patents, and equipment also make it difficult for companies to shift away from existing materials and processes.

Thin-film semiconductor compounds are obtained from metals derived from by-products of mining and smelting operations. Common primary metals used in thin-films include copper and zinc, while secondary metals include cadmium, tellurium, gallium, indium, selenium, and molybdenum, which may be derived from primary operations seeking copper, tin, zinc, and bauxite. Metals are refined and compounded into p-type, n-type, and conductive materials that are deposited as thin films onto a substrate such as glass, plastic, or foil. As each layer of semiconductor is applied, lines are scribed with chemical, laser, or heat treatments to isolate rows of cells that harvest and convert photons into electrons and deliver them to circuits. The result is a product made mostly of glass, with relatively low feedstock requirements and energy inputs compared to crystalline silicon photovoltaics.

Research scientists at the national energy labs were already saying what Silicon Valley and the blogosphere turned their attention toward by 2007—the promise of cheaper and more sustainable photovoltaics.[10] Thin-film photovoltaic manufacturing dates back to the 1950s, with early explorers including General Electric, Kodak, Monosolar, Matsushita, and AMETEK. Federal investments through the National Renewable Energy Laboratory (NREL) Thin-Film PV Partnership in the 1990s advanced the technology with matching investments from companies like British Petroleum's BP Solar and Solar Cells, Inc. (the latter would become First Solar).[11] Numerous materials and substrates were experimented with, and several thin-film manufacturing techniques evolved through this program. The $100 million investment in the thin-film partnership over the course of fifteen years yielded a considerable return on investment through taxes recouped from sales and employment taxes, not to mention the income to wage earners throughout the manufacturing and installation phases.[12] First Solar's revenues since inception would be measured in the tens of billions.[13] NREL has since retained several thin-film patents available for private companies to license.

Rapid growth in thin-film photovoltaics began in the mid-2000s, drawing on technological advancements in the manufacturing of flat-panel displays, hard drives, and other computer components—processes technically called chemical baths, vapor deposition, and sputtering. Around this time, many startup firms emerged on the Silicon Valley scene and received venture capital from prominent clean-tech investment firms. Thin-film photovoltaics had the potential to be a game-changing technology that could displace crystalline silicon solar cells. They were attractive to investors, scientists, and startups because of the possibility of lower energy and materials use and lower manufacturing costs. One major advantage was faster manufacturing throughput; crystalline silicon photovoltaics, which are made in batch processes, take longer to make than thin-films, which are made via a continuous process. The semiconductor layers in thin-films are two orders of magnitude thinner than conventional solar, because they typically have a good absorption coefficient. They will continue to get thinner over time. Thin-film production results in less waste and completely avoids the complicated melting, drawing, and slicing steps used to make traditional silicon cells. Since damage at the various stages of crystalline silicon manufacture can lead to low yields, thin-films offer opportunities to make significant improvements in yield as well. There are several well-documented research papers arguing for the sustainability merits of thin-films compared to crystalline silicon.[14]

CdTe thin-films date back through several lineages, but many of the major innovations began to transform the technology when Harold McMaster, one of the world's foremost experts on tempered glass, founded Solar Cells, Inc., in 1990. McMaster's team experimented with new techniques to apply thin-film semiconductors to glass. As the technology progressed, True North Partners (an investment arm owned by the son of Sam Walton, founder of Walmart) bought the firm and changed its name to First Solar. First Solar's stock experienced a meteoric rise, an order of magnitude higher than any other publicly traded photovoltaic manufacturer, allowing it to vertically integrate. It had a low-cost technology that worked well enough to compete with and in many cases beat silicon-based technologies on cost in solar power plant projects. To facilitate its growth, it vertically integrated by acquiring companies with development experience and project pipelines that included numerous land deals and rights-of-way on public lands. It acquired OptiSolar and Ted Turner's Turner Renewable Energy, both firms having investments in land for renewable energy development.[15]

First Solar rapidly ascended to be the largest thin-film manufacturer, and soon after, the largest overall in the solar energy industry. It built the world's largest utility-scale solar power plants. But several regulatory and governance issues were raised regarding environmental sustainability issues related to cadmium compounds and management options for safe and responsible disposal at the product's end of life. The controversies described next illustrate some of the challenges of balancing regulatory protections for the environment and workers with innovations in emerging technologies. Regulators are faced with the challenge of protecting workers and the environment while not stifling technological innovation, or in this case specifically solar deployment. These kinds of trade-offs will continue to be raised as low-carbon energy technologies are deployed for climate action and clearer air.

DEBATES OVER THE USE OF CADMIUM
IN PHOTOVOLTAICS

Some thin-film photovoltaics contain cadmium compounds, which are heavy metals known to pose environmental, health, and safety risks with high levels of exposure.[16] Acute exposure to elemental and soluble forms of elemental cadmium and cadmium compounds is known to cause kidney disease, bone weakening, birth defects, infertility, renal failure, severe pulmonary inflammation, and pulmonary fibrosis.[17] Cadmium is a known carcinogen (cancer-causing), mutagen (mutation-causing), and genotoxin (genetic information-damaging).[18] The most widely cited epidemiological case of cadmium exposure is from Japan in the mid-twentieth century, where *itai-itai* (ouch-ouch) disease caused severe bone damage downstream from a zinc mine whose cadmium pollution had found its way into the rice eaten by the community.[19] Long used in pigments and as an anti-corrosive, cadmium is no longer used in many products, especially those which could eventually release it to the environment, such as paints. Its use in photovoltaic thin-films is one of the few areas where demand for cadmium compounds is growing.

The bluish metal cadmium and its compounds, which can range from black to yellow, is one of four heavy metals prohibited in any products sold in the European Union by the Restriction of Hazardous Substances (RoHS) Directive—a Europe-wide regulation based on the precautionary principle that could block the sale of photovoltaics that contain them. Since August 2005, companies selling a broad range of electrical goods in European markets must conform to RoHS and the Waste Electrical and Electronic Equipment (WEEE) Directive. WEEE is designed to reduce the amount of electrical and electronic equipment waste going to final disposal, while RoHS aims to reduce the amount of hazardous substances found in electrical and electronic equipment. The two directives represent an expanded focus from reducing the environmental impacts of production processes to reducing the environmental impacts of both production and disposal of manufactured products.

RoHS prohibits products containing 0.01% cadmium by mass in a homogeneous material. The prohibition of cadmium compounds posed a market barrier to CdTe thin-film manufacturers, unless it is granted an exemption, because the modules intrinsically contain cadmium.[20] A typical CdTe module can contain 7 grams of cadmium, which the Brookhaven National Laboratory of the U.S. Department of Energy (DOE) describes as about as much as in a "C-sized" nickel–cadmium battery.[21] The prohibition of such products would exclude any manufacturer of cadmium-based thin-films from the market, so a life cycle management strategy—extended producer responsibility—was pursued to manage this tension.[22] Europe at the time was the leading market for photovoltaics, with three-quarters of the global demand throughout the

2000s. Selling into the European market, particularly Germany, Italy, and Spain, where there were generous feed-in-tariffs, was critical to success in the industry during this time. Solar growth would taper off several years later in Germany and particularly Spain, but thin-films and photovoltaics in general were able to move in if their technology was mature. This raised the stakes in this regulatory controversy. Investors in the clean-tech blogosphere were already noting by 2007 that CdTe manufacturers could be shut out of Europe with the RoHS recast.[23] In a Securities and Exchange Commission filing to report investor risks, CdTe manufacturer First Solar disclosed:

> Currently, photovoltaic solar modules in general are not subject to the WEEE or RoHS Directives; however, these directives allow for future amendments subjecting additional products to their requirements and the scope, applicability and the products included in the WEEE and RoHS Directives are currently being considered and may change. If, in the future, our solar modules become subject to requirements such as these, we may be required to apply for an exemption. If we were unable to obtain an exemption, we would be required to redesign our solar modules in order to continue to offer them for sale within the European Union, which would be impractical. Failure to comply with these directives could result in the imposition of fines and penalties, the inability to sell our solar modules in the European Union, competitive disadvantages and loss of net sales, all of which could have a material adverse effect on our business, financial condition and results of operations.[24]

Since RoHS became binding in 2006, photovoltaics have been outside its scope. Most photovoltaic manufacturers were against the inclusion of photovoltaics in the scope of RoHS. They argued that the purpose of the directive was to keep toxic products out of landfills and prevent improper disposal. This meant keeping toxics out of *consumer* products, and therefore RoHS should apply only to household e-waste. Photovoltaic modules, on the other hand, are specialized electrical equipment, requiring specially trained employees to install, and represent a kind of managed waste product.

By the time of the RoHS debate, there was already a niche research literature on the environmental impacts of metals in thin-films, specifically CdTe. The Brookhaven National Laboratory explored the environmental impacts of photovoltaics, including thin-films, in the Photovoltaic Environmental Research Center from the early 1980s to the 2000s.[25] The researchers published reports on the environmental, health, and safety (EHS) aspects of production of all kinds of photovoltaics. The topics included hazard analyses and best practices to ensure safe handling of silane, a pyrophoric chemical used in amorphous silicon and crystalline silicon manufacturing, and heavy metals used in thin-films.[26] They also produced scores of life-cycle assessments (LCAs), focused on identifying hazards and the cumulative effects of particular material and energy flows of different material combinations. While important within their academic fields, professional organizations, and science-industry networks, the EHS literature on photovoltaics remained relatively hidden from public view and restricted to an isolated technical debate from the late 1970s through the mid-2000s. So when these issues were first publicly raised around 2007, many policymakers, solar industry professionals, energy utilities, and members of the public were hearing of toxics in photovoltaics for the first time.

Early research suggested that cadmium compounds were the most toxic inputs used in commercially available thin-film photovoltaic technologies. Of the numerous heavy metal compounds used in thin-films, cadmium and tellurium are considered more toxic than indium or gallium, and indium is more toxic than gallium.[27] Exposure to these toxics in the workplace can be controlled with robust and responsive industrial hygiene monitoring, engineering controls, and protective equipment. There are no emissions during operation, and even studies of broken modules and modules destroyed in the lab by fire show minimal leaching of cadmium to the environment.[28]

When photovoltaics reach their end-of-life, proper management strategies are required to ensure that the modules do not end up in landfills or other places where metals can be released to the environment. CdTe decomposes under acidic and anaerobic conditions, releasing elemental cadmium.[29] This could occur if end-of-life modules are landfilled, though it would presumably require very large volumes to have any appreciable impact on concentrations of heavy metals in landfill leachate.[30]

Several manufacturers and researchers of thin-films over the years have emphasized that CdTe thin-films do not contain cadmium per se but the compound cadmium telluride, which has different properties. And cadmium telluride may be less toxic than other cadmium compounds, because it has "low vapor pressure, high boiling and melting points, and low solubility."[31] The high boiling point suggests less risk to living organisms.[32] Only a handful of studies have evaluated the toxicological properties of CdTe because it is an uncommon compound, so few data support claims about CdTe toxicity either way. A dossier of the physical, chemical, and toxicological properties of CdTe was registered with the European Chemicals Agency to comply with the

European Union's REACH law.[33] For example, there are few publicly available longitudinal studies of worker health and safety in the peer-reviewed research literature, despite calls for better EHS understanding of CdTe photovoltaic modules long before they were widely commercially available.[34] Studies of CdTe quantum dots suggest damage to cell membranes and mitochondria.[35] But this may not be generalizable to CdTe because nanoscale quantum effects may confound the results. Researchers on the risks of nanotechnology argue that a different set of laws of physics dominate at this scale, making conclusions more challenging, as the nano-scale effect may be influencing the outcome.

The largest occupational and public health risks from cadmium compounds across the CdTe photovoltaic product life cycle occur at the sites of mining, smelting, and end-of-life management activities. CdTe may be a safer material to handle than other cadmium compounds, but it is human-made; it is a compound that does not occur in the ores of Earth's crust. Irrespective of the CdTe's safety, exposure to cadmium compounds may occur deeper in the supply chain at mines and smelters, which in some cases have been linked to cancer clusters, though these sites emit numerous heavy metals.[36] Cadmium is produced from residues of zinc and from lead and zinc smelting.[37] Cadmium and tellurium are usually found in lesser quantities in the ores of other metals, such as zinc, lead, and copper. Because cadmium is not a primary mining activity, some argue that cadmium exposure in mining would occur whether it is destined for market or stored onsite as tailings. In fact, more cadmium is produced by smelters than is sold into commercial activities. The emissions associated with smelters will vary based on the technology used at the facility and regulatory rules.

Several other, less hazardous, cadmium compounds are used in thin-film manufacturing. These include cadmium stannate, cadmium sulfide, and cadmium chloride—none of which are present in the final device. One study of workers at the cadmium chloride station along CdTe manufacturing lines found higher levels of cadmium in their blood, although well below the threshold for concern and below employees who smoke tobacco, who have much more cadmium.[38] A study of five years of exposures from production at a First Solar factory in Malaysia found average blood and urine levels below occupational limits for cadmium compounds and confirmed that smoking is the predominant exposure to cadmium.[39]

The U.S. Occupational Safety and Health Administration (OSHA) requires that workers exposed to cadmium compounds be monitored to ensure that indoor air concentrations are kept below threshold action levels.[40] OSHA requires continuous monitoring if facilities are unable to keep cadmium dust or fumes below a threshold (the permissible exposure limit time weighted average) for cadmium of 0.005 milligrams per cubic meter. These rules do not distinguish between CdTe and other cadmium compounds, even though their physical properties could differ.

In the late 2000s, there were rumors that the European Parliament would propose another exemption from RoHS for photovoltaics. This got the attention of several European environmental groups considered about toxics in products. The Non-Toxic Solar Alliance (NTSA), formed in December 2009 and headed by a well-respected European research scientist, produced a white paper supporting the inclusion of photovoltaics in RoHS. Jan Kallmorgen of NTSA argued that "the use of hazardous and harmful materials in a green industry such as photovoltaics damages the whole renewable energy sector, distorts competition and slows down the emergence of innovative growth."[41] Two other European NGOs, Bellona Europa and ChemSec, echoed NTSA's position on hazardous materials in photovoltaics. Bellona Europa is a group founded to keep tabs on Norway's oil and gas sector, while ChemSec is a group that has long worked to make the world free of hazardous chemicals.

The organizations circulated strongly worded statements against cadmium-based thin-films. "Since photovoltaic panels can also be produced without containing the toxics regulated by RoHS, such as silicon, European politicians should create incentives for the solar industry to develop and deploy the cleaner technologies," argued Tone Knudsen of Bellona Europe in a press release.[42] These statements neglected the lead compounds often found in crystalline silicon photovoltaics. As the statements spread, numerous reports suggested and eventually verified that NTSA was underwritten by the private fortune of SolarWorld's founder and chairman, flamboyant German billionaire Frank Asbeck.[43] SolarWorld, one of the top ten photovoltaic manufacturers, was then manufacturing crystalline silicon photovoltaics in Freiberg, Germany, Camarillo, California, and Hillsboro, Oregon. Asbeck commented, "Clearly these are products that bring my good name [SolarWorld] into disrepute. The industry should remain clean . . . Cadmium is unnecessary and hazardous." He said that he wanted to "protect his children and other children from cadmium emissions."[44] A series of news articles around this time implicated CdTe modules in several house and barn fires in France and Germany and described the hazardous waste left behind from melted modules (ignoring all the other hazardous waste in the home). Another SolarWorld company spokesperson said, "Ultimately, the market will be divided into sustainable and non-sustainable products—and the customer will decide. Already today, we are seeing that this point will gain in importance on the market."[45]

Jürgen Werner, a solar energy research physicist in the Institute of Physical Electronics at the University of Stuttgart, advocated exclusion of CdTe from the widely heralded feed-in tariff that led to the dramatic growth of solar in Germany in the late 2000s. But he did not want photovoltaic modules excluded. "It's a scandal," he added, "the whole thing is being sold under the cover of green technology."[46] He later joined several other solar and materials scientists to publicly argue against an RoHS exemption for CdTe. "The only way to rule out the risks associated with the use of cadmium in photovoltaics," explained the institute, "is to refrain from using cadmium in the first place. This requires nontoxic substitutes to be readily available, which they are (e.g., silicon-based photovoltaics)."[47]

Crystalline silicon manufacturers, most likely more concerned with thin-film technologies occupying increasing market share, also targeted cadmium. The rise of CdTe manufacturer First Solar to number one in 2010 market share drove many competitors to take aim at cadmium. They pointed to the risk of damaging the public perception of the solar industry. One silicon manufacturer worried that "cadtel will attract a dark cloud over the entire solar energy industry."[48] To illustrate the toxicity of cadmium-based thin-films, NTSA purchased some of First Solar's CdTe thin-film modules and sent them to an independent analytical lab to conduct a hazardous waste determination test. The lab performed a very sensitive test used to determine hazardous waste in California and found that the modules would be considered hazardous under state waste laws (although they are deemed nonhazardous by the federal EPA). The tests would be submitted to the committee reviewing the RoHS recast, and later to other public hearings on hazardous waste laws, such as one in California to discuss end-of-life management classification options for photovoltaics.

At the 2010 Solar Power International Conference in Los Angeles, an executive for crystalline photovoltaic manufacturer Sharp remarked, "We have a premium for not being carcinogenic."[49] Sharp, a pioneer in making modules since the early 1960s, had just the year before been unseated as the top global photovoltaic manufacturer by First Solar, so the quote made it seem that Sharp would take an aggressive position on cadmium, questioning the environmental attributes of CdTe. Making similar claims were other crystalline silicon photovoltaic manufacturers, and even a few CIGS manufacturers that did not use cadmium sulfide. The CEO of German crystalline photovoltaic manufacturer SolarWorld made clear his opposition to cadmium-based thin-films: "The world cannot live on cadmium—I hope it will not, due to its toxicity."[50] Crystalline silicon companies argued that consumers would associate all photovoltaic modules with hazardous materials and that the use of cadmium would give the whole industry a black eye. Yet, these same manufacturers admitted to having lead, a known toxic heavy metal, in their own modules. Only a few crystalline silicon companies had phased lead out by 2018.

As discussions about photovoltaic regulations and RoHS were underway in 2010, two reports by the Norwegian Geotechnical Institute (and financed by a group of crystalline silicon photovoltaic manufacturers) detailed new research on leaching levels for CdTe. It had also purchased First Solar CdTe photovoltaic modules and sent them to an analytical lab for testing. The lab found that leaching would exceed the limit for ordinary landfills.[51] On the other hand, cadmium compounds were not likely to appear in landfill leachate—the hazardous waste fluid collected at the bottom of landfill liners—unless the loading of end-of-life modules at the landfill was substantial.[52] Ultimately, while the NTSA and Norwegian Geotechnical Institute reports led to more debate, photovoltaics remained outside the scope of RoHS through 2019. Two CdTe thin-film manufacturers in particular would be influential in making this argument, based on a strong program supported by scientific evidence, thereby shaping RoHS.

Arizona-based First Solar built a factory in Frankfurt Oder, a city near the German border with Poland. And Calyxo, a German startup owned by the top crystalline silicon solar cell maker at the time, Q-Cells, built a small production line in Germany's Solar Valley, a few hours south of Berlin.[53] First Solar put together a sustainability program that strongly demonstrated that cadmium could be handled responsibly through a detailed end-of-life recycling stewardship plan for its modules and manufacturing sites starting in 2005. One financial analyst suggested that RoHS was influenced by lobbying by First Solar's public relations and crisis management company, Burson-Marsteller, a global firm that infamously worked to restore the image of Dow after the horrendous leak of methyl isocyanate gas in Bhopal, India, and Philip Morris at the peak of the secondhand-smoke controversy in the 1980s. If such a firm could repair the images of those two notorious companies, surely it would have no problem convincing the public about a closed-loop solution to the cadmium issue. At the time, CdTe modules were exempt from mandatory end-of-life product responsibility regulations under WEEE. Some experts contend that this was in part because the primary manufacturer, First Solar, implemented a prepaid extended producer responsibility (EPR) program for every photovoltaic module sold worldwide (a program ended for non-European customers in 2013). First Solar set aside money in a restricted investment account to ensure that even if the company became insolvent, there would be sufficient funds to offer collection services. All photovoltaic technologies became subject to WEEE in February 2014. In the U.S. the EPR program is no longer prepaid but is offered as a service to customers. LCA research on the First Solar recycling process suggests that the total impacts are reduced when recycling, including the metrics acidification potential, eutrophication potential, global warming potential, and photochemical ozone potential.[54]

There are a few interesting nuances in this debate on the toxicity and sustainability of thin-films. CdTe photovoltaic manufacturers argue that the use of cadmium is sustainable because cadmium is a by-product of zinc and copper mining and they are turning a waste product into photovoltaic modules.[55] Cadmium has a supply greater than its market, and it may otherwise be left behind as mining slag or at zinc and lead smelters if there is no market to extract it.[56] It was not uncommon to hear proponents claim that CdTe technologies remediate mining and smelter sites by removing and safely encapsulating the waste. There is little information about exposures to workers involved with cadmium at earlier stages in the life cycle of the metal compounds, aside from a handful of studies of cadmium's impact on the health of workers working in and communities living near lead and zinc smelters.[57] Environmental exposures occur in the zinc mines where much of the cadmium originates, through the zinc and lead smelting process that purifies cadmium and turns it into semiconductors.

Cadmium-based products use a waste product of other industrial activities, resonating with themes of the "circular economy." One German manufacturer, Calyxo, claimed in its promotional materials that its CdTe thin-films represented "next generation sustainability."[58] Calyxo at the time was owned by crystalline silicon giant Q-Cells, the world's largest cell supplier in the late 2000s. The two companies shared a campus in Germany's Solar Valley, near Bitterfeld-Wolfen. One of its advertisements claimed, "By using [cadmium], we are relieving strain on the environment."[59] This claim merits serious consideration, since one of the key concepts of industrial ecology is reusing waste products from other industries. CdTe photovoltaics used an input in low demand in other industries and that would otherwise be left behind in the tailings at mines or in smelter waste. Some CdTe promoters even claim that cadmium thin-films safely sequester cadmium waste.[60] At a solar energy trade show and symposium in San Francisco in 2009, CdTe manufacturer First Solar's director of device development, Benny Buller, argued, "We are taking a waste product and turning it into a green product."[61]

A startup named PrimeStar, acquired by General Electric in 2010, similarly claimed "our technology prevents cadmium from entering the environment" on its website.[62] PrimeStar was led by one of the esteemed researchers from Brookhaven National Labs, Ken Zweibel, who had prepared many of the statements, comparisons, and calculations showing that concerns about cadmium exposure in CdTe thin-film manufacturing were greatly exaggerated.[63] According to this view, cadmium not used in photovoltaics still poses a threat to the environment because it would remain at smelter and mine sites. Its use eliminates these disposal pathways.

Research using LCA frameworks has found environmental benefits from directing cadmium flows toward photovoltaics.[64] Given the success of industry leader First Solar and the favorable band gap of cadmium compounds, it is likely that cadmium thin-films will be part of the industry for at least the next decade if not longer. Ensuring that workers' exposure is limited and that end-of-life photovoltaics are managed responsibly will maximize the environmental benefits of CdTe photovoltaics across their life cycle.

Several manufacturers of CIGS thin-films phased out cadmium in the buffer layer, replacing it with zinc, claiming concerns about hazardous materials regulation. The Japanese manufacturer Solar Frontier (formerly Showa Shell, a manufacturer with origins in the 1970s) phased out cadmium from the CdS layer in their CIGS modules and claimed in an interview on a clean-technology blog that "our energy solution has no toxic cadmium in it."[65] Japan has very strict rules regarding the use of cadmium in production, and given the market for photovoltaics in Japan that emerged after the Fukushima nuclear accident, there was genuine interest in eliminating toxic materials from products. Germany's Avancis similarly eliminated cadmium from its CIGS production, claiming, "we are phasing out cadmium because it is not compatible with our sustainability ethic."[66] Incidentally, SolarWorld, Solar Frontier, and Avancis were all spun out of the oil major Royal Dutch Shell at various points in the 1980s.

While cadmium compounds present environmental management challenges, thin-film solar cells do not have many of the environmental and safety hazards of crystalline silicon manufacturing. There is no need for certain problematic high-volume chemicals—no hydrofluoric acid, no hydrochloric acid, and no use of silane, which is the most hazardous chemical used in the industry, according to industry statistics.[67]

European regulators required that thin-film companies have EPR plans in place as a condition for entering the market. The fact that the European Union had prioritized renewables development was a leading factor, as was the idea that CdTe photovoltaics were consistent with ideas in industrial ecology because they incorporated a waste product. Industrial ecology aims to remake industrial systems as ecosystems, where one manufacturer's waste is another's feedstock. The consensus of experts across the space suggests that CdTe thin-film leader First Solar is an exemplar in its treatment of EHS and sustainability issues when it comes to manufacturing its product and being an excellent product steward. This should pave the way for other manufacturers of thin-films that use cadmium compounds or other heavy metals, like the emerging perovskite solar technologies. But the behavior of one firm does not predict all future cadmium-based thin-film manufacturers. Solar energy transition research will have to monitor the use of heavy metals and other chemical compounds in photovoltaics for any such changes to ensure chemical and product stewardship.

ARRA'S BIG INVESTMENTS IN THIN-FILMS

By 2009, First Solar was the largest photovoltaic manufacturer in the world. It received high praise from the clean-tech and financial press, for example *Fast Company* and *Forbes*.[68] Its success in capital markets lent credence to the "low risk" claim attached to over $4 billion in ARRA investments (through the DOE loan guarantee, investment tax credit, and Treasury 1603 programs) to build several utility-scale thin-film power plants across the American West. First Solar produced modules at factories in Ohio, Germany, France, and Malaysia, with the Asian country hosting the bulk of operations. The Malaysian facility, in the city of Kulim, employed as many as 3,700 locally hired workers in 2015.[69] Table 7 lists thin-film versus crystalline silicon projects. Thin-

TABLE 7

Thin-film technologies

Project	Developer/owner	Loan (millions)
200 MW CIGS/CdS PV manufacturing	Solyndra	$535
180 MW CdTe/CdS PV manufacturing	Abound	$400
150 MW CIGS/CdS PV manufacturing	SoloPower	$197
290 MW CdTe PV Agua Caliente Solar Farm	First Solar / NRG, MidAmerican	$967
242 MW Antelope Valley Solar Ranch	First Solar / Exelon	$646
170 MW Mesquite Solar	First Solar/Sempra	$337
550 MW Desert Sunlight	First Solar / NextEra	$1,460
Total		$4,205

Crystalline silicon technologies

Project	Developer/owner	Loan (millions)
250 MW PV manufacturing / supply chain	1366 Technologies	$150
250 MW California Valley Solar Ranch	SunPower	$1,237
Total		$1,387

films ultimately received DOE loan guarantees at a three-to-one ratio over crystalline silicon. Loans were offered to several other solar farm projects, but were turned down when other sources of financing became available or timelines for development did not match ARRA groundbreaking and construction mileposts.

ARRA investments were made in three major solar thin-film manufacturing facilities—Abound Solar, Solyndra, and SoloPower—and three utility-scale power plant projects, all by First Solar. Not all of the loan monies allocated were disbursed. For example, Abound and SoloPower never took the full loan amount to build factories. First Solar's loans for solar farms had been paid as of 2017. The lone investment in crystalline silicon technology was 1366 Technologies. It would go on to make important contributions to the efficiency of modules sold by Hanwha Q-Cells, a South Korean company with a research network based in Germany (because of its acquisition of the one-time solar cell giant Q-Cells struggling by 2012—Q-Cells struggles are one of the outcomes of the rapid ascent of China in the photovoltaic space).

Capitalizing on the advances in CdTe thin-film manufacturing was Abound Solar, a startup incubated at NREL and spun out of Colorado State University. Abound Solar made CdTe thin-films with a CdTe and

CdS semiconductor formula similar to First Solar's. With financing from ARRA, Abound Solar leveraged a $400 million loan with private capital to build an 890 MW capacity manufacturing facility out of an old Chrysler factory in Indiana. CdTe thin-film modules are widely used in utility-scale projects, but Abound found little success in securing such large-scale contracts, even though competitors like First Solar, using the same technologies, were building the largest utility-scale solar farms in the world (mostly through subsidiary LLCs.)

By the time Abound was ready to deliver the volume of modules to support larger-scale projects, the renewable portfolio standards for utilities were well past filled. For example, according to data in California, the most lucrative market in which to sell electricity, Michael Picker, a senior adviser to governor Jerry Brown, said that more than double the quota (which was 33% at the time in California) had been proposed for development by 2010. Without vertical integration with a land management and acquisition firm, it was hard to compete. Facing manufacturing challenges as well, Abound ultimately declared bankruptcy in 2012. (Only $60 million of the $400 million loan was lost to the federal treasury.)

One lesson learned from the bankruptcy of Abound is that companies without plans in place to minimize the costs of waste and hazardous materials disposal can be riskier investments. Abound was later ordered to encase thousands of pallets of photovoltaic modules deemed hazardous in cement for permanent burial. Its Longmont, Colorado, facility left behind thousands of gallons of cadmium-based liquids, costing taxpayers over $2.2 million to clean up and dispose of, and four other sites were also ordered cleaned up.[70] It was widely reported that First Solar recycled a large portion of Abound's defective and unsold modules. The purpose of this would be to recover the valuable tellurium and return it to feedstock, but also to ensure that public perception of CdTe photovoltaics remained positive.[71] Reserves of tellurium are around 24,000 tons, but annual refinery production was only around 110 tons in 2016, so recycling this metal helps stretch supplies further.[72]

Two thin-film companies, Solyndra and SoloPower, also received DOE loans (supported by ARRA) to build manufacturing facilities. Both made CIGS thin-films with a layer of CIGS and a second semiconductor layer of CdS. Solyndra built a factory across from the Tesla Motors factory in Fremont, California (also built with ARRA support, under the Advanced Technology Vehicle Manufacturing Program) and adjacent to its pilot facility. SoloPower would grow out of its San Jose pilot facility, used prior to 2010, and move to Oregon to scale up production with its ARRA grant.

While the loan was being considered, the Silicon Valley Toxics Coalition submitted a letter to the DOE, the lead agency for the environmental assessment, documenting some of the impacts of manufacturing with cadmium. The early warning signs that Solyndra was headed for trouble were related to the use of cadmium. First, it was learned that Solyndra's photovoltaic modules did not pass California's or the EPA's toxic waste characterization tests. Anyone de-installing Solyndra's modules would become a hazardous waste generator in the state, or anywhere in the U.S., which would add significantly to the cost of end-of-life disposal. Second, other more commercially mature CIGS companies, like Solar Frontier (formerly Showa Shell) and Avancis (also at one point formerly Shell), had already phased out the cadmium in its buffer layer, replacing it with zinc. This would be attractive to investors looking at the strict rules on cadmium in markets in Japan and Europe, despite RoHS.

A California Public Records Act request was leveraged to obtain toxic waste numbers from Solyndra kept by the California Department of Toxic Substances Control. As a thought experiment, conducted with the Associated Press reporter who obtained the data, we assumed that Solyndra had manufactured 500,000 photovoltaic modules, or 100 MW, as claimed on its website. These half-million modules would generate about 3,300 GWh over their 20-year life. At 6.2 million pounds of chemical waste from that facility, that amounted to about 1,800 pounds of chemical waste per GWh produced. Previous studies had measured pollution in grams, not pounds. The total amount of hazardous waste produced at Solyndra's facilities was 12.5 million pounds, when including waste generated during the development of its manufacturing process. Most of the hazardous waste was water tainted with cadmium, and most was shipped out of state from 2007 to 2011, until the company declared bankruptcy. (It is worth noting that water is very heavy.) An investigative report by the Associated Press found that most of this waste was trucked across the country to Arkansas, Mississippi, and a handful of East Coast states.[73] The data mainly reflect an early startup manufacturer, which was experiencing production problems and selling only a few modules. They also reflect the early development of an industry, as waste flows in a more fully developed chain of custody should be shorter.

Solyndra warehoused small amounts of cadmium waste at a facility in Milpitas, California, a town neighboring its Fremont plant. This facility was not financed by the loan, and when Solyndra filed for bankruptcy, it abandoned the storage facility. For several days it headlined the local news in the San Francisco Bay Area, with the public linking the national story about potential cronyism in the loan program with these claims of hazardous waste abandonment. Solyndra's cadmium-based modules played politically into the hands of critics of the Obama administration's policy. Southern California congressman Darrell Issa led the opposition to the loan program, with his office holding hearings and preparing reports critical of the loan program and the DOE.[74]

SoloPower never scaled to the production capacity it had intended for its Portland, Oregon, facility. Two new manufacturing lines in its factory were planned after ARRA support was awarded, but it used very little of its loan guarantee. To summarize the results of the loan program, two thin-film manufacturers of CIGS and CdTe photovoltaic semiconductors were bankrupt by 2012, while the third is still in limited production, as it has been since 2010. Loan program leaders would later testify before Congress that China's illegal dumping may have played a role in the struggles of the three thin-film companies that received ARRA support (Abound Solar, SoloPower, and Solyndra) to build manufacturing facilities.[75]

Solyndra defaulted on its loan in August 2011, before it could sell many modules. The fallout was accompanied by claims of cadmium pollution, including at the facility in Milpitas where cadmium waste had been abandoned.[76] Solyndra left behind dozens of drums of hazardous waste, as well as manufacturing equipment that was contaminated with semiconductor materials. The incident was not yet causing pollution in the community and would turn out to be relatively easy to clean up, but it poured more fuel on the bad news and created more negative headlines for ARRA.

Manufacturers using cadmium in photovoltaic manufacturing, like First Solar, have a strong record of protecting workers from cadmium exposure during manufacturing. They have a track record of keeping emissions and effluents below local regulatory limits. But startups and less scrupulous companies could be less careful. Most regulatory institutions have decided that the benefits of CdTe and CIGS technologies warrant relaxing or exempting them from rules and requirements that might be applied to consumer products. These companies had to endure much criticism and opposition to the use of cadmium, requiring that they collaborate with research scientists or contract for independent research over the years to reaffirm that the renewable energy benefits outweigh the toxic issues.[77]

USING LIFE CYCLE ASSESSMENT TO RESOLVE
ENVIRONMENTAL CONTROVERSY

CdTe thin-films did not receive similar scrutiny from U.S. regulatory institutions as they did from RoHS in Europe, in part because there is no similar federal directive governing toxics in electronics or rules for e-waste. A few environmental organizations raised concern about the use of cadmium in solar manufacturing in the late 1990s. In 2002, Greenpeace asked the California Public Utilities Commission whether California ratepayer investments in cadmium were compatible with an ethic on sustainable energy.[78] In one statement, used in a public presentation at the Brookhaven National Laboratory, Greenpeace compared cadmium in CdTe thin-films to cadmium in coal, arguing that investments in toxic photovoltaic modules were not worth the impact on the environment.[79]

> Greenpeace is deeply concerned with the possibility of the CPA [California Power Authority] choosing to purchase solar modules that contain toxic metals. . . . Current CdTe panels contain approximately 6 grams per square meter, resulting in cadmium emissions of 0.5 grams per GWh, equivalent to that of a coal fired power plant. The majority of these emissions (77%) result from mining and utilization of the modules, therefore a comprehensive collection and recycling program would not reduce the environmental impacts of these panels. Greenpeace believes that the [California Power Authority's] commitment to sustainability would be compromised if you were to purchase CdTe panels. The state should not spend taxpayer money on technologies that will ultimately end up polluting the environment. There are many solar photovoltaic technologies that have fewer adverse effects on the environment that will satisfy the needs of the California Power Authority's request for bids.[80]

To respond to the claims about cadmium pollution, researchers from the Brookhaven National Laboratory and NREL used LCA to evaluate cadmium emissions across all phases of production from various electricity generation sources, including CdTe thin-films.[81] LCA is a quantitative environmental assessment framework that looks at the cradle-to-grave impacts of commodity production. The use of LCA is increasingly popular in corporate social responsibility efforts, eco-labeling, and regulatory decision-making and public policy, where quantifying the net impacts of a product's life cycle, from raw material extraction through manufacturing, use, and disposal, is of interest.

LCA is emerging as productive framework for environmental knowledge production and synthesis by the expert engineers, scientists, and computer programmers that do this kind of work. Practitioners have developed guidance with the help of the International Organization for Standardization (ISO) on how LCA should be conducted and what kinds of claims manufacturers and experts can make based on data quality and other interpretive criteria. In working groups with the most prominent LCA practitioners, the International Energy Agency has developed more detailed guidance on how to conduct this research on photovoltaics.[82] In the peer-reviewed literature, LCA is extensively used to compare energy systems. The systematic and quantitative approach of LCA readily lends itself to accounting for the environmental impacts of energy because energy resources can be easily made commensurable. Energy is a derived demand, meaning that the demand for energy services mediates the demand for energy resources. The electricity grid requires managing a complex system of power plants and power demands drawing from various resources, making such comparisons of electricity resources not only possible but also meaningful for environmental and climate policy. People want electric outlets to provide electricity and automobiles to provide mobility; they do not care as much what resources make this possible. This means that energy lends itself to commensurable functional units (all energy can be converted smoothly between units: kilowatt-hours, megajoules, barrels of oil equivalent, etc.). There are more LCAs written in English about energy systems than any other industry or sector. Database searches in April 2016 produced 104,000 results for "life cycle assessment" plus "energy" in Google Scholar; the same search in Science Direct yielded 115,710. Similar searches for "life cycle assessment" and "food" (89,160 in Science Direct), and "agriculture" (38,900 in Google Scholar, 31,185 in Science Direct) yielded similar results. There is arguably no comparable set of industry-specific LCAs undergoing extensive harmonization—a process for standardizing the rules for conducting LCAs so their findings are more easily compared. LCA is also becoming embedded in many regulatory instruments, which in the U.S. includes LCA-based performance standards set by the EPA, DOE, and state agencies such as California's Air Resources Control Board.

Conducting an LCA requires proceeding along several major stages, though in practice these are iterative processes, with constant tunings, adjustments, and quests for more representative data and model treatments. The first stage is to establish the goal and scope of the LCA. This is where the practitioner or client determines the major parameters of the life cycle, which ensures that the approach meets the project objectives (to compare, for design, etc.) by confirming the functional unit, setting system boundaries, and identifying environmental impact categories. These impact categories may include global warming potential, eutrophication, acidification, toxicity, ground-level ozone/smog formation, stratospheric ozone depletion, natural resource depletion, water pollution, water use, eco-toxicity, and human toxicity. There are a few organizations that standardize impact category data, including the EPA's Tool for Reduction and Assessment of Chemicals and Other Environmental Impacts (TRACI). A review of 150 published LCAs found over 90 different metrics reported in publications from 1976 to 2011. The proliferation of metrics has led to calls for greater standardization and harmonization. NREL's harmonization project produced a meta-analysis of energy-related LCAs.[83]

Environmental metrics have discursive power because they appear as impartial representations of technologies. Because it relies primarily on quantitative data, LCA appears objective, overcoming problems of distance and distrust.[84] LCA is conducted in four phases: goal and scope; inventory; assessment; and interpretation. Note that though it is described as a quantitative tool, two of these four phases are deeply qualitative and therefore mediated by human subjectivity. Even where objective accounting of inventories is done, the process of translation require commensuration—reducing complex socio-ecological interactions to metrics. Commensuration has a long history as "an instrument of social thought, and as a mode of power."[85] Commensuration occurs when the complexities of social life and exchange are reduced to metrics.

The first LCA of photovoltaics, in 1976, found that "energy payback times" for photovoltaics—the amount of time photovoltaics must generate electricity to offset the amount of energy invested—were decades long, meaning that many would never pay off the energy investment before they were de-installed.[86] In the subsequent forty years, energy payback times have shrunk considerably, to a few years or even less than a year, and environmental performance has improved with scaling-up of production.[87] Governance and corporate practice are increasingly using performance metrics and tools like LCA for purposes of investments and auditing.[88] LCA can be used as a framework to help make decisions, differentiate technologies, improve production processes, or explore the interconnected aspects of environmental impacts across the commodity chain.[89] LCA results are commonly compiled and published as technical reports and research articles that describe the environmental impacts of different combinations of flows of materials and energy.[90]

The complexity of production processes is reduced to material flows, often lacking the context for community and worker exposures.

The Brookhaven research found that cumulative flows of cadmium to the environment are higher from coal-fired electricity than from CdTe photovoltaics per unit of energy.[91] This put to rest the warning from Greenpeace. When electricity from CdTe devices displaces electricity from coal, overall cadmium emissions to the environment decrease considerably. This study further suggested that there were more cadmium emissions from crystalline silicon photovoltaics, which themselves contain no cadmium and produce no cadmium by-products. The reason is that crystalline silicon requires higher energy inputs, and that energy comes largely from coal, which releases cadmium during combustion.

In some back-of-the-envelope estimates intended to educate the public on this issue, Brookhaven's scientists, Vasilis Fthenakis and Ken Zweibel, also compared the amount of cadmium in various household items. One calculation favorably compared CdTe photovoltaic modules to the larger amount of cadmium found in nickel–cadmium batteries, commonly used in household flashlights at the time (they have since fallen out of favor in commercial and consumer applications).[92] They emphasized that the scientific evidence shows that most exposure to cadmium comes from phosphate fertilizers and fossil fuel combustion.[93] In Germany, for example, most cadmium emissions from energy production are from coal-fired power plants, and this amount would decrease if all this coal were replaced with cadmium-based thin-films.

Another Brookhaven study considered the effect on cadmium emissions of replacing the CdS semiconductor in CIGS photovoltaic modules with zinc sulfide.[94] It found that the zinc-based version would result in higher cadmium emissions, even though it did not contain cadmium. As in the case mentioned earlier, the higher cadmium emissions are from upstream coal-fired electricity. This leads to the seemingly paradoxical claim that a crystalline silicon photovoltaic module with no cadmium in it can generate more cadmium pollution than a comparable cadmium-based photovoltaic module. (This research assumed module and ingot production using electricity from the coal-intensive grid of Europe in the mid-2000s.)

How do we interpret these LCA results? In what ways are they meaningful? Numerical data sometimes require careful interpretation—the last and perhaps more important, yet subjective element of LCA—to become actionable information. The study provides insight and context on emissions of cadmium compounds to the environment. The overall emissions from energy technologies are important to know. But this research does not model the distribution of risks to workers along the commodity chain, or spatially, in ways that map actual facility emissions to existing communities. There are no human lives in these LCAs, only flows of materials and energy.[95] These too are important information in regard to sustainability decisions. But even if they identify where and how emissions occur, it is difficult to identify opportunities to reduce emissions and effluents, or where EHS efforts are needed.

These studies were also limited in what they could say about the occupational health and safety of workers upstream or downstream—upstream, where cadmium exposure occurs prior to semiconductor preparation, or downstream, at end-of-life. In some cases, the supply chains of metals are unknown; they can be difficult to trace. Metals may be sourced from multiple production streams. Aluminum, for example, can be sourced from recycled supplies or bauxite smelters. Limited information on worker exposure can be drawn from regulatory standards. Regulatory exposure thresholds are generally overestimates of actual exposure, but serve as a good proxy for understanding safe levels of exposure. OSHA took action on exposure to cadmium in the 1980s.[96] The acceptable concentration of cadmium in an industrial facility is 2,000–50,000 nanograms per cubic meter, three to five orders of magnitude higher than ambient cadmium concentrations typically measured near industrial parks centered around coal-fired power plants (15–150 nanograms per cubic meter).[97] Worker exposures to cadmium are permitted to be an order of magnitude higher in a facility handling primary metals than product manufacturing.[98] But these workers are also exposed to many other heavy metals, such as lead and arsenic.

In the spirit of using LCA to inform environmental justice and EHS topics, it is helpful to shed light on the results for workers exposed to cadmium compounds in the workplace. While it is important to know that coal causes more cadmium pollution than thin-film photovoltaics per unit of energy, the kinds and routes of exposure are very different. Perhaps in some places, as the paper suggests, there are more cadmium emissions to the environment from crystalline silicon photovoltaics. It is also true that workers manufacturing crystalline silicon photovoltaics will never be exposed cadmium in the workplace (though they may be exposed to other hazards). Understanding the risks of chemical use for worker and community safety requires focusing on workplace exposure and emissions directly to the community, not necessarily the full life-cycle environmental releases.

THE POWER OF NUMBERS

These LCA and EHS studies of thin-films were influential beyond the initial controversy with Greenpeace. The Silicon Valley Toxics Coalition used the reports to choose a position of technology agnosticism in its Clean and Just Solar Energy Campaign. It weighed the opinions of Greenpeace against the technical evidence presented in the research, as well as conversations with manufacturers, to take the position that the risks from heavy metals could be reduced with strong EPR and manufacturing best practices. Several controversial utility-scale solar projects, including several ARRA projects that used First Solar's CdTe thin-films, submitted copies of these LCA studies to the public record in proceedings required under the National Environmental Policy Act. These submissions helped defuse cadmium toxicity concerns by showing that detailed investigations had already found minimal impacts from cadmium-compound exposure, including issues with leaching and even fires. These peer-reviewed studies showed minimal if any risks. Some of the studies appeared as appendices to official environmental impact statements. With this expertise brought into the public and decision-making dialogue, concerns about cadmium compounds were minimized, and claims of cadmium pollution were easy to dismiss as missing the broader public health benefits of solar power.

Numbers are powerful because they appear as objective representations of reality.[99] They have the property of ordering and ranking, which makes numbers more active participants in decision-making than we realize. Translating social and environmental phenomena into numbers can redistribute agency typically attributed to humans because the numbers themselves tell the stories and act as arbiters.[100] Representing objects as numbers puts them on a relative scale, and in doing so mechanizes decision-making. Numbers force the hands of decision-makers, who often rely on formulaic responses; it is difficult to refute quantitative evidence without quantitative counter-evidence.

Research at the intersection of geographical thought and science and technology studies shows how the social processes involved in the construction of metrics make them less objective than they appear.[101] Metrics are produced through the social activities of abstraction (making material things into numbers), commensuration (making things comparable), and reification (making the numbers back into "things" or representations that appear objective).[102] This makes metric construction and use subject to political and cultural forces.[103] As numbers increasingly represent environmental impacts through the more widespread use of LCA, the political power to talk about environmental impact shifts toward experts. Quantified environmental claims tend to carry more weight with decision-makers and even the public. These trends suggest that LCA experts will increasingly be interlocutors who weigh in and define the multiple narratives that travel with the resource and technologies they describe.

LCAs aggregate information about environmental impacts across various stages of production, and this can obscure who is exposed to heavy metals, how, and by how much. Acquiring this latter information about exposures accounts for risk—testing for heavy metal levels in blood, urine, or ambient air is common practice where such materials are present in industrial and manufacturing settings.[104] These impacts are the specific and locally situated differences that LCA approaches can erase when data are aggregated into material and energy flows and represented as bar graphs. Whereas inputs and emissions are relatively easy to report, occupational or epidemiological data is far more burdensome to collect, interpret, and analyze—not to mention raising all sorts of implications for liability and regulation.

Numbers have power to shape public discourse.[105] Environmental impacts and material flows are transformed into numbers that purport to be objective but are shaped by social activities, mediated by values and decisions about what and how to count. Constructing and using metrics moves matters to the realm of technical expertise, and out of the realm of political debate. Environmental metrics shift claims about pollution into the technical sphere, where expertise is required to weigh in or challenge. Political concerns become more muted when debates shift to the technical sphere. "The apparent objectivity of numbers, and of those who fabricate and manipulate them, helps configure the respective boundaries of the political and the technical."[106] Metrics are technical instruments subject to their own politics and human subjectivities and errors. The activities of classifying and compiling metrics create new categories, and "each standard and category valorizes some point of view and silences another."[107] Numbers can discipline and delimit the conversation about the impacts of technology, tilting the scales in comparisons, or offering insights into future possibilities or counterfactuals.

For energy transitions, it is important to know that leading companies are more than adequately practicing chemical and product stewardship today. But it should not be assumed that the most responsible companies will always be making decisions because of the geography and governance of commodity chains. If there are new entrants into the market with less competent EHS practices, or if manufacturing happens in places with poor regulatory standards, perhaps industrial accidents and worker safety problems will occur more frequently. There also are legitimate concerns about cadmium compounds that are not well characterized in the supply chain. The consensus is that photovoltaics are among the least impactful sources of energy available, but careful EHS management is required to maximize their benefits.[108] Numbers can represent future states of the world, and should be read as only a partial view of a complex global environmental and social change. Further complicating these matters is whether criticisms of technology are based on competitive posturing or other motivations, or whether genuine issues are at stake. For example, much of the criticism of thin-film photovoltaics has been linked to proponents or producers of crystalline silicon photovoltaics.

Solar energy innovations focus on driving down costs by working through the challenges associated with innovations in policy, materials science, manufacturing, and the balance of materials. Making photovoltaics cheaper is the top priority for programs that received support such as the DOE's SunShot Initiative and ARPA-E. The SunShot Initiative, managed by the DOE's Solar Energy Technologies Program, established price goals of $1 per watt for utility-scale solar, $1.25 per watt for commercial, and $1.50 per watt for residential photovoltaics. Projects funded by ARPA-E have led to scientific discoveries that reduce costs by increasing output and efficiency, advance materials science, and make the total system costs lower. While these material innovations promise cheaper solar, many rely on nanotechnologies such as quantum dots and nanospears, and emerging materials where little is known about health or environmental effects. Driving down costs opens the door to questions about environmental justice, when innovations use hazardous materials or affect workers' exposure to hazardous materials, or changes to workers' rights and standards occur through geographical shifts in production. These too are manageable EHS issues, but warrant early investigation to ensure that best practices are available at the time of commercialization or scaling up.

The tensions between justice and innovation are further illustrated at the frontiers of solar technology research. With the collapse of Solyndra in 2011, enthusiasm shifted from innovations in first-generation thin-films (CdTe, CIGS, amorphous silicon) to second-generation organic, dye-sensitized, and perovskite solar cells. The laboratory efficiency of perovskite solar cells already exceeds the commercial efficiency of crystalline silicon.[109] Perovskite semiconductors are made of a methylammonium lead or tin halide and an inorganic–organic hybrid lead compound. This innovation in materials science offers opportunities for significant cost reductions, because the materials are abundant and cheap and processing is relatively straightforward. Most current configurations of these solar cells contain lead. But efforts are already underway to eliminate the toxic heavy metal. Companies like First Solar have shown it is possible to manage the risks from heavy metals with strict EHS standards and product stewardship. Companies pursuing photovoltaics based on heavy metals will require similar best practices to ensure just transitions for all workers and communities, if this technology is brought to the scale needed for meaningful climate solutions. For now, physicists and materials scientists are working to find the most stable compounds to make commercially available perovskite solar cells, which most likely will enter the market combined with crystalline silicon.

CHAPTER 4

Recycling and Product Stewardship

Recycling is an imperative for a successful, sustainable photovoltaic industry.

—Karsten Wambach, engineer for Sunicon, a former subsidiary of SolarWorld[1]

FORECASTING PHOTOVOLTAIC WASTE

The photovoltaic industry's remarkable growth will eventually result in huge quantities of retired modules that have become waste in need of disposal. A report from the International Renewable Energy Association estimates 80 million metric tons by 2050.[2] Assuming that each gigawatt of photovoltaic capacity represents somewhere around ten million modules, there are billions of photovoltaic modules installed worldwide today. The U.S. has installed about 50 GW of distributed and utility-scale photovoltaics as of 2018, so roughly half a billion modules. The Energy Information Agency of the Department of Energy (DOE) estimates that residential rooftop photovoltaics will continue to grow at around 7% each year through 2040.[3] By 2030, cumulative global installed photovoltaics will be around 8 terawatts.[4] This amounts to hundreds of billions of photovoltaic modules installed in utility-scale, distributed, and off-grid applications. Where will all this photovoltaic material go after the end of its useful life?

There are tremendous benefits coming online with these photovoltaic systems. Electricity from photovoltaics displaces the dirtiest energy sources, the fossil fuel peaker plants that provide electricity during peak hours of the day. Replacing peaker plants with solar power means we breathe cleaner air, fewer nitrogen oxide emissions that cause photochemical smog, and less particulate-matter pollution. There are also unquestionable benefits for our climate. The greenhouse gas emissions related to the manufacturing of photovoltaic modules are negligible compared to the lifetime emissions they save after installation.[5] Photovoltaics are one of the few energy technologies that can generate electricity close by healthy communities. They do not cause air pollution, and can be integrated with the built environment.

However, photovoltaics could present disposal challenges when large amounts start to enter the waste stream. Photovoltaic modules degrade in power output over time until they are no longer generating sufficient electricity or meeting the needs of the homeowner or power plant owner. Nearly all companies offer warranties on photovoltaic modules that guarantee a certain power output (commonly 80%) for 20 to 25 years. Because of these long warranties, companies aim to make a product with minimal degradation, and many photovoltaic modules can operate for many decades without fading appreciably. Most "solar panels" on rooftops and in power plants are not going to be decommissioned anytime soon. Still, every photovoltaic module will eventually become photovoltaic waste.

Some photovoltaic module waste comes from defective or broken modules that require disposal before they are installed. Some manufacturers may produce photovoltaic modules that fail quality control or have manufacturing extrusions, creating waste on the factory floor. Other modules may break en route to or at an installation site, or may fail at some point early in operation. These photovoltaic waste streams will represent early waste flows.

The challenge for policymakers is that the steep wave of photovoltaic waste is over the horizon, too far off to spur action, as large volumes will not show up for twenty to forty years, or even later. But as early photovoltaic installations are upgraded and replaced with new modules, the waste stream will quickly rise to very high volumes. If there is no plan in place, it is not clear where they may end up. Where will they be directed? Developing countries? Landfills? Smelters? High-value recycling? Finding a way to close the loop on photovoltaics, shifting the material flows closer to "circular economy" principles, could be one way to manage future photovoltaic waste streams.

Some photovoltaics will become waste much earlier. The so-called bathtub curve shows a high rate of failure at first that slowly diminishes, only to rise again as the expected lifetime approaches. Early failures are often due to mechanical issues or breakage. Modules that are defective will show that right away, if they made it to installation

TABLE 8

	Photovoltaic modules (approx.)	Owner/developer	County
Desert Sunlight Solar Farm	8.9 million	First Solar	Riverside
Topaz Solar Farm	8.9 million	First Solar	San Luis Obispo
Antelope Valley Solar Ranch	2.9 million	First Solar	Los Angeles
California Valley Solar Ranch	1.3 million	SunPower	San Luis Obispo
Mount Signal Solar	1 million	PayneCrest	Imperial

without being caught by quality control. Modules can be broken during transport and installation, or damaged by random or acute events like hail, mice, heat stress at high temperatures, mechanical stress (mounts fastened too tight, for example), bullet holes, even tornadoes. These are all-low probability events, so the volume of middle-aged end-of-life modules remains small.

Another source of end-of-life photovoltaic module waste is discards from manufacturing facilities. These end-of-life modules might be clearly defective or broken, or fail quality control (even for cosmetic reasons) at the fab. A similar stream of end-of-life photovoltaics is from modules that are removed because they are defective or do not meet warranty claims, if the company directly processes warranty reserves. These are modules that might generate electricity for a few years, but not last as long as the rest of the system or power plant.

California will be a sentinel for photovoltaic waste because the state is by far the leader in photovoltaic module installations in the U.S. California also has the most utility-scale photovoltaic power plants. The means that in addition to the distributed flows of photovoltaic waste, there will be places with very high concentrations. In California, many of the rural areas where the largest solar farms are there may have disposal challenges due to the remoteness the of facilities or the lack of disposal infrastructure. However, utility-scale installations are likely to have some kind of waste logistics company involved in case disposal is needed. Table 8 shows the largest installations, developers, and locations of some of the large photovoltaic installations in California.

Photovoltaic module waste flows can be estimated with a few assumptions about when waste will be generated and where end-of-life management opportunities arise. Dr. Vasilis Fthenakis, of Brookhaven National Laboratory and a professor at Columbia University, developed a framework for addressing this question. It begins by asking how many photovoltaic modules are produced by the industry each year. This value is readily available, because numerous industry reports estimate it regularly. Second, how much does each photovoltaic module weigh per rated power output? This is estimated in tons per MW-peak. This is more speculative, because photovoltaic modules are getting more powerful over time, but some assumptions can be made about the power output and mass of panels every year to account for increased output and lighter modules. Third, how much waste is generated in production at photovoltaic manufacturing facilities? All manufacturing results in some kind of waste. Manufacturing extrusions can damage photovoltaic modules during manufacture, and they can be removed for quality control, or downstream if the problem remains undetected before it leaves the factory gate. Fourth, what percentage of photovoltaic modules are damaged or defective en route or during operation? Installers receive broken photovoltaic modules, and others may become broken or damaged in the field. This includes damage from external factors such as extreme weather, falling debris, and so on. Finally, how long will the modules operate? When will customers start replacing them, and when will they begin to show up for disposal? Some photovoltaic modules might be sent for reuse, since many decommissioned photovoltaic modules are still capable of delivering power.

RARE, VALUABLE, AND PRECIOUS METALS IN PHOTOVOLTAIC WASTE

Photovoltaic modules contain a number of valuable or rare materials, and recycling will ensure that these rare materials are recovered for reuse.[6] Scarce and precious metals are now even geopolitical issues. In 2011, China blocked exports of rare earth elements (the lanthanides on chemistry's periodic table) such as terbium, europium, cerium, and neodymium to several countries for strategic reasons. China controlled 97% of global rare earth production that year. The announcement drove prices for these commodities up over 750% within a year.[7] Other rare inputs have similar volatility. A looming limitation for decarbonization technologies (renewable energy, energy efficiency) is the availability of a handful of rare metals, including rare earth elements, precious metals, and other metals, which are used in products such as electric vehicles, hydrogen fuel cells, photovoltaics, and next-generation LED lights. The only rare earth element associated with the photovoltaic industry has been cerium, which is believed to have been used in a filtration process, but only by one known manufacturer over time.

Almost all photovoltaic modules require inputs that are relatively scarce, precious, or subject to price volatility: silver, tellurium, and indium. The price and availability of rare and precious metal can influence costs of electronics and photovoltaics significantly. The importance of ensuring adequate supplies of critical elements has long been recognized by the Department of Defense.[8] The Department of Energy's SunShot Initiative recognized this too and tracks four metals (silver, indium, gallium, and tellurium) that can influence the cost of photovoltaics and possibly even hinder cost reductions.[9]

Crystalline silicon photovoltaic modules contain a number of valuable materials that can be recovered and reused, such as copper, aluminum, and glass. The precious metal silver is also used in these modules, in the electrical contacts (look closely at a module and notice thin the lines of metal running across the blue, grey, or black photovoltaic material). From 2008 to 2009, silver prices rose while many other costs fell for photovoltaic manufacturers as investors moved out of securities and other risky financial products to the safe refuges of precious metal commodities. In 2011, the solar industry consumed 11% of the global silver supply.[10] By 2018, that number was over 20%.[11] Silver is commonly recovered in mines where there are also ores associated with other metals, such as lead, gold, copper, zinc, and cobalt. The dominant producers include Mexico, Peru, Australia, the U.S., and Chile. Smelters that produce silver can be designed to recover other metals, but only about 32% of that silver is recovered, and the rest disposed of as slag.[12]

Few materials from end-of-life photovoltaics can be recycled into high-value glass cullet (recycled glass that has been crushed and is ready to be melted) or "downcycled" into lower-value secondary glass products. A photovoltaic module is mostly glass by weight (75–90%), but recyclers report that much of this glass cannot be made into flat glass again due to impurities. Common problematic impurities in glass cullet include plastics, lead, cadmium, and antimony; where they are present, the glass can only be downcycled. Recycled silicon wafers also have value since a significant energy investment is required to make them.[13] Recovering the silicon from photovoltaic cells can reduce overall energy use because less polysilicon needs to be refined.

The tellurium, indium, and gallium used in thin-film photovoltaic technologies are among the rarest elements in the Earth's crust. Recycling is one way to recover these materials. Modules are first disassembled to recover valuable materials like copper wiring and aluminum (if they have frames). Then they are shredded into small pieces. A process developed by Brookhaven National Laboratory and First Solar uses dilute acids to remove the CdTe from the module.[14] Up to 95% of the cadmium and tellurium is recovered from filter cake, and reprocessed into new semiconductor material by a third party. To ensure that CdTe modules are recycled safely and responsibly, ambient cadmium emissions are monitored in real time to ensure that occupational exposure is minimized.[15]

Tellurium occurs in the earth's crust at the rate of only one part per billion. It is about a thousand times as scarce as the rare earth elements that are the subject of ongoing trade disputes among the U.S., China, Malaysia, and Japan. Annual global production is on the order of 200 to 1,000 tons. The availability of and market for tellurium have interesting implications for the photovoltaic industry, in part because one photovoltaic manufacturer dominates the market. First Solar, of Tempe, Arizona, currently purchases 40% of the total volume of tellurium sold.[16] In 2010 First Solar agreed to buy a significant portion of high-purity tellurium from Apollo Solar Energy, a Chinese tellurium supplier operating open-pit mines on its Dashuigou property in Sichuan Province.[17] This is the only mine in the world where tellurium is the primary product. Most other tellurium supplies are secondary or tertiary products of copper or gold. Recovery is largely driven by the price; as the price goes up mines are more willing to put in the effort to recover tellurium from ores. The DOE expects that by 2031 there will need to be additional main-product tellurium mines and an ample secondary supply of recycled CdTe photovoltaic modules.[18] In 2011, financial analysts observed that First Solar had acquired a gold-tellurium mine in Mexico to secure future cadmium telluride supplies.[19]

First Solar's efforts to secure a tellurium supply are bolstered by the reuse and recovery of tellurium from manufacturing scrap and end-of-life photovoltaic modules. It builds recycling operations into its factories, and ships the resulting filter cake to its supplier, which returns it as high-value semiconductor feedstock. It is also making its semiconductor layers thinner and more efficient over time, yielding more energy per module, with less tellurium. Its current line of photovoltaic modules contain semiconductor layers that are about three microns thick, and there is potential to reach two microns in the not-too-distant future.

Indium is a key input to some thin-film photovoltaic technologies and is also relatively rare. Competition for indium for use in flat-screen televisions could ultimately restrict its availability for commercial CIGS photovoltaic production.[20] The need to recover indium should therefore encourage the recycling of CIGS, and this is seen as an important issue for large-scale commercialization and deployment. The recycling of amorphous silicon is said to be one of the cleaner recycling processes of all photovoltaic technologies. Though most of the literature mentions this, none mentions that indium tin oxide has been linked to occupational lung problems in flat-panel television recycling facilities.[21] Amorphous silicon manufacturing has been marred by several silane accidents, and the technology suffers from low efficiency.

The major indium producers are China, France, Canada, and Japan. Indium is mainly recovered from zinc production at smelters. Indium is also recovered from waste flat-panel displays. The price of indium has been far more volatile in recent years than that of tellurium, even though it is more widely available and more widespread in the Earth's crust (about one part per million in the Earth's crust). There is no production of indium in the U.S.

Gallium is one of the four scarce metals tracked by the DOE, and its scarcity is a potential barrier to low-cost photovoltaics.[22] Gallium is used in CIGS photovoltaics, as well as multijunction and dual-junction solar cells used in satellites and space craft. The metal is widely distributed, and no country dominates the supply, though the primary global suppliers are Australia, Guinea, Brazil, and Jamaica.

Some advanced photovoltaic technologies currently being explored in scientific laboratories, such as dye-sensitized photovoltaic cells, require ruthenium, a platinum-group metal of which only about 12 tons is mined annually.[23] Molybdenum is used as a contact layer in some kinds of photovoltaics. The presence of these valuable materials suggests that recycling would be both economically and environmentally beneficial in the long term. In fact, many studies show that the limited availability of some of these key inputs caps the total amount of photovoltaic module production possible without materials recovery.[24] Other valuable materials used in thin-film modules include copper, aluminum, and glass (Table 9).

Today, very few of the components of a photovoltaic module are recovered for reuse as feedstock in similar products. Even where photovoltaic modules are safely and responsibly recycled at facilities renowned for worker health and safety, not all the components of photovoltaic modules are recovered. At ECS Refining, a San José company locally recognized as a leader in product stewardship and recycling advocacy, photovoltaic modules result in only smelter flux, sold off in bulk with other glass and e-waste products.

TABLE 9

Material	Content by weight	Use
Glass	~90%	Substrate, weatherization
Steel	0–50%	Frame and mounting equipment
Aluminum	0–5%	Frame and mounting equipment
Copper	~2%	Wires, electrical equipment
Plastics	5%	Junction boxes
Silver	<1%	Metallization frit paste in c-Si
Tellurium	1%	Semiconductor in CdTe photovoltaics
Silicon	2%	Silicon wafers
Cadmium	1%	Semiconductor in CdTe photovoltaics
Lead	<1%	Solder; also found in glass
Antimony	trace	Solar glass
Gallium arsenide	5%	Semiconductor in multijunction cells

Almost all photovoltaic modules contain plastics of some kind. For example, all must have a junction box—the unit that connects the solar cell to the wiring needed to deliver electricity. Though most of the thermoplastic materials used for junction boxes are "halogen free" (meaning they contain no bromine or chlorine), some module junction boxes may contain brominated flame retardants. The EU and California both have restrictions on selling products containing brominated flame retardants. While it is critical that photovoltaic modules use the most fire-resistant materials possible for electrical equipment and wiring, there are many such materials available that do not contain carcinogenic bromine-based materials. More recently, there have been efforts to remove halogens from backsheet, laminates, and wires and cables.

PHOTOVOLTAIC MODULE TOXICITY

Where photovoltaic modules are disposed of, they can present hazardous-waste disposal issues. One concern raised with end-of-life photovoltaics is that they may enter into the global e-waste trade in end-of-life electronics. Photovoltaics have the recipe for such e-waste: toxic materials are embedded in valuable ones. The heavy metals of concern in photovoltaics are lead and cadmium, depending on the design (a handful of modules have no toxic metals).

Regulated metals in end-of-life CIGS include cadmium, copper, and selenium, while in end-of-life CdTe, cadmium compounds are the primary metal of exposure concern. California regulators debated how to classify end-of-life photovoltaic modules, with focus on whether photovoltaics would burden local governments and transfer stations. In 2011, CdTe thin-film manufacturers First Solar and Abound asked the California Department of Toxic Substances Control (DTSC) to deregulate end-of-life photovoltaic modules that otherwise would be considered hazardous waste under California law. Other thin-film producers, such as SoloPower and Solyndra, participated as well. This new rule would eliminate costly and time-consuming manifests and other paperwork. At issue is whether photovoltaic modules must pass a hazardous waste characterization test designed to estimate the amount of toxic material that might leach from landfilled material.

Because crystalline silicon is the most common photovoltaic technology, lead compounds are the most widespread toxic materials in photovoltaics. The amounts of lead in crystalline silicon photovoltaics vary from zero to several hundred grams per module. According to an annual survey of photovoltaic manufacturers by the Silicon Valley Toxics Coalition (SVTC), several companies are able to make modules without lead for several or all of the module types they offer, and more plan to, though only a handful do so as of 2018.

Two thin-film CdTe manufacturers with projects in California—First Solar and Abound Solar—made modules that passed the U.S. Environmental Protection Agency waste determination test, but failed California's more stringent test, meaning that they would be considered hazardous in the state. Both manufacturers asked for an exemption from DTSC rules for the disposal of end-of-life modules. Project developers at the time were unclear of the decommissioning costs of projects, so there was interest in confirming and clarifying the rules. The rules would treat photovoltaic waste as "universal waste" (widely produced household hazardous waste, like compact fluorescent bulbs or mercury thermostats), which required that the state develop management programs, often paid for by companies or consumers. The agency finalized a rule in 2012, but owing to rules limiting administrative procedures, the rule required legislation, which was introduced by state senator William Monning and passed in 2014.[25]

Some have argued that CdTe has different physical properties than elemental Cd and could be less toxic. For example, CdTe melts at a much higher temperature (1041 °C) than cadmium (321 °C), and is less soluble in water. CdTe appears to be less toxic than elemental cadmium in terms of acute exposure, but the highly reactive oxidizing surface of CdTe quantum dots can damage cell membranes, mitochondria, and cell nuclei, though this study was not able to disentangle the nano-scale effects from the effects of cadmium compounds.[26] There are a few studies of the toxicology of CdTe, but most do not accurately capture the primary exposure route for CdTe because they rely on ingestion, not inhalation, as the primary pathway.[27]

The global e-waste trade includes advanced automated facilities, but also facilities with lax labor laws and low occupational health standards, and a separate informal processing sector that uses especially crude tools and equipment. Strong and enforced environmental health and worker protection standards for recycling can help minimize the toxic exposure and human rights abuses that currently plague the trade.

Possible destinations for end-of-life photovoltaics are landfills, waste transfer stations, and recycling facilities. Incinerators operate at temperatures high enough to volatize the CdTe into the air, although such incinerators should have pollution-control equipment to prevent significant emissions, and most of the cadmium will be dissolved into the molten glass.[28] Discarding CdTe photovoltaic modules in landfills can pose risks because the cadmium could leach into groundwater.[29] However, one study found that the amount would not violate the drinking water standard in Germany.[30] A report from the Norwegian Geotechnical Institute argued that leaching of cadmium into the environment could occur even in slightly acidic water, suggesting it could end up in landfill leachate.[31]

There has been much debate among policymakers about whether photovoltaic modules that fail hazardous waste determination tests should be considered hazardous waste under the U.S. federal Resource Conservation and Recovery Act of 1976. The act sets certain national standards, but states are free to set stricter rules. This means that some modules might be hazardous waste in some states, but not others. So whether or not a photovoltaic module is considered hazardous waste largely depends on the thresholds set by regulators. As mentioned, some modules pass the EPA tests but fail the stricter California tests. This has economic implications, because hazardous waste can be more costly to transport and dispose of. Many local waste managers feared that California and its major cities will bear many of the costs associated with end-of-life photovoltaic module disposal. Estimates of photovoltaic module disposal costs depend on the waste classification and the distance to recycling and disposal facilities. These estimates range from $0.04 to $0.13 per watt installed, somewhere in the range of $5 to $10 per module, depending on the size.[32] On an analyst call with First Solar, the only company with any recycling infrastructure at its manufacturing facilities, the company said the cost was about $0.04 per watt.[33]

California's hazardous waste laws are codified in the California Code of Regulations at Title 22, Environmental Health Standards for the Management of Hazardous Waste.[34] This law authorizes the DTSC to implement a system of registration and permitting for the transport and disposal of hazardous waste.

Following Photovoltaic Waste

On a rainy winter day in San José, California, a team from SVTC visited an e-waste recycling facility, ECS Refining. SVTC has made photovoltaic recycling the core of its Green Jobs Platform for Solar and its Solar Scorecard since 2008. SVTC's goal is to keep photovoltaics out of global e-waste streams by encouraging extended producer responsibility (EPR), preventing e-waste exports, and ensuring that no prison labor is used to disassemble modules.

ECS has long offered tours to SVTC to help members understand e-waste recycling, and learned from SVTC about global e-waste problems. The company earned the esteemed e-Stewards certification from the Basel Action Network for its recycling practices. The volume of e-waste generated in the Bay Area is staggering. Our guide noted that a room 150 feet across and 100 feet wide, and filled to the ceiling with old cathode-ray-tube television sets, turned over several times a day. Workers wearing masks wielded hammers, breaking the plastic off the sets and sending the cathode ray tubes along conveyor belts toward a hammermill, where they would be broken into smaller bits. The facility was a simple, yet effective, means of separating the various elements of electronics for secondary uses.

On the north side of the building was an outdoor area where large containers held scrap materials. One area of the facility was a room for hazardous liquids built over a containment liner. The liner prevented leakage into the soil of the toxic photo-processing chemicals that were stored at these facilities before the era of digital cameras. Several dumpsters full of broken or off-spec Solyndra photovoltaic modules were parked in the center of this room. The tubes were unmistakable, as no other company was exploring any similar form factor. Many were still clear, meaning the semiconductor layers had not yet been applied. The reason they needed to be stored in a place prepared for a chemical leak was that the tubes contained a mineral-oil-like substance that was supposed to enhance the photovoltaic effect and long-term stability of the layers. Should the tubes break, the oily mess might be difficult to clean up. Hence, it was not the presence of cadmium but the liquid form and the inconvenience of handling spills that drove the infrastructure needs for these particular end-of-life photovoltaic modules.

Nearby were several more dumpsters of modules from a competing thin-film CIGS manufacturer, MiaSolé. Its semiconductor cocktail was an n-layer of CIGS and a p-layer of cadmium sulfide. Since the cadmium compounds were present only as a solid, they did not require the same levels of infrastructure and containment. MiaSolé, too, was seeking a DOE loan guarantee. It had gone so far as to seek out a life cycle assessment (LCA) expert to help it prepare the documentation and analysis required by the DOE background check, which requires that the project over its lifetime leads to greenhouse gas reductions. A photovoltaic module facility could claim that the modules made there would lead to returns on energy investments. The energy invested during facility construction would be returned or "paid back" by the electricity generated from modules made there. Energy payback time is a very common metric used in LCA of photovoltaics. A survey of LCA of photovoltaic modules shows extensive use of this metric in one form or another.

Like another local CIGS manufacturer, Nanosolar, MiaSolé saw interest in its technology dwindle with numerous delays, and technical and market developments that led to steep price declines in crystalline silicon photovoltaics. MiaSolé was eventually purchased by the Chinese energy giant Hanergy as that company went on a buying spree of CIGS manufacturers to complement its existing line of crystalline silicon modules. The company is owned by one of the richest men in China and began to have trouble in 2015, when the stock was the most shorted of any on the Hong Kong stock exchange.

Near Dresden, Germany, nestled among rolling hills, is a small town called Freiburg. It is famous as Germany's "solar city," and it exports to neighboring cities four times the amount of energy it uses, thanks to photovoltaics installed on the city's rooftops.[35] The city is also a headquarters for SolarWorld, a major manufacturer of crystalline silicon photovoltaics with operations in several countries, including the U.S. and Germany. It was here that SolarWorld had spun off a company called Sunicon, which was a project to investigate recycling solutions for crystalline silicon photovoltaics. The project began during the peak of the polysilicon shortage of 2008 and continued until about 2013. Its lead engineer, Karsten Wambach, was instrumental in the design and implementation of the pilot facility. Yet the success of the facility hinged, not on good engineering and design, but on automation and improving the economics of recycling these high-value materials. Wambach was instrumental in the development and founding of PV Cycle in 2005, the first industry-wide EPR program for photovoltaics.

Extended Producer Responsibility

The key to effective long-term management of end-of-life photovoltaic waste is the establishment of recycling infrastructure based on EPR, a "polluter-pays" policy framework aiming to ensure that consumer products are safely disposed or recycled. One approach to limiting the amount of end-of-life photovoltaic waste entering the environment is to employ a lifecycle management strategy based on EPR. The programs usually involve some kind of collection scheme to ensure that money is available to collect and or even recycle the product. EPR is a widely used policy instrument for products as varied as electronics, paint, carpet, batteries, and even pharmaceuticals. The rise of EPR in electronics is in part a reaction to the concerns raised by activists and government environmental agencies about e-waste. This e-waste is regarded as a serious environmental justice issue, as the development of informal sites of valuable-metal recovery operations has occurred in places with high rates of poverty in Ghana and China. These informal operations have few if any safeguards to protect people from the toxic materials to which these valuable materials are bound. Estimates vary, but somewhere on the order of fifty million tons of e-waste is produced annually.[36]

There are multiple benefits from recycling end-of-life photovoltaics, including energy and resource savings, green jobs creation, and landfill diversion. Once this was recognized as an important issue in Europe, the first major site of growth for photovoltaics in the past decade, an organization named PV Cycle worked to develop a take-back and collection scheme in 2007. Several interview informants speculated that the industry was acting to get ahead of any proposed regulation under the directive on Waste Electrical and Electronic Equipment (WEEE). Recycling of electronics, especially photovoltaics, is recognized as a critical issue for the sustainability of photovoltaics by researchers at government labs and agencies in the U.S., such as DOE's National Renewable Energy Laboratory. Photovoltaic modules contain many toxic materials, so to protect workers and communities many safeguards are required, akin to those warranted for other electronic products. Many environmental organizations and some policymakers believe that strong EPR policies can even create incentives for product designs that are safer, easier, more environmentally friendly, and cheaper to recycle.

The German Federal Institute for Materials Research and Testing, in cooperation with First Solar and Deutsche Solar—the two photovoltaic companies that operate recycling processes for end-of-life thin-films—as well as the Universities of Utrecht and Miskolc, conducted LCAs of recycling processes in a project called RESOLVED (Recovery of Solar Valuable Materials, Enrichment, and Decontamination). The project found broad benefits from photovoltaic recycling, including for materials availability, energy recovery, and environmental emissions. The findings from that project offered a proof-of-concept for recycling processes that a handful companies would adopt as they built recycling infrastructure.

Abound Solar, a CdTe manufacturer that also received ARRA support in the form of a loan guarantee, also planned to develop a recycling program. It signed an agreement with CdTe powder refiner 5N Plus to operate a recycling facility in Wisconsin. The life of the facility was brief, however, and the company closed the plant only a few years later. After Abound Solar declared bankruptcy it worked with First Solar to recycle the end-of-life modules left behind in its shuttered factory.

Companies operating in European Union member states are required to meet the legislated requirements of the local state. These vary from one EU country to the next, because the legal basis for WEEE allows each member state to interpret WEEE's scope differently, setting slightly different objectives, benchmarks, and goals. Rules, requirements, and even procedures can be different in different markets. WEEE sets minimum thresholds for end-of-life e-waste recovery. Each member state has authority to go beyond these. Costs and recycling markets differ from state to state.

The first cast of WEEE included ten product categories in the scope of electrical and electronic equipment, including large household appliances, lighting equipment, information technology devices, and many more. This first cast of WEEE did not create a product category for photovoltaic modules to be included in the directive. In the scoping documentation for WEEE in 2002, photovoltaic modules are discussed in a separate article that specifically deals with WEEE's adaptation to scientific and technical progress over time. Most photovoltaic modules manufactured to date are still far from end-of-life. The EU determined that it did not yet need specific WEEE regulations for photovoltaic modules. Looking forward, this issue will rise to prominence if not planned for and anticipated properly. Text from an article in the 2002 Directive of the European Parliament briefly mentions "solar panels" and leaves open the possibility that they could be included in a future recast of WEEE:

> Article 13: Adaptation to scientific and technical progress. Any amendments which are necessary in order to adapt Article 7(3), Annex IB, (in particular with a view to possibly adding luminaires in households, filament bulbs and photovoltaic products, i.e. solar panels).[37]

Several leading photovoltaic manufacturers, the European Photovoltaic Industry Association (EPIA), and the German Solar Business Association joined together to launch PV Cycle, "a European-wide collection, recycling and recovery system."[38] In anticipation of the possibility that WEEE could include photovoltaic modules in its recast, the photovoltaic industry in 2007 developed an initiative to implement a take-back system to ensure that defective and used photovoltaic solar panels are properly recycled and their hazardous materials safely removed. The initiative was largely led by European industry actors and policy-makers but included global photovoltaic companies, because Europe was by far the largest market for photovoltaics. At the time, 80% of photovoltaic modules sold were installed in Europe, led by Germany, Italy, Spain, the Czech Republic, and Greece.

PV Cycle's founding photovoltaic manufacturers were Avancis, Isofo-ton, Conenergy, Schott Solar, SolarWorld, and Sulfur Cell. Membership ballooned to fifty members in two years, and over five hundred members in five short years. PV Cycle was shorthand for European Association for the Voluntary Take-Back and Recycling of Photovoltaic Modules. The association developed a framework based on "self-control and reporting." The organization would establish the broad framework for a central management system for end-of-life photovoltaics, either as a clearinghouse or a complete disposal service. WEEE mandates "insolvency insured guarantees" to ensure that even companies that go out of business will take back end-of-life modules. Under PV Cycle's scheme, businesses must submit evidence that a financial guarantee is in place for each new module brought to market, though there were still unresolved questions about the value of the financial guarantee, and how performance is overseen. PV Cycle also defined uniform quality and technical standards for collection and recycling.

The group announced that the scheme would be in operation by 2008 and cover about 90% of photovoltaic waste by 2015. They established benchmarks for recovery rates that would improve over time. They also laid the groundwork for a system for developing and sharing best practices for end-of-life photovoltaic waste handling, recycling and reuse projects, and for minimizing the overall waste in the design phase. Marie Latour of the EPIA said, "There is currently little need for specific measures for recycling and waste control for solar panels as most of the solar panels are yet to reach the end of their first life cycles." However, she added, the "industry is already preparing for the solar panel waste issues likely to emerge over the next 15 to 20 years."[39]

Early on, the organization promoted a voluntary framework. The threat that WEEE could be extended to include solar panels in a future revision was believed to be enough to motivate the industry to act. EPIA and individual companies argued that the PV Cycle scheme would be more effective at increasing recycling rates than a mandatory scheme. "An inclusion of the photovoltaic sector in WEEE would result in 27 differently designed recycling systems, with inherent administrative procedures and costs," noted a European Commission Report.[40] This would offer an opportunity for companies to participate, but not require them to comply with rules that could differ by jurisdiction.

PV Cycle announced, "The take-back system proposed by PV Cycle will instead create a coherent EU-wide recycling system that will enable efficient and economically viable management of waste from the photovoltaic sector. EPIA together with PV Cycle is therefore urging the Commission . . . not to include photovoltaic products within the future scope of the revised WEEE Directive."[41] The voluntary efforts by the industry aimed to get ahead of European regulators, showing that voluntary regulation works and that the EU should continue to exclude photovoltaics from WEEE.

PV Cycle would contract with Ökopol in 2008 to research different schemes for a take-back and recovery system for photovoltaic products. One major conclusion of Ökopol's report was that a mandatory system could have high regulatory costs that would be a financial burden on an immature, young industry. The Ökopol study also found that a voluntary program would have higher collection and recycling rates than a WEEE-type scheme. The voluntary framework would have high initial costs for manufacturers, but unlike in the WEEE scenario, those costs would go down over time, as a consequence of allowing companies to develop their own opportunities for innovation.

At a 2009 meeting in Brussels with SVTC, the leadership of PV Cycle and the EPIA proudly spoke of the PV Cycle effort and the "doubly green" benefits of the recycling policy. They enthusiastically shared a recent video of a de-installation, set to a high-quality musical score. The founding CEO of PV Cycle is Jan Clynke, an engineer from the e-waste industry, who brought intimate knowledge of recycling e-waste to the team tasked with designing an end-of-life management scheme for photovoltaics. Clynke said that instead of regulation the industry would rely on "self-control and reporting" to ensure that it was meeting targets. However, the possibility that WEEE would regulate photovoltaics remained.

By 2011, new rumblings of a recast of WEEE spurred PV Cycle to greater action. Many details remained to be decided about how the take-back, collection, and recycling system would be organized. Would it be based on direct reverse logistics, where photovoltaic installers would provide the services? Would customers have to drop them off at big box retailers or other pickup points? One key sticking point in the early debates was whether photovoltaic manufacturers making modules that were considered hazardous waste would have to pay a larger set-aside per panel, or per watt. Eventually, PV Cycle's progress on developing the take-back and recycling scheme began to slow.

A key area of disagreement centered on whether the scheme would be paid for per photovoltaic module manufactured or as photovoltaic modules arrived at collection centers. How would it be financed? One scheme is a pre-paid approach, where funds are set aside in a restricted investment account. The second is a pay-as-you-go approach. The problem with the latter is that there is no guarantee that companies will be solvent—modules could arrive at recycling centers after a company is in bankruptcy. Given the turnover in the industry and the long time between installation and end of life, some modules would be "orphaned" in a pay-as-you-go model.

The solar industry set collection and recycling targets that were much higher than those for many other consumer electronics devices. Collection targets refer to the amount of waste collected, out of the total sold. The recycling target is the amount of waste that is recycled. As of 2015, the target collection rate was 85%, and the target recycling rate 90%.[42] One might expect higher numbers if manufacturing discards were covered under WEEE. Recyclers suggested that this would also enhance the business model for photovoltaic waste haulers and reserve logistics companies. Today no remnant of the word "voluntary" appears in PV Cycle's websites or brochures, or in formal WEEE documentation.

Strong EPR policies can even create incentives for product designs that are safer, easier, and less expensive to recycle. To ensure safe and responsible recycling of modules sold in the U.S., more domestic recycling networks and infrastructure will need to be developed. However, more collaboration and collective action is needed across the photovoltaic industry to encourage strong recycling policies and best practices, and to encourage design for environment and recycling. In an annual survey conducted since 2010, 75% of photovoltaic module manufacturers reportedly support mandatory take-backs and a responsible recycling policy.[43]

Private initiatives are being pursued also. In 2012, SVTC collaborated with the Basel Action Network to develop language to include photovoltaics in its widely used e-Stewards certification process. This is a certification standard for electronics recyclers who adopt certain best practices for material recovery and worker health and safety. The process included bringing together photovoltaic industry experts, electronics recycling industry representatives, scientists, and policymakers to develop a list of best practices for handling end-of-life photovoltaics.

The welcome rapid growth of photovoltaic installations means there is a limited window of opportunity to establish recycling policies and practices to manage end-of-life photovoltaic waste. If waste issues are not preemptively addressed now, the industry risks repeating the disastrous environmental mistakes of the electronics industry. Toxic e-waste—made up of discarded computers and other electronics—is shipped to developing nations like India, China, and Ghana for manual disassembly, exposing workers and communities to highly hazardous chemicals.

A mandatory EPR and recycling law would achieve several objectives that are critical to a just and sustainable solar energy industry. First, responsible take-back and disposal can help prevent end-of-life photovoltaic waste from adding to the already burdensome global flows of e-waste. Second, there are real material limitations on the availability of some critical inputs to photovoltaics, which cap the total production of photovoltaics containing certain materials, including tellurium, silver, gallium, and indium. Third, and relatedly, a supply of these inputs from recycled sources could help stabilize the costs and price volatility of inputs that are subject to price fluctuations, such as indium and tellurium. Fourth, purifying materials from ores and minerals requires far more energy than recovering them from recycled materials, meaning less energy is needed, and thus less greenhouse gases are emitted, when sourcing recycled materials. Finally, the major burdens of end-of-life photovoltaics will be borne by local governments who operate and own landfills, waste transfer stations, and recycling facilities and that will have to find final disposal places.

The long time between when a module is made and when it becomes waste suggests that mandatory recycling will be required. Many interviews with manufacturers point to the need to eliminate free riders—companies that take advantage of recycling services, but do not pay for them. The Basel Action Network and SVTC identified several best practices for end-of-life photovoltaic module management, which aid safe de-installation and handling, encourage reuse, protect vulnerable communities in developing countries, protect workers, and minimize emissions from recycling facilities. The lack of recycling in the U.S. can be partly explained by the lack of volume of manufacturing scrap, the small volume of waste arriving from the field, the lack of consideration of the value of materials recovery, and the lack of interest and funding by the government.[44] Ultimately, a successful take-back, collection, and high-value-materials recycling system will depend on governance and cooperation throughout the industry.

At the urging of several photovoltaic manufacturers, including First Solar and Abound Solar, California's DTSC, part of the California EPA, engaged in several efforts from 2010 to 2014 to clarify the hazardous waste laws in California with regard to photovoltaics, as several hundred million modules were being installed. California legislator William Monning, whose district includes the Santa Clara/Silicon Valley but also areas further south, in San Luis Obispo, where large utility-scale solar energy projects have been built, took an interest in photovoltaic recycling and authored legislation that would declare end-of-life photovoltaics a type of universal waste.[45] Universal waste is a classification of materials that relaxes certain requirements. The state senator's staff said that this would open the door for a more comprehensive EPR law in the future. Washington State became the first state in the country to require manufacturers to have (by 2020) an EPR program in order to sell in the state.

Product stewardship requires a policy framework that takes a cradle-to-grave approach and better labeling and identification schemes to ensure proper disposal and treatment of end-of-life photovoltaic modules. But it is not the only way to improve end-of-life impacts. Design efforts could use the tools and principles of green chemistry to phase out toxic materials.[46] Many of the concerns about environmental and worker protection standards to minimize the toxic exposure and human rights abuses that currently plague the global e-waste trade would be obviated by the absence of toxic heavy metals. While green chemistry is discussed in the solar industry in general, the latest technological innovations of interest in solar are perovskite solar cells, which can contain soluble lead compounds.[47] This suggests that some recycling and take-back scheme will be necessary to prevent heavy metals from entering the environment, even in emerging technologies. End-of-life photovoltaic management will offer the most benefits if recycling services are offered domestically and locally, with minimized transport requirements, and an added emphasis on recycling and recovery of high-value materials. Some argue that the best way to make sure that photovoltaics stay out of landfills is to make them last longer. Products made to last longer would delay the need to find management strategies and solutions. But this alone will not safeguard the long-term sustainability of the photovoltaic industry. Some form of EPR will best ensure that there are no e-waste crises, that worker health and safety will take priority, and that a long-term supply of critical metals will be recovered, augmenting future supply chains.

CHAPTER 5

Green Civil War

They say that we want renewable energy, but we don't want
you to put it anywhere. I mean, if we cannot put solar power
plants in the Mojave Desert, I don't know where the hell we
can put it.
—California governor Arnold Schwarzenegger, Yale
University, 2009[1]

The persistent strong winds and strong insolation of the Western deserts
have reemerged as beacons for a new resource rush, this time for land on
which to harness renewable energy. In the California deserts, earlier
rushes brought mining and other extractive industries—gold and silver
mining, boron harvesting, grazing, and later, various recreational uses. As
renewables began to take a foothold here in the 1980s, the wind industry
built extensive wind farms at desert sites in the San Gorgonio and Tehachapi Passes, and a company called Lux built nine utility-scale solar
thermal power plants. Together, these projects would help make California the global leader in renewable energy throughout the 1980s and
1990s. And they were built out with relatively little controversy.[2]

Later efforts to deploy solar energy in America's southwestern
deserts, where the most solar energy is present, quickly encountered
significant resistance, mainly due to their placement on ecologically and
culturally sensitive land. Utility-scale solar energy (USSE) facilities on
the order of square miles in size, some with over a gigawatt of power
capacity and capable of providing power to tens of thousands of homes,
were proposed, which would have required clearing tens of thousands
of acres of vegetation and grading land with heavy equipment. Numerous projects slated for solar energy development on public lands would
be on quality habitat for the Agassiz's desert tortoise (*Gopherus agassizii*), a federally threatened species.[3]

Conservationists worried about the wholesale transformation of vast stretches of public lands and open desert into industrial zones of renewable energy production. Since most public lands are in conservation by default, owing to never having been put to work under Homestead Act provisions, mainly due to a lack of water, this meant that the lands offered for solar development were bound to raise concerns for ecosystem and cultural resource conservation, particularly land managed under the California Desert Protection Act. So much high-quality desert habitat coming under development pressure was alarming to environmental groups as well as to Native American tribes.

As many states pursued ambitious climate change and renewable energy goals, USSE projects were proposed across public lands and the deserts of the American Southwest. The Bureau of Land Management (BLM) manages 15.2 million acres of public lands in California (this is about 6% of the total land area it manages across the United States).[4] The agency's policies and practices affect biodiversity conservation on over 250 million acres across the contiguous U.S. The BLM's conservation management agenda must be balanced with other agency missions, because it has manages such varied activities as energy development, grazing, geologic exploration, and off-road vehicle recreation. These opposing commitments can pit endangered species conservation against renewable energy development. The BLM is typically being pulled in different directions by multiple missions and stakeholder concerns.[5]

Solar developers were attracted to California because it is the largest renewables market by far and offers higher electricity prices, which translate into more profitable contracts to purchase electricity. Voters, and customers of electric utilities, show widespread support for renewables.[6] Renewable energy projects have bipartisan support, according to most polls and surveys of public opinion, especially in California.[7] USSE development is largely driven by public support for regulation, legislative action, or utility investments that aim to meet environmental or climate goals, such as state renewable portfolio standards, which were responsible for many early projects. One study of the politics of the solar renaissance in the Mojave Desert pointed to the role of negative public attitudes to deserts in mobilizing public support for renewables on public lands and support for action on climate change as reasons that the projects were not opposed by major environmental organizations.[8] A telephone survey of residents of the California deserts suggests that proximity to project development does influence the degree of opposition.[9] This chapter aims to illustrate how many of these projects came into conflict over land-use and ecosystem impacts, particularly projects proposed on public lands managed by the BLM. To understand why these land-use conflicts emerged when and where they did requires a deeper investigation of the recent conservation history of the region.

The BLM's solar policy sparked a debate about using public lands for low-carbon energy generation versus ecosystem conservation and wilderness preservation. The landmark California Desert Protection Act of 1994 fundamentally altered the management of public lands across 25 million acres of Southern California desert. In the California Desert Conservation Area (CDCA), public lands designated by the Congress in 1976 would be managed for multiple use, sustained yield, and environmental quality. The 25 million acres of CDCA lands contain 11 million acres managed by BLM, 5.3 million acres of national parks, 4.2 million acres of private land, 3.2 million acres of military bases, and 1.1 million acres owned by the state of California, with the rest being Indian lands and wildlife refuges. Championed by U.S. senator Diane Feinstein (D-California), the bill designated 3.6 million acres of BLM lands as wilderness and off-limits to energy development, as well as transferring millions of acres of public lands to the National Park Service.[10] The initial applications to the BLM included dozens of proposals to build USSE projects within the CDCA. This tension between ensuring that areas of high concern in the California deserts remain conserved and the availability of land for lease for solar development to slow carbon pollution led to inevitable conflict, particularly in this region of California.

USSE power plant construction entails transformation of the desert landscape, with surface impacts comparable to those of agriculture. Solar farms commonly require extensive surface scraping, grading, and bulldozing across thousands of acres of flora and fauna. Projects can also include dozens of miles of roads. With some solar power plants measuring up to eight kilometers on a side, a single project can have serious implications for wildlife. On public lands in the West that were under a strong stewardship regime for more than a century, rural and wild desert landscapes are transformed into industrial ones. Critics argued that solar power plants were never land uses considered in the scientific evaluation that informed the CDCA planning and that a careful approach to development needed to balance the interests of solar energy developers and environmental conservation.

When public lands across the American Southwest were drafted into the fight against climate change, it set off local controversies that pitted vulnerable ecosystems against renewable energy development. Solar developments on public lands managed by the BLM became entangled in controversies over impacts on biodiversity and threatened or endangered species, particularly the desert tortoise. The intensive land-use impacts of USSE development raised many questions about local ecosystems, embroiling projects in controversy and even causing fissures between issues of climate and biodiversity in the "green" community.

This conflict between solar projects and wilderness preservation led the editors of the *New York Times* to call it a "green civil war."[11] Despite the ecological problems with many of the proposed projects, advocates of strong climate change policies chastised those concerned about tortoises, birds, cultural artifacts, and public lands for being shortsighted. Arnold Schwarzenegger—an outspoken solar advocate during his tenure as California's governor—leveled public criticisms at those raising concerns about the Mojave Desert's ecosystems and inhabitants. In a public talk at Yale University, Schwarzenegger described opponents as NIMBYists and the desert as a wasteland with no better use.[12] These views highlighted the urgency of reducing emissions to ward off climate catastrophe and stressed the need for solar energy innovation and deployment, but also often lacked any sensitivity to ecological impacts that might follow. To renewable energy advocates, these land and finance policies would help foster innovation in the solar energy industry and make important contributions to decarbonization and climate security. Desert wilderness organizations—the Wildlands Conservancy, Wilderness Society, Center for Biological Diversity, and Defenders of Wildlife, to name a few—argued that alternative sites, including abandoned agricultural land, brownfields, abandoned mines, and other disturbed lands, were more appropriate for solar energy development. These organizations questioned the wisdom of using vulnerable ecosystems for energy development when more appropriate sites were available.

VIRTUAL PRIVATIZATION AND FAST-TRACKING OF PUBLIC LANDS IN THE AMERICAN SOUTHWEST

The history of renewable energy development on public lands owes much more to the Energy Policy Act of 2005, along with a series of subsequent secretarial orders making 22.5 million acres of BLM lands available for solar development. Championed by then Colorado senator Ken Salazar, the bipartisan Energy Policy Act encouraged renewable energy projects on public lands by 2015. Often referred to as the BLM's "renewable energy mandate," the language specifically says:

> It is the sense of the Congress that the Secretary of the Interior should, before
> the end of the 10-year period beginning on the date of enactment of this Act,
> seek to have approved non-hydropower renewable energy projects located
> on the public lands with a generation capacity of at least 10,000 megawatts
> of electricity.[13]

Since a production ramp-up in the 1970s, public lands have become important sites of domestic energy production. In 2008, nearly 10% of the federal mineral estate, including the subsurface, was under lease for oil and gas development.[14] As the new secretary of the interior, in 2009 Salazar signaled a new era in the management of energy development on public lands with Secretarial Order 3285. The BLM made relatively inexpensive and contiguous public lands available for lease to solar developers. The public land was attractive to solar developers because it was cheaper than private land. Land prices can be hard to predict when negotiating deals among multiple owners. Developers were also aware that the BLM had authority to "fast-track" environmental and cultural resource reviews for selected projects eligible for DOE loan guarantees. To maintain eligibility for finance, companies had to demonstrate that their projects were "shovel ready" and economically viable.[15]

The pace and scale of proposed landscape transformation was unprecedented in the American Southwest. Nearly a tenth of the total land that BLM manages was made available on a first-come, first-served basis for solar development. If granted, this would be the largest transfer of public lands to private use in U.S. history. The mandate to develop renewables on public lands made renewable energy development "one of the Department's highest priorities," instructing agencies to fast-track new applications and remove impediments to permitting, siting, and development of renewable energy projects seeking ARRA support.[16] It allowed proposed renewable energy facilities, some of which had been in queue since 2005, to proceed with applications for right-of-way (ROW) permits on public lands. ROW authorizations grant specific rights to individuals and companies to use public lands for projects for a specified time. The public lands were not privatized exactly, but virtually privatized for some time into the future (before returning to federal ownership). Developers would have all of the authority and rights of a private landowner without actually retaining the rights to land.

Companies of all sorts and sizes—from the largest multinational energy corporations and financial services firms to startup venture capital firms with little more equipment than rented sport utility vehicles—

were enticed to the desert by the promise of profits in the clean-energy race and the West's latest land and resource rush.

The BLM used the ROW application process to offer land leases to developers through a process required under its guiding "organic act," the 1976 Federal Land Policy and Management Act (FLPMA, often pronounced "flip-ma"). FLPMA requires that the BLM administer land on a "multiple use" basis, accounting for the views of stakeholders. Under new provisions in FLPMA, in the Energy Policy Act of 2005, and in several executive and administrative orders, the BLM was authorized to grant ROW permits for renewable energy projects on public lands. "We're open for business with respect to renewable energy on public lands," Interior Secretary Salazar would declare in 2009.[17]

BLM had historically authorized ROWs for water and gas pipelines, water storage, roads, oil and gas leases, and "systems for generation, transmission, and distribution of electric energy," making the approach consistent with the leasing practices for grazing permits and other resource and recreational uses where formal land rights are not customary.[18] Land is leased to solar energy developers at rates that are based on a combination of power capacity and total acreage. The BLM earns Interior tens of millions of dollars from ROW leases to USSE power plants.[19] The rates adopted by the BLM aim to reflect market costs at the same time as becoming an important source of agency revenue. The cost of land is a relatively small portion of the overall cost of USSE facilities, but it still a cost that can be saved on. The lease option allows solar developers to avoid some up-front costs, making the project more economical. Other benefits of using public lands include favorable lease rates, the fast-tracking of environmental and cultural resource reviews, and the benefit of negotiating with a single land manager.

In June 2009, flanked by Senate majority leader Harry Reid, Secretary Salazar proclaimed, "We are putting a bull's-eye on the development of solar energy on our public lands."[20] Desert conservation experts and activists saw a "land rush" and "virtual privatization of public lands."[21] The adjective "virtual" is used because the lands are not taken out of the public domain, but they are removed from public access.[22] The announcement by Salazar that public lands would be available for renewable energy brought interest from developers, entrepreneurs, and speculators, including major Wall Street investment banks, investor-owned utilities, and venture capital firms. By the end of 2009, there were ROW applications proposing solar power plants on over a million acres of BLM-managed public lands—more than the public lands used for bedrock and coal mining and oil and gas development since the passage of the Mining Act of 1872.

Companies with varying business interests, from investment houses like Goldman Sachs to startup photovoltaic manufacturer OptiSolar, sought ROW applications, each on hundreds of thousands of acres. In aggregate, the proposed projects represented several times the amount of renewable electricity that California investor owned-utilities must buy to satisfy RPS obligations and far more power than needed to fulfill the BLM's own 10 GW renewable energy mandate. Only a portion of the projects would have had guaranteed markets for their renewable electricity, so the BLM was processing applications knowing that a portion would never be built or permitted.

From 2005 through 2017, the BLM received over 400 ROW applications for solar projects. By 2010, active applications covered 1.2 million acres across the American Southwest.[23] California alone had 79 solar energy facility applications representing 569,802 acres of BLM lands and estimated at 48 GW of power (Figure 6). The California RPS would require about 20 GW by 2020, so there was already over twice the amount of power officially seeking permits as would be required to fulfill the first RPS targets.[24] Developers' ambitions were larger on public than on private lands, averaging over 700 MW per project. This average is larger than the largest USSE actually built in California (Desert Sunlight, in Desert Center, and Topaz, in California Valley, are rated at 550 MW each). One project proposed in 2006 was a 4 GW solar farm, six times larger than the largest solar plant in operation ten years later. The 648 MW array in Kamuthi, Tamil Nadu, India is the largest USSE facility in the world as of 2017.[25] By 2017 only 32 of the 400 projects proposed for USSE on public lands had been built, and only a handful are still under development.

ROW authorizations are subject to provisions in FLPMA for public participation and to the 1969 National Environmental Policy Act (NEPA). NEPA requires that all federal actions "significantly affecting the quality of the human environment" undergo a thorough environmental review.[26] Lead agencies must disclose the potential environmental impacts of a project and several alternative versions to potentially mitigate them.[27] ROW authorizations, and the environmental review entailed therein, became flashpoints in conflicts over solar energy development on public lands. Several loan-guarantee applicants were deeply concerned that NEPA compliance would delay their ability to break ground or meet required spending mileposts.[28] Congress debated exempting all stimulus spending from NEPA, but decided against it.[29]

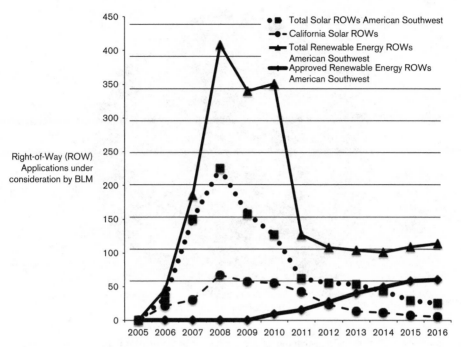

FIGURE 6. Bureau of Land Management right-of-way applications, 2005 to 2016.

Ultimately, California senator Barbara Boxer added an amendment that became Section 1609 of ARRA, requiring that federal agencies devote adequate resources to expeditiously complete environmental reviews.[30]

The volume of ROW applications for projects on public lands overwhelmed some local BLM district offices. At the office in New Mexico, near a proposed "low-conflict" Solar Energy Zone, one official complained of a lack of time to prepare an adequate assessment of solar energy development on public lands they manage.[31] USSE proposals were different from the other kinds of land-use proposals, such as transmission line development, grazing, hunting, and recreation activities, that the agency's experts were used to evaluating. In areas where solar applications were landing, many district offices did not have time to properly evaluate proposed solar projects. They were also unsure of what solar development entails specifically. Does it keep vegetation intact, or do developers leave sites barren of topsoil? How many roads are needed? Are groundwater wells required? Although no agency official would go on record owing to the controversial nature of the question, there was a sense of dissent among agency staff, particularly among

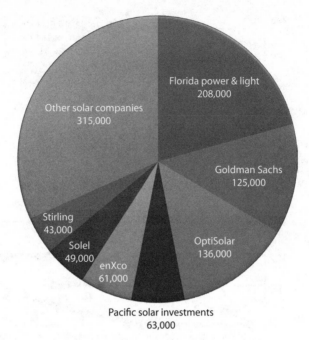

FIGURE 7. A handful of firms had most of the right-of-way applications (by acreage) for public lands in 2008.

employees responsible for conservation-related issues at the BLM such as habitat management plans or other land stewardship issues. One California BLM scientist complained of lack of resources to conduct long-overdue resource conservation plans with staff time dedicated to processing ROW applications and providing input to everyone from counties to the federal government regarding solar power plant development. Figure 6 shows ROW applications over time for renewable across all public lands, USSE projects, and USSE projects in California. Figure 7 shows the relative share of ROW acreage applied for by several firms.

One difference between grazing and energy leasing (development of solar, oil and gas, coal, etc.) is how the projects change access to public lands. Industrial facilities like utility-scale solar power plants require fencing and private security to ensure public safety and reliable operation. This means historic access is restricted, often with cyclone fences, sometimes topped with razor wire. Oil and gas equipment has a smaller footprint, and often the equipment does not restrict site access at all, there are simply signs that say, "Enter at your own risk." This helps explains the opposition to solar power plants by many hunters and even libertarian activists who are concerned about public access. In fact, the land at the center of the controversy between Cliven Bundy and the BLM that started in 2014—a long-standing dispute over unpaid grazing fees—was rumored to be an inaccurate "fake news" story claiming grazing lands were being sold off for solar and wind development.[32]

Public lands have historically required administrative review through a lengthy environmental review process involving local, state, and federal agencies, stakeholders, and the public. Executive Order 13212, signed by president George W. Bush in 2001, directed federal agencies to expand and expedite environmental reviews of oil, gas, and coal production on public lands, a move suggested by the Cheney Energy Task Force. The order opens with statement that "the increased production and transmission of energy in a safe and environmentally sound manner is essential."[33] The order also instructed federal agencies to take all actions possible to expedite the permitting and construction of projects that would "increase energy production, transmission, or conservation of energy . . . while maintaining safety, public health, and environmental protections."[34] This became the legal basis for fast-tracking solar energy ROW applications funded by ARRA.

In October 2010, the Department of the Interior authorized six fast-tracked USSE facilities covering 21,324 acres.[35] To ensure that ARRA milestones could be achieved, these projects were fast-tracked so they could "be reasonably built before the ARRA funds expire."[36] This became the so-called "shovel-ready" provision, a fast-track status for expedited environmental impact assessment.[37] Over the next 25 days, several more would be announced, with 3 GW of cumulative nameplate capacity.[38] These approvals made several projects eligible for DOE loan guarantees. Solar Millennium, for example, was approved two months later for a $2.1 billion ARRA loan guarantee to build the 1 GW Blythe solar power plant. But the loan was never finalized before Solar Millennium filed for bankruptcy just fourteen months later.[39] Fourteen USSE projects were identified as fast-track projects—twelve in California, and two in Arizona.[40] For projects built on public lands this meant that groundbreaking would have to commence before the end of 2011, just over a year away.

The selection of fast-track USSE projects in California was through a memorandum of understanding between the California governor and the interior secretary.[41] This Renewable Energy Policy Group worked with the Renewable Energy Action Team, composed of representatives of the

California Energy Commission, the BLM, the California Department of Fish and Wildlife, and the U.S. Fish and Wildlife Service (USFWS). Michael Picker led this effort as both Arnold Schwarzenegger's and Jerry Brown's senior adviser to the governor for renewable energy facilities. In this early phase of development, local BLM district offices did not provide solar developers or the national BLM any guidance about appropriate sites to minimize environmental impacts. The main information given to developers was from the Renewable Energy Action Team, with input from the California governor's office and the Department of the Interior. Desert conservation organizations were readily able to differentiate state and federal officials who appeared to be making decisions at the BLM (national) level instead of using the district office agency staff, "real people, the individuals working on the ground."[42]

Environmental groups were not sure which of the 1.2 million acres of projects in California would be ultimately fast-tracked, so they found it difficult to keep up with the "land rush and feeding-frenzy mentality." This made it challenging to legally intervene in environmental review processes. Though the landscape was full of proposed projects, it was not until 2010 that any were approved. "The BLM permitting system set up to process solar applications is basically broken . . . that is why after five years there still hasn't been one project approved. A lot have been speculative projects."[43] This speculation referred to companies that applied for ROW permits, not because they intended to develop projects, but because they planned to sell their permits to other companies seeking land for projects.

Fast-tracking was intended to benefit solar developers by signaling to the investment community that there would be no regulatory barriers in the siting process. One month after the spate of fast-track approvals in 2010, one company, Tessera, which had not one but *two* of the six fast-track projects under development and approved by BLM—the Imperial Valley solar project and the Calico project—went bankrupt. Tessera sold its development pipeline of planned projects for an undisclosed sum to K Road Power, a group of Goldman Sachs executives, who planned to use photovoltaic modules instead of Tessera's novel but troubled SunCatcher Stirling dish design.[44] The projects would continue to advance through permitting, even though there was little information about K Road or the photovoltaic technology or layout it planned. Table 10 shows the projects selected for fast-track status.

Opponents argued that by weakening environmental review with fast-track authority, the agency was exposing itself to litigation where projects encountered threatened or endangered species such as the iconic Agassiz's desert tortoise. Environmental organizations specifically working on issues related to the tortoise became particularly concerned about solar development at Iron Mountain, Chuckwalla Valley, McCoy Wash, Pisgah Crater, and the Ivanpah Valley. The BLM estimated 161,943 ha of Agassiz's desert tortoise habitat would be directly impacted by USSE development across the American Southwest. The Sierra Club, Natural Resources Defense Council (NRDC), and Defenders of Wildlife sued to stop development of K Road's 4,000-acre Calico Solar Project, located on the controversial Catellus lands, after failure of a formal protest to BLM director Bob Abbey about a controversial desert tortoise relocation plan for the project. "If this project moves forward at this location, Calico will irreversibly harm the sensitive Pisgah Valley and the desert tortoise," argued Kim Delfino, a lawyer for Defenders of Wildlife.[45]

Several fast-tracked solar projects were targeted by litigation. A lawsuit by the Quechan tribe and the La Cuna de Aztlan Sacred Sites Protection Circle over six utility-scale solar projects went before the Southern District Court of California alleging that the BLM had failed to consult with tribes as required by a memorandum of understanding between them.[46] The plaintiffs maintained that the BLM had violated the National Historic Preservation Act and the Native American Graves Protection and Repatriation Act by inadequately consulting with Native American tribes regarding the siting of the projects and potential resource conflicts. The courts eventually dismissed the case.

The fast-track process that expedited the environmental review of numerous projects on western public lands is one of the primary drivers of these environmental and cultural resource controversies. Fast-tracking ensured that ARRA support could be applied to particular projects, fostering the innovation and green job growth that ARRA projects were intended to create alongside low-carbon electricity. Ultimately, the BLM fast-tracked 31 projects slated for ARRA support.[47] These same institutional machinations undergirding BLM land and California climate policy put the lack of a robust environmental review on a collision course with controversies over desert biodiversity.

Procedural justice issues are interwoven with the solar siting controversies and may continue to challenge solar energy transitions moving forward. The American West is unique in many ways owing to the great species diversity and the fragility of western habitats compared to deciduous or tropical environments, which also face siting challenges

TABLE 10

Project name	Developer	Technology type	MW rating (proposed)	Public lands (total)	DOE loan?	Status[a]
Ivanpah Solar Electric	BrightSource Energy	Power tower	370 (400)	3,472 ac. (4,073)	$1.6 billion	Fast-tracked, approved, operating since 2014
Blythe Solar Power Project	Solar Millenium & Chevron Energy	Parabolic trough	968	6,300 ac. (7,025)	$2.1 billion (declined)	Fast-tracked, approved,[b] operating since 2017
Desert Sunlight Solar Project	First Solar Development	Photovoltaic	550	4,144 ac.	$1.46 billion	Fast-tracked, approved, operating since 2015
Genesis / Ford Dry Lake Solar Energy Project	Next Era Energy Resources	Parabolic trough	250	1,950 ac.	$852 million	Fast-tracked, approved, operating since 2015
Stirling Energy Solar One	Stirling Energy Systems	Stirling engine	850	8,230 ac.	No	Fast-tracked, approved, declared bankruptcy
Stirling Energy Solar Two	Stirling Energy Systems	Stirling engine			No	Fast-tracked, approved, declared bankruptcy
Calico Solar Project	K Road Power	Photovoltaic & Stirling engine	663.5	4,604 ac.	No	Permit denied
Imperial Valley Solar Project	Stirling Energy Systems	Stirling engine	709	6,360 ac.	No	Withdrawn
Desert Harvest Solar Farm	EDF Energy (UK)	Photovoltaic	100	930 ac.	No	Operating since 2018
Palen Solar Power Project	Solar Millennium / Chevron Energy	Parabolic trough	484	3,075 ac. (5,176)	No	Approved in 2017, under development
McCoy Solar Energy Project	Next Era Energy Resources	Photovoltaic	750	7,700 ac.	No	Operating since 2015

Project	Developer	Technology	MW	Acres		Status
Lucerne Valley Solar	Chevron Energy Solutions	Photovoltaic	45	516 ac.	No	Fast-tracked, approved, but never constructed
Stateline Solar Farm	First Solar Development	Photovoltaic	300	2,000 ac.	No	Operating since 2015
Ridgecrest Solar Power Project	Solar Millennium	Parabolic trough	250	3,920 ac.	No	Proposed but withdrawn
Rio Mesa Solar Electric Facility	BrightSource Energy	Power tower	750	unknown	No	Proposed but withdrawn

SOURCE: Bureau of Land Management (2012a, 2012b).

NOTE: There were an additional five fast-tracked projects sited on private land which have applied to run transmission lines and/or access roads across public lands, totaling 1,126 MW and crossing 425 acres of public lands. Data compiled from various BLM sources.

[a] Status as of March, 2015. "Approved" means that the project is ready to begin construction; "in permitting," that a draft EIS is being prepared; "proposed," that a notice of intent has been filed but nothing more.

[b] Solar Millennium is in insolvency proceedings. The project was sold to German developer Solar Hybrid, which intends to use First Solar PV modules. The project was halted because of a possible prehistoric Native American settlement. The site was approved again for a new developer in 2015.

but have more rapid rates of recovery. These impacts are very geographically specific and need to be understood on a case-by-case basis before drawing conclusions about environmental impacts. The BLM solar development policy came under scrutiny by environmental organizations because it lacked sensitivity to habitat, cultural resources, and land-use suitability beyond some general information about solar insolation, slope, and distance to electricity transmission infrastructure.

SOLAR PROJECTS VERSUS PRESERVATION AND THE SOCIAL GAP IN RENEWABLE ENERGY

Researchers of socio-technical change find that the public is a political actor that can shape the outcomes of infrastructure projects, and thus technological transitions more generally.[48] Research on social resistance to renewable energy facilities is beginning to identify frictions and means to lessen controversies or mitigate impacts.[49] USSE projects are far less represented in this literature, owing in part to their low profile and visibility in most places. A key concept used in the study of social resistance to renewable energy facilities is the "social gap" in renewable energy—strong, consistent support in general for renewables alongside local resistance to specific projects. Surveys routinely show the American Southwest to be overwhelmingly in support of the growth of USSE.[50] Yet, numerous renewable energy projects throughout the region have faced stiff social opposition.

One of four explanations offered for persistence of the social gap—the "democratic deficit" hypothesis—suggests that local stakeholders oppose projects because they are far removed from the decision-making locus.[51] Project developers often take a "decide-announce-defend" approach, where they show up having already completed the project financing and other key steps and then go through a process of environmental review. These are more likely to find local resistance. Processes that are perceived as undemocratic can elicit negative attitudes irrespective of other attributes. Community groups, organizations, and citizens resist local developments because of inadequate public participation. The developer-led or top-down approach is one source of conflict in siting issues because it lacks a credible means for public participation. Public involvement in project design, planning, and siting efforts reduced frictions over development when divergent views on a proposed rural wind farm development were collected and incorporated into decision-making.[52] An extensive literature suggests that the kinds of public participation used in NEPA public comment processes are insufficient for effective controversy extenuation.[53] The National Research Council, at the urging of the Environmental Protection Agency, DOE, and other federal agencies, convened an expert panel to assess whether and under what conditions public participation achieves the desired outcomes.[54] Despite these recommendations, public participation was complicated by the need to fast-track USSE projects to access stimulus funds for cash grants and loan guarantees. Instead of the two-to-three-year review process, environmental impact statement (EIS) processes were expedited to eight to twelve months by the invocation of the fast-tracking executive order.

A second explanation—the "qualified support" hypothesis—suggests that when people offer support for renewables, they do so with qualifications. They support renewables, but not without knowing some particulars about the project in question.[55] Actual support for renewable energy is typically qualified in ways that elude surveys of social attitudes. Such qualifications might include ecological and community impacts, although this information may be difficult to ascertain without detailed personal interviews, and such arguments could be raised for self-interested factors. Previous research on wind farm controversies at San Gorgonio Pass, near Palm Springs, California, suggests that opposition to wind farms decreases over time, with qualifications including support for local economic benefits such as job creation and expanding the local tax base.[56] There is also evidence for reduced community friction when benefits are realized, such as increased economic activity and tax abatements.[57]

Third, socio-political acceptance also often hinges on the insider–outsider frame. Framing solar energy developers as "big solar" evokes the tendencies of capitalist companies to act like the very powerful "big banks" and "big oil." "Big wind" positions developers as outsiders not sensitive to local concerns.[58]

Finally, in self-interest explanations, opposition stems from a project's impact on an individual's interests, property, or otherwise. This last explanation is encapsulated by the acronym NIMBY, for "not in my backyard." But scholarship on the social gap in renewable energy deployment shows that NIMBY-based explanations are by and large unsatisfactory here.[59] "The [NIMBY] syndrome really exists, but . . . we must conclude that its significance remains very limited."[60] NIMBY explanations fail to account for the complex motivations of various stakeholders and the role of political, cultural, and institutional factors that better explain social resistance.[61] This logic contends that it is at root a collective-action problem: even if they support renewables, individuals may have no incentive to support *local* projects because they can free-ride on renewable energy developments elsewhere. But if NIMBYism is more complex and local opposition more nuanced than a collective carbon free-rider problem, what explains the social resistance to local projects?

One challenge to NIMBY explanations in the California deserts is that very few people actually have backyards there. Local residents were represented in public comments submitted on projects, but outside organizations were also well represented. Labor unions bussed out the rank and file to advocate projects, and environmental organizations sent people to oppose them. As public lands, they are a shared resource that benefits the public in general. But public lands and renewables would also be fed into the politics of federal land management, which has been an object of political mobilization for over 100 years—the contemporary period starting with the Sagebrush Rebellion. Without understanding the cross-section of federal land politics, one might miss this important factor shaping public attitudes to solar siting.

Aside from findings that emphasize the need for participation and stakeholder engagement, some research suggests that the social gap is best explained by how a project fits into its regional context. Early research on opposition to renewable energy projects in the California desert suggested that wind energy landscapes were the most contentious.[62] With the rapid acceleration of wind developments associated with the new markets for renewable energy, opposition to siting wind turbines is still an important part of the politics of the American West.[63] But as the costs of solar energy technologies fall, conflicts are rapidly growing. Key reasons for this include the intensive land use of solar facilities compared to wind turbines, which have a smaller direct footprint on the land. In addition, solar energy facilities have by and large relied on public lands, while wind energy developments are sited on both public and private lands. Use of public lands arouses public opposition, first because there are more opportunities to comment on projects, so negative representations have a visible public platform. The federal nexus here also made projects easier to oppose because there were more legal opportunities to intervene.

Some researchers posit that local resistance to renewable energy is due to a failure of local groups to recognize the imperative of climate change. Sound social science can help overcome barriers and obstacles to renewable energy development.[64] A lack of familiarity with renewable energy facilities may be at the core of the problem, which would suggest educating the public about the benefits of renewable energy and how it works, though this contradicts the claims about the benefits of collaboration and participation from other research.[65] Research on opposition to renewable energy projects has found that many opponents are articulate and well reasoned.[66] It may be a firmly held psychological idea that makes groups hold on to the local in the face of a global, distant problem like climate change.[67]

Public referenda and collaborative planning may help fill the social gap, as poor communication and mistrust are primary points of conflict in opposition to renewable energy.[68] Incorporating local knowledge, experiential learning, and access to information into project proposals could reduce social opposition.[69] Visual simulations of impacts may offer opportunities to mediate conflicts. Redistributing benefits and providing a sense of ownership to community members have also been shown to reduce social-gap frictions; community involvement reduces resistance to projects compared to communities where there was no community involvement.[70] Tolerance maps and decision support systems may help minimize conflicts where resolving aesthetic issues could help resolve such controversies. Public involvement in planning can foster a more collaborative spirit around wind farm proposals, suggesting possibilities for community collaboration to find acceptable outcomes even if the sides are not in agreement with the final results entirely; such processes are key to satisfying "wind justice."[71] Where opponents base their judgments on a sense of landscape justice, a process that respects multiple landscape valuations may help ensure equitable and fair outcomes.[72]

SLEEPING BEAUTY AND THE CATELLUS LANDS CONTROVERSY

With the announcement of the BLM solar development program, desert conservation groups immediately began to question the conservation implications of fast-tracked USSE projects. Some believed that solar developers were receiving a "green halo" for projects that otherwise would be criticized for their ecological impacts and even attract lawsuits or other legal interventions.[73] Most organizations were unequivocal that the siting of many proposed solar developments was at odds with habitat and species conservation, and even climate mitigations for wildlife. Fast-tracking allowed too little time for review and would compromise sound scientific judgment by preventing robust, science-based impact assessments. Basin and Range Watch, one of the sentinels observing solar developments in the American West, remarked on its blog, below an image of ten large Caterpillar scrapers preparing land for a solar project, "Can't we slow down a little and put together a coherent energy policy?"[74]

The Wildlands Conservancy is an environmental organization that raises support to purchase lands of conservation significance and donates them to land management organizations. Its executive director, David Myers, was near Anza-Borrego Desert State Park, east of San Diego, when he first heard about the proposed solar projects in the California deserts. Myers read in the *San Diego Union-Tribune* that a startup company named Stirling Energy had requested that the California Energy Commission expedite the environmental impact review of the proposed Imperial Valley solar project near Plaster City, California, south of the Salton Sea and about ten miles north of the U.S.–Mexico border.[75] The developers had asked that a review process that usually takes two years be reduced to ninety days.[76]

Complicating matters for the fast-tracked environmental review request for the Imperial Valley solar project were roughly 5,000 flat-tail horned lizards (*Phrynosoma mcallii*). If development occurred, these reptiles would need to be transferred off-site to mitigation lands. The flat-tail horned lizard was under consideration for listing under the California Endangered Species Act.[77] Designing that element of the project planning alone would take considerable time and money. Meyers said,

> They wanted to shorten the review period for their environmental impact statement, but . . . they had to relocate 5,000 flat-tailed horned lizards from their site, which is a candidate for a federally endangered species . . . and here you have a technology that has never been proven on a large scale. They [solar developers] say they are going to have to get their power engines made on an automobile assembly line to get their costs down 90%. But here you have the U.S. government offering them a 30% grant; and on top of that 30% to cover the cost of the project, they are offering them another 50% in a guaranteed loan for an unproven technology, for a foreign company where all the profits are gonna go outside of the United States . . . where we have PV like First Solar coming down to a dollar a watt. There is just no way Stirling can compete with PV, except for in the venue of lobbying congressmen and lobbying this administration to give them low-interest loans and 30% grants for their projects. That's what is most frustrating for us. The administration, in its haste, not unlike its haste with offshore oil drilling . . . the biggest thing is, he [Interior Secretary Salazar] has no experience. And that's what we are seeing. Ken Salazar, with no experience in energy, just opening up the California desert for companies on a first-come, first-served basis. With no competing

bids for who has the best technology, who has the proven technology, who is actually going to be able to meet their power purchase agreement, who is capitalized, who is speculating. I mean this is just a free-for-all, and it's terribly naive to think we are going to have anything less than a lot of dead dinosaurs that have destroyed 6,000 acres of land here and there, scattered throughout the California desert, when all is said and done.[78]

Meyers spoke highly of solar in general, proudly mentioning a home he had built with ARCO M70 photovoltaic modules in 1984; the modules still operate at a high output (ARCO's legacy is currently SolarWorld). He also pointed to a trend where the industry and government seemed to be favoring a solar deployment strategy based on utility-scale solar and wind farms connected through expanded and new transmission corridors. He pointed to projects by the Los Angeles Department of Water and Power and projects that cut through two desert wilderness reserves. "Governments shouldn't be picking sites," he said.[79] The Imperial Valley solar project would be connected to metropolitan San Diego through the Sunrise Powerlink, a high-voltage transmission line connecting wind, solar, geothermal, and natural gas power plants in the Imperial Valley westward, over the mountains. The proposed transmission line was mentioned numerous times in public comments on many of the first wave of USSE applications on public lands in the Colorado Desert.

Myers pointed out that there was no evidence that the Stirling-engine technology could be cost-effective. A Stirling engine is an external combustion engine that absorbs solar thermal energy to create temperature differences inside a piston chamber; the moving piston spins a generator, creating electricity. The environmental controversy around the Imperial Valley solar project was a harbinger of things to come. No Stirling-engine technology would be proposed over the next ten years as the costs of photovoltaics fell and concentrated solar power (CSP) fell into disfavor among investors.

The spark for this environmental flareup came with solar developments proposed for the Catellus lands in southeastern California, along a lonely stretch of the iconic route 66, near Pisgah Crater, at the center of the Mojave Desert. The Mojave Desert is in southeastern California and southern Nevada, bounded by the southern tip of the Sierra Nevada and the Great Basin to the north, the Tehachapi Mountains to the west, and to the south by Mount Baldy and where Joshua Tree National Park drops down to a lower elevation, where it transitions to the Colorado Desert. Other scientists suggest that it and other deserts of the American Southwest will be impacted by climate change more than any other place in North America south of the Arctic.[80] The Mojave Desert is considered one of the most ecologically intact high desert ecosystem in the world, with 86% of the area documented as having high conservation value.[81] At the same time, it sits on the edge of California's largest population center, which is ripe for renewable energy development. The "desert scrub" term used by some to describe the Mojave Desert ecosystem actually represents several community types, including creosote bush scrub, saltbush scrub, shadscale scrub, blackbush scrub, and Joshua Tree woodland. Somewhere on the order of 80–90% of species in the Mojave are endemic, so the region has unique flora and fauna. The Mojave Desert has faced numerous ecological challenges from westward expansion and particularly the development of Los Angeles and Las Vegas. The major stressors on the Mojave Desert include urbanization, intensive livestock grazing, off-road vehicle recreation, road construction, military operations, invasive species, and mining. These same lands now face pressure from USSE proposals. The desert has a tremendous diversity of geological features, but also reveals two centuries of extractive industries.

After learning of the Imperial Valley solar project, Myers and April Sall, also of the Wildlands Conservancy, received word that several new projects were being proposed on the Catellus lands. The Catellus lands are 600,000 acres of former railroad land purchased by the Wildlands Conservancy and donated to the BLM for long-term conservation. The lands were a legacy of the 1864 Homestead Act. The Catellus Development Corporation is the real estate arm of the Santa Fe Pacific Corporation. The Catellus lands are scenic lands with interesting geologic formations—basin and range, cinder cones and lava flows—and viewsheds that offer views from horizon to horizon. They were also oddly configured, owing to their Homestead Act origins, which granted the railroad every other parcel in a 20-mile wide checkerboard of one square mile blocks for fifty miles from the Colorado River west to Barstow, California. The rest were public land, managed by the BLM.

Starting in the 1980s and carrying through to the 1990s, the Wildlands Conservancy raised $45 million in private and $18 million in federal support to protect 600,000 acres of desert wilderness around Joshua Tree National Park. In 1999, the conservancy completed the largest land acquisition ever donated to the federal government. The Southern California NGO purchased 587,000 acres of Catellus land and donated these parcels to the BLM and the National Park Service; the latter were added to Joshua Tree National Park and the Mojave

National Preserve in 1999 during the William Clinton administration. Railroad explorers in the 1850s knew the first rail line through the Mojave as the 35th Parallel Route. If chosen as the first transcontinental railroad route, it could have rewritten the history of the west coast of North America. Union Pacific's eventual route would spawn the Mojave Desert towns of Kelso, Cima, and Nipton.

Wilderness organizations claimed that the public lands were under assault. Solar power plant proposals were popping up everywhere. Environmental groups chastised the BLM for processing permits to solar developers without adequately knowing the conservation status of particular parcels. One company proposed a 19,000-acre solar farm, nearly 100 times larger than any existing solar generation facility. Anthem Solar proposed a project of 10,000 acres. Sall noted 19 projects inside an area that was proposed in 2008 to become two national monuments with the help of Feinstein. President Obama used the Antiquities Act in 2016 to proclaim the monument designations. The 94,100-acre Mojave Trails National Monument now connects Joshua Tree and the Mojave Preserve units of the National Park Service, while the 134,000-acre Sand to Snow National Monument connects the highest peaks of the California desert to dry lakes and basins.

The most controversial of the Catellus projects was a 5,130-acre CSP tower project proposed by BrightSource Energy for Broadwell Dry Lake, in Sleeping Beauty Valley (Figure 8). Located outside the western edge of the Mojave National Preserve and adjacent to the Kelso Dunes Wilderness Area, a bit north of Ludlow (and near the Calico Project mentioned earlier), it is one of the most remote USSE projects ever proposed in the California desert. The area is populated by bighorn sheep (*Ovis canadensis*).

BrightSource was a venture-capital startup based in Oakland, California, and owned by BrightSource Industries Israel. BrightSource's heliostat and power tower design was the brainchild of Arnold Goldman, an experienced solar power plant designer who had completed a smaller prototype in the Negev Desert in Israel in 2008. He received the 2009 World Economic Forum's Technology Pioneers Award. Goldman, BrightSource's founder, also founded Luz International, a developer that built nine solar energy generation systems near Barstow, California, in the 1980s, but was bankrupt by 1991. Goldman had a particular utopian idea for solar energy development, with energy systems surrounded by communities, culminating in a mythical city he called Luz, where angels climb a ladder to heaven.[82]

FIGURE 8. Numerous solar projects were proposed for the Sleeping Beauty Wilderness, which is now part of Mojave Trails National Monument.

With solar developments threatening to undermine the promised conservation of these ecosystems, the Wildlands Conservancy voiced concern about a breach of trust with the BLM. With the support of David Gelbaum, the conservancy's largest donor, a clean-tech investor and venture capitalist and a personal friend of California senator Diane Feinstein, a plan to block the solar development were set into motion. The plan would establish two new national monuments via federal legislation. All told, BrightSource filed 19 proposals for CSP towers on public lands, and complained about the proposed monuments. Robert F. Kennedy Jr., whose venture capital firm, VantagePoint Capital Partners, was heavily invested in BrightSource, was quoted in the *New York Times* chastising the senator. "This is arguably the best solar land in the world," he said, "and Senator Feinstein shouldn't be allowed to take this land off the table without a proper and scientific environmental review."[83] At the time of the conflict between BrightSource and Wildlands, this firm was the only visible member of the group of investors that would invest in BrightSource. Kennedy continued, "I respect the belief that it's all local. But they're putting the democratic process and sound scientific judgment on hold to jeopardize the energy future of our country. . . . Harnessing the sun's energy will be paramount to addressing climate change and protecting our natural heritage. Proven and cost-effective technologies like BrightSource Energy's solar thermal systems exist today and are ready to be implemented. The time to act is now."[84]

By now BrightSource was receiving negative press from popular venues such as *Forbes,* the *New York Times,* and the *Los Angeles Times.* Environmental groups asked the developers to consider more appropriate sites. In a press release, NRDC senior attorney Johanna Wald clarified that they had "tried very hard to avoid litigation and filed this suit as the last resort. We have focused instead on consensus building to improve as many large-scale solar projects as possible to transition our nation to clean energy sources while protecting wild lands and wildlife."[85] The courts disagreed, and the Calico project was eventually approved. But by 2013, K Road had withdrawn its application for the site and sold off its development pipeline.

Ironically, Kennedy four years earlier had penned an op-ed opposing the Cape Wind project off Nantucket Sound. "As an environmentalist, I support wind power, including wind power on the high seas. I am also involved in siting wind farms in appropriate landscapes, of which there are many. But I do believe that some places should be off limits to any sort of industrial development. I wouldn't build a wind farm in Yosemite National Park. Nor would I build one on Nantucket Sound, which is exactly what the company Energy Management is trying to do with its Cape Wind project."[86] The proposed wind turbines would have stood less than 10 miles from the Kennedy family compound. One activist quipped, "BrightSource [is pursuing] the worst projects in the worst locations, but they have the best PR firm, because Robert Kennedy is involved."[87] The activist was referring to the fact that Kennedy is an outspoken environmentalist and his opinion carries weight in the environmental community.

Shortly after the public flare-up in 2008, BrightSource withdrew its application. All told, six companies would withdraw their applications, and thirteen more would have been blocked by the proposed national monuments. Other companies, such as Congentrix Energy, a subsidiary of Goldman Sachs, also canceled their projects. Congentrix's vice president said, "When we attended the onsite desert meeting with Senator Feinstein, it was clear she was very serious about this."[88] Buried deep in the legislation were rules authorizing the BLM to expedite USSE projects outside the monument area, while protecting the lands contained by the monuments. The projects proposed for Catellus lands were removed, including the controversial Calico project and the BrightSource project. One-third of the Calico project area overlapped with the range of the entire species of white-margined penstemon, a plant the California Native Plant Society is petitioning to be listed under the Endangered Species Act (plants are notoriously difficult to get listed).[89]

There remained salient questions about the effectiveness of the BLM's approach. Sall noted, "I have to say I think that the BLM process is a broken process—it's geared for conflict."[90] The Catellus lands would later be included in the boundaries of two new national monuments in California, initially proposed by Feinstein in 2008 and added by President Obama via the Antiquities Act in 2016. The lands received permanent protection from solar development in a planning process that led to the Western Solar Plan.[91] Some of these public lands eventually would become the Sand to Snow and Mojave Trails National Monuments.

CULTURAL RESOURCE ISSUES AND TRIBES

Fast-tracking hastened solar deployment but also exacerbated some environmental justice tensions. Across the American West there is evidence of past peoples and civilizations. The Fort Mohave, Chemehuevi, and Quechan are just some of the Colorado River tribes that expect prior consultation from developers and the BLM, and failing to do so could lead to costly legal actions or project delays. The public lands where solar development is focused must comply with the National Historic Preservation Act. Early projects approved by the BLM were a model of how not to consult with tribes (cursory consultation, failing to keep tribal leaders informed), and the BLM has revised its practices of consultation considerably since 2010. Figure 9 shows utility-scale solar projects across the American Southwest.

Several Colorado River Native American tribes consider this region a sacred ancestral home. Numerous lawsuits from 2010 to 2012 claim that the BLM failed to adequately consult tribes on cultural resource issues, raising a question of procedural justice. The La Cuna de Aztlan Sacred Sites Protection Circle filed suit against six solar power plants on BLM lands shortly after the interior secretary approved them. The suits claimed that the BLM did not take this and other Native American concerns into account when evaluating the impacts of fast-track projects. The Indian tribes spoke of a long history of Europeans exploiting and unfairly displacing tribes. Arrow-weed admonished, "They seem to want to do it at the price of destroying history. . . . It's an assault. They've already wiped out a lot of things and now they want to wipe out the desert and any evidence of our past."[92]

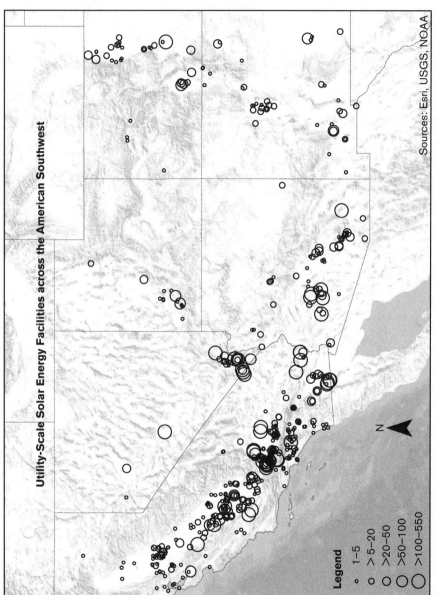

FIGURE 9. Map of fast-tracked solar energy projects on public lands (project size in megawatts).

FIGURE 10. Geoglyph in desert pavement near the Blythe Solar Power Project.

At one site, tribes argued that a project would harm the fringe-toed lizard, a species that is central to their creation story.[93] Harm to these organisms was seen as damage to Quechan culture. Another project, the Genesis Solar Project, was temporarily halted after the discovery of human remains in a suspected prehistoric cremation site.[94] A third project damaged several geoglyphs near Blythe, California (Figure 10).

The cultural resources found in the lower Colorado River Valley, which spans Arizona and California, contain unique features known as geoglyphs or intaglios, which are sixty to one-hundred-foot figures depicting humans, animals, and spirits. They were made one to three thousand years ago by Native Americans, by turning over dark stones so that their lighter bottom sides are visible, and were not rediscovered until the advent of airplanes. They form part of the spiritual basis for the religion of the

FIGURE 11. Alfredo Acosta Figueroa, who once worked to organize farmworkers with Cesar Chavez, led efforts by tribes to increase scrutiny of solar projects slated for public lands.

Colorado River tribes.[95] According to Native American elder Alfredo Figueroa (Figure 11), developers bulldozed a geoglyph of Kokopilli (a deity of fertility that looks like a flute player with wild hair) and a sun geoglyph on public lands shortly they went bankrupt, leaving the spot vacant until construction commenced on a solar farm in 2014. These geoglyphs are made of desert pavement—small pebbles firmly settled atop desert soils—and tribes call them sacred places, according to filmmaker Robert Lundahl, who made a documentary on the conflict. Elsewhere, clear patterns of anthropogenic origin scattered across the region in the form of giant intaglios, figures tens of meters across, can be seen in the desert pavement. Many intaglios are thought to be up to 10,000 years old. Native Americans have occupied this region for thousands of years, and it is believed to have been an important area for settlement by the ancestors of the Aztec, Mayan, and Incan civilizations, as humans migrated across the Bering Strait and toward Central and South America.

My interview with the Chemehuevi Indian leader began in his living room, where he pointed to the answering machine. Under a memorandum of understanding between tribes and the BLM, tribes are supposed to have a prior consultation. He noted that the BLM office that was required to seek his input did not leave a voice message. Alfredo contended that a call leaving no voice message did not constitute an effort to seek prior input. Figueroa is a historian by training and organized the La Cuna de Aztlan Sacred Sites Protection Circle, which seeks to protect lands along the Colorado River for their spiritual significance. Aztlan is the mythical place of origin in the creation story of many tribes in the Americas, and Figueroa points to evidence that the lower Colorado River region is the birthplace of the Aztec and Mayan systems of belief.

The tribes also maintained that there are significant gravesites in the project areas of some of the developments. Alfredo Acosta Figueroa is a long-time social justice activist. Great-grandfather Figueroa is Cheme-huevi Indian (a branch of the Southern Piute, based near Lake Havasu on the Colorado River) and a longtime resident of Blythe, California, which was the epicenter of solar development in the state in terms of total acreage. In an interview in 2011, Figueroa showed pictures of his work with Cesar Chavez organizing farmworkers in the 1960s and 1970s. He marched alongside Chavez with farmworkers, and worked to block nuclear waste dumps in the 1980s. More recently, he turned his organizing toward the damage being done to cultural resources in the Colorado River Valley as a founding member of La Cuna, but also a member of Californians for Renewable Energy. He opposed several solar projects that threatened cultural resources, but also led campaigns against natural gas peaker plants and the proposed (but never built) Sundesert Nuclear Power Plant, and today was pointing to intaglios that had been destroyed at a nearby site called the Blythe solar power plant, originally under development by Solar Millennium (a company based in Cologne, Germany) before it declared bankruptcy in 2011.

During a visit to the site of the Blythe project site, Figueroa and his two grandchildren spoke of the importance of the horned toad and the desert tortoise.[96] He noted that "the tortoise is at the center of the Aztec sunstone calendar. . . . This represents more of the creation story than any other relic."[97] Figueroa made reference numerous times to the role that Western peoples have long played in exterminating the cosmic tradition of Native American communities and the sacred nature of desert wilderness for Uto-Aztecan language speakers. To Figueroa, solar energy development in the desert wilderness is "antithetical to the sacred sites' purpose and appears to be intended to essentially trap the Creator Quetzalcoatl as the deity descends at sundown."[98] This theme of the continued genocide of Native Americans by Western culture was echoed in public comments submitted to several other renewable energy projects, including Ivanpah, the Palen Solar Project, the Ocotillo Wind Energy Facility, the Panoche Valley Solar Project, and the Genesis Solar Energy Project.

During my interview with Figueroa, standing at the foot of the McCoy Mountains, he pointed toward the Palen Mountains and noted that the ancient watering hole near the site almost certainly bore the remains of his ancestors, as they would frequently convene and reside near water sources in this arid stretch that connects the Colorado River to the Pacific Ocean. After climbing a hill to an overlook viewing the neighboring valleys, where the McCoy, Genesis, and Palen USSE projects were planned, Figueroa noted the sacred importance of historic trails. Chemehuevi means "people who play with fish" in Mojave, and the name seems ironic until one considers the plentiful fish in the Colorado River flowing out to the Pacific Ocean, places that were connected by Native American trails, some of which are officially registered as national historic trails. "Our people lived near springs along the trail; our ancestors are buried there," he said.[99] There are numerous other historic trails passing through many of these sites, including those used as supply points for early settlers and overland stage routes that were part of the great western migration.

In 2009, a subsidiary of Next Era Energy Resources proposed building a parabolic trough solar power plant in the Colorado Desert, close to the Colorado River boundary near Blythe, California, on Ford Dry Lake. The $1 billion Genesis Solar Energy Project would generate 250 MW of power on BLM lands. While the project was much heralded for its low-carbon electricity, after being approved, it was almost immediately embroiled in controversy.

Early in the public review process, Native American elders warned that the plant was sited near a desert watering hole along an ancient trail connecting the Colorado River to the Pacific Ocean. There were concerns raised about the cultural resources of Native American tribes. The project is near Ford Dry Lake, which is widely held by several Native American tribes to be a significant cultural heritage site and important ecological site for a reptile of special spiritual significance.

The next year, in 2011, at the site for the Genesis solar power project site Figueroa referred to, grinding stones and a layer of charcoal believed by Colorado River tribes to be an ancient cremation site were uncovered during construction, resulting in a lengthy delay. The tribes demanded that 80 hectares be taken out of the proposed development. One tribal elder was quoted in the *Los Angeles Times* as saying that the project "disrupted the peace of our ancestors and our relationship with the land. There is no mitigation for such a loss."[100] The special place of Native Americans in the socio-ecology of North America warrants their influence on the direction of socio-ecological change as society determines the pathways for responding to global climate change. These considerations extend beyond the deserts of the American West, as lands rich in solar resources are facing pressures across the arid parts of the world, including dispossession and socio-ecological change.[101]

The tribes justly claim that there was no prior consultation with them on the site, a requirement for public lands through a special arrangement with the U.S. government. Figueroa spoke specifically of the problematic nature of the fast-track process, noting that the land was taken from them when their numbers were small, over a century ago. He believes that much of the land belonged to the tribes as reservation land until it came under federal control and the reservation size was reduced. Figueroa claims that the BLM only called once and did not leave a message, abrogating its responsibility, something the BLM denies.

David Myers and April Sall, the two leaders of the Wildlands Conservancy, expressed notable disdain for the big environmental NGOs, like the Sierra Club, NRDC, Defenders of Wildlife, and Wilderness Society. They suggested that these organizations were prioritizing renewable energy development as a climate change imperative no matter how much development might harm biodiversity. "We got dragged into this because the big groups were standing on the sidelines and we were watching this big conservation legacy practically go under a bulldozer," said Sall, the organization's conservation director. "We said, 'we can't be silent anymore.' The Sierra Club and the NRDC—their mission is to work on climate change above all else," Sall said. "We refuse to compromise on that level."[102] Even the Sierra Club remained split at local and national levels. At a 2007 meeting of the Sierra Club's California/Nevada Desert Committee, a senior representative of the national organization said that local concerns about siting were less important than getting projects implemented and developed quickly.

Projects on public lands faced stiff social resistance, through litigation or public protest. At the same time, a number of ARRA-supported projects were being developed on private lands and facing much less scrutiny. The Agua Caliente Solar Farm in Arizona, built by First Solar, received no visible public opposition because the private land had been used for conventional agriculture. The private lands that did receive scrutiny were usually those in easements for conservation or grazing, so from an ecological perspective they were similar to the public lands offered by BLM. Though not explicitly public lands, lands in easement receive some pubic support through property tax breaks.

IVANPAH VALLEY, SOLAR POWER, AND AGASSIZ'S DESERT TORTOISE

BrightSource withdrew its Sleeping Beauty Valley project on the Catellus lands in 2009. The spotlight quickly shifted to another BrightSource project that was advancing toward approval. The Ivanpah Solar Electric Generating Stations were the subject of evidentiary hearings before the California Energy Commission, the state agency that permits the siting of thermal power plants, including CSP—photovoltaic solar farms do not need permits from the commission. BrightSource proposed to build three CSP towers, with 173,500 computer-controlled heliostats directed to deliver solar energy to boilers (preheated with natural gas) to make steam to drive a turbine to generate electricity.[103] For the proposed 400 MW Solar Electric Generating Stations, BrightSource Energy requested a permit for 4,055 acres (about six square miles) of public lands, which would require translocation of hundreds of desert tortoises, a federally protected and threated species (Figure 12). This ancient species of the Mojave evolved when the region was more tropical. The desert tortoise can live over a hundred years and can tolerate extreme temperatures and drought.

The Ivanpah Valley is on the California side of the California–Nevada border in the eastern Mojave Desert. Most travelers find themselves here en route to Los Angeles or Las Vegas, as Interstate 15 runs through the valley parallel to Ivanpah Dry Lake and the Union Pacific rail line. Imagine a long valley split by an interstate highway and a parallel rail line, with a dry lake at the low point of the valley. Two towns stand thirty miles apart at opposite ends of the valley, which is bounded by the New York Mountains to the south, Clark Mountain and the Ivanpah Mountains to the west, the Spring Range to the north, and the McCollough Mountains to the east. To the north, Primm, Nevada, is a town that resembles a rest stop, with a thousand people, about the total capacity of the town's three hotels. The unique desert town has a small

FIGURE 12. Agassiz's desert tortoise is a threatened species in the California Desert.

amusement park, with a roller coaster, Ferris wheel, casinos, and shopping outlets connected by a monorail. Every square foot of the town, up to the boundary with public lands, is developed; Primm's parking lots extend to within an inch of the California–Nevada border.

Nipton, California, is much smaller, with a population of 60. Settled as a nineteenth-century Southern Pacific railway stop, the town now sits on the northern edge of the Mojave National Preserve, with a bar, cafe, trailer park, motel, and general store scattered among a handful of residences. The towns themselves are developed to different degrees, Primm nearly a square mile of urban space and pavement and Nipton resembling a square mile of rural ghost town. In the landscape of the Ivanpah Valley, the towns barely occupy any land at all. More than 80% of the valley is considered "ecologically core" or "ecologically intact" habitat by the Nature Conservancy, based on a regional assessment of the Mojave Desert in 2008 in preparation for renewables development in California's deserts.[104]

The remaining expanse of the Ivanpah Valley is 200 square miles of Mojave Desert open space covered in creosote bush scrub, old-growth barrel cacti, and Mojave yucca trees and containing the desert tortoise and the rare white-margined penstemon flower. There is a natural gas power plant looming in the distance above Primm, near a cement factory, and a 36-hole golf course sits at the bottom of the valley, near the dry lake. The southern portion of Ivanpah Valley is protected in the Mojave National Preserve, and a portion of the valley on the California side has been designated critical habitat for the desert tortoise by the USFWS. Though the BLM lands proposed for solar development here did not have any special protections, the landscape is considered wild enough to be a popular place for researchers to study the behavior of wild populations of desert tortoise and collect rare plants.[105] The valley has seen its fair share of other kinds of development proposals; an international airport for Las Vegas was proposed but later withdrawn in the 2000s, and there have been numerous proposals for high-speed rail to connect Los Angeles to Las Vegas.

The Ivanpah Valley's future as an industrial solar zone would be sealed with the announcement of BrightSource's project. On paper, the Ivanpah Solar Electric Generating Stations consisted of four limited-liability corporations, Solar Partners 1, 2, 4, and 8. In the public eye, the company was BrightSource Energy, a venture capital startup backed by Silicon Valley and clean-tech investors and angels, including environmentalist Robert Kennedy Jr. BrightSource was already a controversial name in the environmental community because of the controversy at Sleeping Beauty Valley with the Wildlands Conservancy.

BrightSource's investors at Ivanpah were revealed to be Chevron, Google, CalPERS, the California State Teachers' Retirement System, BP, Morgan Stanley, and VantagePoint Capital Partners, which collectively raised $615 million in private capital, mostly from NRG, a Princeton, New Jersey–based electricity generator and owner of many merchant power plants, which are power plants that sell electricity to utilities in electricity markets. For Google it was its flagship investment in RE<C— its low-carbon technology initiative, an effort to make renewable energy (RE) cheaper than (<) coal (C). BrightSource and its partners borrowed another $1.6 billion from the DOE loan guarantee program with money from Treasury's Federal Financing Bank, the conduit for ARRA. Once complete, the energy conglomerate NRG would own and operate the $2.2-billion solar power towers. Two California investor-owned utilities agreed to purchase Ivanpah's electricity at around $0.12 per kilowatt-hour to comply with state RPSs: Pacific Gas and Electric, serving central and northern California, and Southern California Edison, serving further south. Bechtel led the construction project. It was no stranger to building energy projects in economic hard times, having constructed nearby Hoover Dam during the Great Depression with the labor of the Industrial Workers of the World. Bechtel's strength as a renowned global engineering firm no doubt assuaged some of the worries of the private investment community and likely influenced the financial metrics that awarded them the federal loan guarantee.

Constructing Ivanpah led to the translocation of over 150 desert tortoises. They epitomize the vulnerability of species experiencing severe decline, as they are considered the most likely of higher-order animals to face extinction in the next century.[106] The desert tortoise has experienced increasing cumulative impacts range-wide, such that the Desert Tortoise Council has recommended uplisting the species from federally threatened to endangered. The species is already one of the top recipients of federal spending for an endangered or threatened species. U.S. state and federal agencies spent $93 million on conservation for this desert reptile from 1996 to 2006, more than for the gray wolf, grizzly bear, or bald eagle.[107] After a century of habitat transformation and disturbance, the desert tortoises now face the twin pressures of climate change and land-use changes.[108] The subpopulation west of the Colorado River was federally listed as threatened in 1990, having lost 90% of its population over the previous fifty years.[109] Land-use change and habitat loss is the primary threat to desert tortoise survival and remains the primary driver of extirpation of some genetic subpopulations. Healthy genetic subpopulations and gene flow between them may be critical to the species's resilience and survival. One biologist interviewed for this research noted that "the genetic diversity of the desert tortoise is important to maintain for species health; these power plants could impede gene flow."[110] Species like the tortoise require modest levels of gene flow to prevent genetic problems associated with inbreeding and susceptibility to disease. Through direct individual loss and by posing obstacles to gene flow, USSE could further erode biodiversity in the Eastern Mojave. Land managers play a critical role in desert tortoise protection, as 80% of the desert tortoise population is found on public lands.[111]

The Ivanpah project was described many times by Obama and his cabinet in political speeches on climate change and economic recovery. Obama enthusiastically described the project in 2010 in one of his weekly Saturday morning addresses. The project created 1,000 jobs during construction, and 86 permanent jobs, with an agreement with local unions.

This month, in the Mojave Desert, a company called BrightSource plans to break ground on a revolutionary new type of solar power plant. It's going to put about a thousand people to work building a state-of-the-art facility. And when it's complete, it will turn sunlight into the energy that will power up to 140,000 homes—the largest such plant in the world. Not in China. Not in India. But in California.[112]

President Obama continued to describe how the loan guarantee would facilitate research and development in BrightSource's home of Oakland. It was touted as a job-generating innovation—the Sputnik moment needed to revitalize national economic competitiveness. Yet the high-tech jobs and the patents for this technology are in Israel, not Oakland. Key hardware would be sourced more locally. A company based in Arizona would provide the steel for the heliostats. The gearboxes that allow heliostats to track the sun were made by Cone Drive Gearing Solutions a subsidiary of defense contractor David Brown Group.[113] The same technology used to point munitions at people could target heliostats at the sun. This collection of large corporations, unions, and major investors yielded significant political power in pushing the project forward. Even Kennedy's brother-in-law and renewable energy advocate California governor Schwarzenegger asked Obama and Salazar in a letter to fast-track Ivanpah.[114]

A small wilderness advocacy group, Basin and Range Watch, was one of the earliest intervenors, bringing other groups to the site to make the case for an intervention. Environmental groups such as the Sierra Club, Center for Biological Diversity, and Defenders of Wildlife eventually intervened in the case, alongside citizens and local groups expressing concerns about the project's impact on a relatively large intact ecosystem connecting parts of the Mojave National Preserve. Public comments from environmental organizations raised questions about the impact the solar power towers would have on desert wildlife such as raptors and other avian wildlife, rare plants, and bighorn sheep. Most notably, environmental groups pointed out that the agency decision would be contrary to the principles outlined in desert tortoise recovery plans. These plans emphasized the importance of connectivity and gene flow between tortoise subpopulations around the Ivanpah Valley, which is within the Northeastern Mojave Recovery Unit, one of six evolutionarily significant populations of tortoise designated in the CDCA.[115] The habitats are important to the desert tortoise, given other threats—mining, grazing, urbanization—and climate change is likely to shift the tortoise's habitat range as temperature and precipitation patterns change.[116]

A public comment from the Center for Biological Diversity pointed out the valley's importance to the BLM's bioregional planning efforts. It complained that "the lack of prior planning by BLM for siting of this proposed project and others could undermine the conservation goals of the CDCA plan as a whole, [and] create a de facto industrial solar zone in the Ivanpah Valley, undermining recovery of the desert tortoise in this area."[117] Environmental organizations expressed considerable sympathy for climate action, but pleaded for more appropriate sites for renewable energy development.

The possible conversion of the Ivanpah Valley into an industrial solar zone prompted the Audubon Society to worry about impacts on migratory and raptor bird species.[118] Solar flux directed at solar power towers was known to singe passing birds, and waterbirds might mistake the reflections, glare, or polarized light from the heliostats for a lake. The California Native Plant Society questioned the proposed mitigations for damage to the white-margined penstemon. "There are no known techniques to mitigate for the loss of rare plants and their habitat in desert environments," I was told. "Avoidance is the only mitigation that is appropriate."[119] The director of the University of California's Sweeney Granite Mountains Desert Research Center called the Ivanpah Valley a "biological frontier" where several notable discoveries were recently made and said that little effort has gone into fully surveying and documenting the plant biodiversity of the Mojave Desert. These features give scientific credence for including the Ivanpah Valley as an Area of Critical Environmental Concern, a special BLM designation under FLPMA used when lands require special management considerations for ecological, cultural, or scenic resources. The designation was advanced by Basin and Range Watch.[120]

Despite the challenges to Ivanpah by environmental organizations, the BLM proceeded to process the ROW permit. In preparation for a formal EIS, BrightSource hired biologists to survey the site for, among other species, the desert tortoise, which was known to spend 90% of its up to 100-year life in burrows. In this relatively dry year, biologists found only 17 tortoises on site in their surveys.[121] This contradicted opponents' claims that the site was important tortoise habitat, evidence used in the USFWS's Biological Opinion in the review process, which advised that the tortoises be moved to mitigation sites.

When the BLM approved Ivanpah in 2010, no mainstream environmental organization took legal action to stop or influence the project. This was puzzling, given the attention that the project had attracted in popular venues such as *Forbes*, the *New York Times*, and the *Los Angeles Times*. National environmental groups had a difficult time opposing renewable energy—*any* renewable energy, according to some wilderness advocates—because many large funders thought fighting climate change, at any cost, was more important. An "all-of-the-above" renewable energy policy has dominated mainstream environmental groups, without nuanced consideration of impacts on biodiversity and natural resources. This has only grown in recent years, with most NGOs taking a hands-off approach to opposing USSE projects, preferring to settle instead.[122]

Only the Western Watershed Project, a watchdog organization focused on "private abuses of public lands," filed a lawsuit, arguing that the USFWS had relied only on the project proponents' paid scientists for a "woefully underestimated" count of tortoises to be impacted.[123] The organization argued that the BLM should not have approved the solar electric generating station because it had only assessed adult tortoises in the population. The case was eventually denied in the U.S. Ninth Circuit Court of Appeals, which upheld a federal judge's earlier ruling that supported the project.[124]

The tortoise count would dramatically rise as Ivanpah's construction proceeded. As land clearing commenced in late 2010, BrightSource quickly exceeded the permit limit of 38 tortoises, triggering a second consultation with the USFWS and a temporary construction delay that threatened the loan guarantee.[125] Since a respiratory disease carried by tortoises can be spread by handling them, the developers agreed that tortoises found on site would overwinter in seclusion pens in quarantine. Over 170 tortoises were kept and raised from 2011, and this will continue until 2020.[126] Tortoises were outfitted with radio-telemetry transmitters and evaluated for disease before being released onto adjacent lands when deemed healthy. Each tortoise would be tracked to assess survival rates from translocation. Tortoise translocations were already controversial in the region and face skepticism in the Mojave Desert because of earlier efforts to translocate nearly 770 tortoises for an expansion of nearby Fort Irwin. The project was suspended when nearly a hundred died within several weeks.[127] In early 2011, BLM scientists issued a report that BrightSource's Ivanpah project would disturb 3,000 desert tortoises, with 700 juveniles killed from construction alone.[128] Tortoise mitigations will cost BrightSource $56 million overall through 2020.[129]

By 2012, the financial viability of BrightSource seemed to be in question. The company found it difficult to get other proposed projects approved in the U.S. and abroad. The clean-tech investment community took note when BrightSource withdrew its $150 million dollar initial public offering on the New York Stock Exchange. The mood on CSP technologies soured in the venture capital community. A successful Silicon Valley clean-tech investor, Nancy Pfund, penned an op-ed, "Donuts, Renewable Energy and What They Say about America," declaring the comparably small investment in renewable energy, the very same week that Dunkin' Donuts raised nearly a billion dollars from Wall Street, a troubling sign.

> Now, there's nothing wrong with a glazed cruller now and then. But at a time when climate change is wreaking havoc on weather patterns and wars are being fought over access to fossil fuels, why do doughnuts trump clean energy as an attractive place for investment? When nations all over the world are investing heavily in clean energy, why are American investors sitting on the sidelines? Last time I checked, while America may run on doughnuts, the rest of the planet runs increasingly on renewable resources. While China has made clean energy one of its strategic industries, we still are hoping we can drill our way to the future.[130]

The next bad news for Ivanpah was that the power plant was delivering less electricity than planned and using far more natural gas. BrightSource and the loan program received severe criticism when Ivanpah was listed as a major *source* of greenhouse gas emissions in the state and had to participate in the cap-and-trade program.[131] In February 2017, the owner, NRG, announced the power plant was finally delivering the designed amount of solar electricity.[132] One environmental attorney described Ivanpah as "just a boondoggle. . . . This isn't about solving an environmental problem or an economic problem. It's corporate welfare."[133] Figure 13 depicts the three solar power towers in the Ivanpah Valley.

The desert tortoise was only the first ecological conflict in the Ivanpah Valley. A USFWS staff biologist interviewed by the Associated Press described a raptor passing through the solar flux of BrightSource's power towers shortly after the power plant was commissioned on February 14, 2014.[134] As the bird passed, he noted, a small plume of smoke could be seen from the bird's feathers as it glided beyond the site perimeter into the adjacent wildlands—the solar flux temperature being far higher than the melting point of feathers. "Streamers" would become a new concept to describe avian mortality. Within three months project monitors confirmed the death of nearly 300 birds at Ivanpah—and only a third of the project site is monitored.[135] The lead biological consultants on the project estimated that the solar power tower was the site of 1,500 bird deaths with known causes in one calendar year (from 2013 to 2014) and another 2,000 with unknown causes.[136]

FIGURE 13. The three solar power towers of the Ivanpah Solar Electric Generating Station, near the California–Nevada border. (Photo: WikiCommons.)

SACRIFICING PUBLIC LANDS

These desert ecosystems would become what Valerie Kuletz called "sacrifice zones" in her book about the nuclear testing in Nevada. Lands proposed for solar farms were viewed as collateral damage in the fight against climate change. Desert ecosystems with renewable energy resource endowments must be sacrificed to save society from climate change, just as they stood in for national security in stories about the Nevada Test Site.[137] Janine Blaeloch of the group Solar Done Right put it this way: "Should we save the desert tortoise, or plow over its habitat to build solar power plants that can help us save ourselves?"[138]

Debate about sacrificing public lands recurs throughout U.S. environmental history. Federal lands were at the center of the Hetch Hetchy Valley controversy in California's Sierra Nevada. This case instructively reveals how sacrifices are shaped by security discourses. As early as 1890, San Francisco's mayor, James Phelan, backed by the city's political elite and later emboldened by the utilitarian conservationist and forester Gifford Pinchot, sought water rights to build a reservoir along the Tuolumne River in Yosemite National Park. John Muir described the Hetch Hetchy Valley in 1873 in his serialized and syndicated travel column as a majestic valley similar in appearance to the more famous

Yosemite Valley, with towering waterfalls pouring over steep granite walls. After the 1906 earthquake set off fires that burned much of San Francisco, arguments for a more robust water supply became more compelling. The Hetch Hetchy Valley, the argument went, must be sacrificed to secure the city against natural disasters. (A more accurate version of history recalls that the cause of the ineffective response to the post-earthquake fires was not a lack of water but a failing delivery system.) Efforts to claim the reservoir were instigated by city leadership once they realized that the limited water resources would only support a small population and they needed more water resources for San Francisco's growth.[139] Despite the protests of John Muir and the Sierra Club and efforts to promote alternatives, the Hetch Hetchy Valley was flooded after the completion of the O'Shaughnessy Dam in 1924. Some argue that this wilderness was sacrificed to build San Francisco.

In the cases of the Ivanpah and Sleeping Beauty Valleys, conservation and wildlife organizations claimed that desert biodiversity was being sacrificed for industrial solar. Some even reluctantly agreed that the climate issue was so paramount that some ecosystems would have to be sacrificed to save others and human civilization, making solar energy development sites "sacrifice zones" where the land uses were contested. Sacrifice zones are spaces offered up for some greater good or purpose. The term is loosely used to describe marginalized places that bear the burdens of the industrial economy, purportedly to benefit society overall as measured by abstract notions of progress or development. For example, chemical production[140] and mountaintop removal for coal[141] have negative local impacts but arguably provide society with cheap plastics and electricity. Yet, used this way, the term remains theoretically underdeveloped, commonly referring simply to a place unequally devastated or lands converted for the sake of development.

The act of sacrifice implies an instrumental, deliberate act with clear underlying motivations. Sacrifice means more than giving away, discarding, or neglecting, because whatever is being sacrificed has some value to some person or community, even intrinsic or symbolic value. Discarding or neglecting a place is not the same as sacrificing it, because often decision-makers are too far removed, or worse, do not value the place. It is a slight but important nuance, and it maps onto the debate about siting solar power plants. Sacrifice is the outcome of a contemplated trade-off, vetted to be acceptable to the one committing the act, even if not acceptable or just to all.

Developing social theory around sacrifice zones to inform social planning for energy transitions requires a more selective application of the term. Sacrifice zone narratives that loosely apply the concept can lack detail on constitutive social processes, or the agents may not be evident. When chemical pollution is highly concentrated in a community, it could be more a consequence of neglect than any conscious attempt at trade-off among different benefits and losses.[142] In some places, these trade-offs are worth the benefits to the majority of local residents and voters. Research in West Virginia's Appalachia describes the mountains there as an environmental sacrifice zone "surrendered to keep power cheap,"[143] but it is not clear these are outcomes of any specific logic other than the metanarratives of capitalist development, globalization, and consumerism.

Shortly after the BLM decision to approve the Ivanpah project, California governor Schwarzenegger remarked, "I applaud the Bureau of Land Management's decision and I look forward to more decisions that will help grow our green economy, promote energy independence and strengthen our national security."[144] Similarly, interior secretary Ken Salazar, who presided over and announced the decision, justified it under the banner of energy security. "Under the leadership of President Obama, the renewable energy world is opening a new frontier. . . . The Department of the Interior is resolute and determined to secure a safer, more sustainable energy future for our nation. We do so because we can't afford to remain so dependent on foreign oil. We do so because we can't afford the risks that our energy dependence creates for national security, economic security, and environmental security."[145]

Desert landscapes have historically been depicted and described as wastelands, invoking inhospitable qualities: extreme heat, scarce water, abrasive winds, and freezing nights. Deserts can be forbidding landscapes, and making lands useful can be challenging. Environmental writer John McPhee recounts numerous failed attempts to cross the Great Basin, the endorheic watershed covering 10,000 square miles of arid western U.S. land roughly surrounding Nevada.[146] Similar stories describe fateful attempts to cross Death Valley by early California pioneers. Today the desert experience is moderated by climate control, reliable automobility, and other modern conveniences, which have turned deserts into cities like Las Vegas, Los Angeles, and Phoenix.

Desolation and emptiness are common frames of reference associated with deserts. How people experience those qualities has changed with time. Views of desolation and emptiness shift from associations with crime, scarcity, and vulnerability to attributes held in higher regard, like privacy, land conservation, and human enlightenment through transcendental experiences of nature. Those seeking refuge from the ills of urban life and humanity can come to desolate landscapes like this. Like many other writers and poets, Edward Abbey found spiritualism in the desert.[147] In *Desert Solitude,* he wrote extensively about arid public lands and their exploitation by human civilization and the industrialization of the West. He recognized the dangers of human ambivalence toward these harsh landscapes, which lack the redwoods, oaks, salmon, and other charismatic megaflora and megafauna that public campaigns could be built around. Appreciation for nature in desert landscapes among the public is more widespread today than during the early desert encounters such as the great western human migration across North America in the nineteenth century, which often was a fight just to survive.

As the railroads rolled west, more Americans came firsthand to see the Grand Canyon and its other remarkable landscapes and geological features. Landscapes such as Yellowstone, Yosemite Valley, and the Grand Canyon were apotheosized in the written word and later by photography, which widely disseminated the magnificence of the great western landscapes, and also tended to erase its human presence.[148] New appreciations for nature and landscapes soon led to protections for land with extraordinary qualities: scenic vistas, unique biophysical features, areas of conservation significance, charismatic megafauna. Many of the first U.S. national parks and monuments are in the West, but these represent only a small portion of federal lands, much of which contain lands of conservation value, but offer more ordinary ecosystem conservation qualities, interconnections, or habitat for species. Public lands (not national parks and monuments) are clearly of conservation importance, but perhaps lack specific extraordinary qualities requiring greater land protections.

Environmental historian Donald Worster distinguishes between the protection of "ordinary nature" and "extraordinary nature" in his biography of John Muir.[149] Ordinary nature consists of functioning and healthy ecosystems that may lack certain aesthetic qualities. Extraordinary nature describes the revered places that humans interpret and celebrate as sacred. Muir, for example, called Yosemite's peaks the cathedrals of the gods. He saw relatively pristine areas of the world, lacking much evidence of human presence (conveniently erasing the people who were already there before the Europeans came). Healthy desert ecosystems might only be ordinary nature if they lack majestic vistas or rock formations. This may partially explain why desert regions may receive less appreciation and advocacy, making them seem worth the sacrifice.

BLM lands have always been tasked to be productive lands. These parcels were originally left over from the Homestead Act. Efforts were made to give these federal lands to individuals and even back to the states, but no one wanted them, with the exception of the occasional mining claim or ranching permit.[150] Today, public lands are in demand by energy developers, recreationalists, ranchers, mining companies, and conservationists.

The multiple-use mission that guides the BLM seeks a balanced approach that puts land to the highest productive use.[151] As established by FLPMA, the BLM takes energy production on public lands to be one of several good uses. The mandate for renewable energy reflects the need to balance out the already widespread use of public lands for fossil fuel and mineral extraction across the West. But even though considered good uses, these activities lead to severe land degradation. The deserts of the Great Basin and Central Asia have already been used as "national sacrifice zones" for militarization and weapons development, with nuclear weapons testing called by some one of the planet's worst ecocides.[152] A hundred and twenty-six nuclear bombs were detonated above ground at the Nevada Atomic Test Site from 1951 to 1963; the site was selected in part because "there's nothing out there."[153] Mike Davis calls the area around the test site the "dead West," affirming that the destiny of the rural American West is as a national dumping ground.[154] Explorer John Wesley Powell pleaded that the scarce resources of the West require a cooperatively managed effort: "Capitalism pure and simple, Powell implied, would destroy the west."[155]

In *Savage Dreams*, Rebecca Solnit remarks that temporality in desert landscapes is experienced on geological scales, which may make them harder to appreciate and value by biological organisms like humans.[156] Even the desert's ecology operates across longer temporal scales, as deserts accumulate biomass very slowly compared to other ecosystems.[157] Desert soil surfaces are easier to damage, particularly the fragile soils covered with cryptobiotic crust, which can take decades and centuries to recover.[158] Lacking appreciation for these subtleties—the slow movement of tectonics and nature—people may underappreciate the things that hold together these ecosystems, Solnit argues. It is perhaps these understated qualities of nature in desert environs that make people more willing to sacrifice desert biodiversity in the fight against climate change. Empty lands are viewed as idle lands, and across other parts of the American West the natural resource beauty contrasts with harsh environs to more thoroughly justify taking the bounty of the otherwise unproductive lands.

Desert landscapes are characterized as barren and neglected wastelands, a point echoed throughout comments submitted to the formal environmental review processes. Every USSE project on public lands required an EIS. These reports are required when federal actions cause significant environmental or social impacts. One area considered in these reports is aesthetic impacts. With Ivanpah and numerous other projects, the images presented in the visual impact assessment corroborate the view that deserts are wastelands. Images of shotgun shells or tires dumped on the side of dirt roads signal that the parcel may have been neglected. Rarely were photographs taken during the wildflower season or after a rainfall caused a burst of desert color; hot, dry, dusty depictions were the norm in EISs.

Public perceptions of deserts as places less deserving of the protections awarded public lands with more extraordinary nature is reinforced by views that desert lands are a homogeneous canvas. In the American West, these landscapes more closely resemble complex ecological matrices of different degrees of habitat quality and conservation value.[159] This region is composed of unique landowners—the federal government's BLM and Forest Service lands, but also military lands, Indian lands, and private lands and inholdings. Many of the latter lands have long been degraded by agriculture, grazing, mining, and vehicular recreation, and some of these parcels no longer harbor the rich species diversity found at sites like the Ivanpah Valley.[160]

Environmental conservation organizations questioned why wild public lands were the first up for sacrifice, given the availability of private land with few or no comparable ecological features. A frequent public meeting participant and scientist with the California Native Plant Society said, "The question that's not being addressed here is basically why are they [BLM and solar developers] going on wild public lands first? Our organization and many others understand why we need renewable energy, and why large-scale utility projects will need to be part of the initial equation. But why put these big-scale projects in the intact wildlands first?"[161] Public officials framed the use of public lands as solutions to the challenges of economic recovery, climate change, and energy security. Other imaginaries envision solar energy deployment much differently, distributed in and around urban areas instead of remote ones.[162] Hence, it is critical to deconstruct the arguments for sacrificing public lands and whether the options presented are the full complement of opportunities and possibilities for solar energy and biodiversity conservation.

Describing the sacrifice zones of the west, Mike Davis and Rebecca Solnit see hope in a global social movement to counter the context of militarization and war.[163] The controversy in the Ivanpah Valley reveals the increasingly difficult green politics of climate change trade-offs. Given the specter of climate change, there may be more green forces calling for increased sacrifice of these areas than defending them. The task of environmental movements might instead be to develop and advocate alternatives to such sacrifice. Some environmental organizations are already trying to decentralize solar energy deployment. The NGO Solar Done Right has campaigned for distributed power generation, producing a series of reports documenting opportunities.[164] The Wilderness Society publicly recognized one solar energy project sited on an abandoned mine.[165] After these early projects, many major NGOs, such as the Sierra Club, Center for Biological Diversity, and NRDC, began to more strongly advocate for distributed solar, particularly as state net metering policies came under attack.[166]

The fate of the desert tortoise and avian biodiversity in the Ivanpah Valley now depends on how well species adapt to the presence of its new industrial solar farms. Wilderness advocates describe the valley as a sacrifice zone for solar energy by an industrial landscape to power the conspicuous consumption of high-energy society. A biologist with Basin and Range Watch asked, "Should we sacrifice public lands to power air conditioners running in empty homes in Los Angeles?"[167] He further asked why more attention and finance was not being directed at energy efficiency and conservation. In 2014, the BLM authorized the construction of two more utility-scale photovoltaic farms in the Ivanpah Valley—First Solar's Stateline Solar and Silver State South projects, each rivaling the scale of BrightSource's Ivanpah project—despite a failed lawsuit from the Defenders of Wildlife that claimed the project jeopardized the fate of a subpopulation of desert tortoises.[168] This brought the total number in the valley to four, with First Solar's Silver State North project. The Stateline project is on the mitigation lands for the Ivanpah project, so many of the same desert tortoises were moved again. The Silver State South project required relocation of 161 desert tortoises from 2014 through 2015, and 21 died.[169]

In a *New York Times Magazine* feature, Rebecca Solnit wrote about the Ivanpah controversy, clearly marking the sacrifice as one that is worthwhile and recapitulating the false dichotomy of biodiversity versus climate.[170] Solnit has written much on the lack of humility and the need for empathy toward the natural world. She captured so eloquently the militarization and industrialization of the Mojave. Solnit observed that "supporters of fossil fuel and deniers of climate change love to trade in stories like the one about Ivanpah, individual tales that make renewable energy seem counterproductive, perverse. Stories cannot so readily capture the far larger avian death toll from coal, gas and nuclear power generation."[171] The clouds of fear that accompany the apocalyptic narratives about climate change perhaps provoke this kind of reasoning, and perhaps it is warranted.

The studies that compare energy sources and avian impacts do reach these conclusions: coal, gas, and nuclear generation all have higher mortality numbers, mainly due to cooling towers. But impacts vary in the types of birds and extent of impacts, and tolls are geographically specific—with birds of concern in some places, and nuisance or invasive birds in other cases, making up large portions of the mortality figures from conventional thermoelectric power plants. While the magnitudes of relative impacts may be generally correct, the comparison relies on the false premise that all avian impacts affect the same species, and that these are real-world trade-offs in a zero-sum game. It also relies on the false narrative that these lands must be developed because there are no other options. And it ignores the fact that this specific project produces 10–15% of its electricity from natural gas. So there is a great deal of irony in a wildlands writer dismissing wilderness organizations' concerns about a proposed area of critical ecological concern becoming an industrial solar zone, while the warehouses of the Inland Empire, much of urban Los Angeles, the Central Valley, and other disturbed or less ecologically valuable landscapes could be used for solar and sited with community support and input.

Moreover, the technological pathway of siting extensive solar farms in and around wildlands is more likely to produce these avian conflicts than projects in urban or agricultural areas. The fact that wilderness organizations like Basin and Range Watch advocate protecting the Ivanpah Valley has far less to do with their inability to process the implications of climate change—they study and advocate conservation on the frontline of climate change, the Mojave Desert. It has more to do with lack of imagination about the different possible locations for solar farms.

Solnit writes:

That one death is a tragedy, a million deaths a statistic, is as true of animals as it is of human beings. It's a lot harder to mourn a potential loss of an entire habitat—as is threatened now for birds like the chestnut-collared longspur—than it is to mourn a golden eagle struck down by a turbine blade, or a warbler scorched in a solar farm. . . . And so we should seek out new kinds of stories—stories that make us more alarmed about our conventional energy sources than the alternatives, that provide context, that show us the future as well as the past, that make us see past the death of a sparrow or a swallow to the systems of survival for whole species and the nature of the planet we leave to the future.[172]

As environmentalism increasingly puts climate at the center of environmental politics, local ecologies and cultures can be erased or subsumed to address this effort. Sacrificing public lands for renewables only seems acceptable because of the potential contribution to holding off the worst consequences of climate change.

As electric utilities are required by law to buy renewable electricity, it is not a matter of solar versus coal or natural gas, but about different configurations of solar generation and electricity demand. A reductionist epistemology of carbon causes Solnit to miss the tremendous violence done to the ecosystems by industrial solar facilities, and to recapitulate the false choices at the surface of the Ivanpah Valley debate, failing to dig deeper, and ignoring the progress articulated in the Western Solar Plan. The question is not whether energy pathways can respond to climate action, but how.

EXPLAINING THE SOCIAL GAP IN RELATION TO SOLAR POWER PLANTS ON PUBLIC LANDS IN THE AMERICAN SOUTHWEST

Conflicts over land resources for solar energy may become increasingly common as energy systems transition from fissile and fossil fuels to renewables. Renewable energy technologies have relatively low power density (power per area). These extensive spatial requirements may clash with efforts to preserve wilderness. This makes it imperative to plan for the impacts of solar energy development, as environmental conflicts are neither desirable nor inevitable.[173] This polarizing dichotomy of solar development versus biodiversity is only a partial truth, because better land selection could drive projects to previously disturbed land, which covers some of the California desert, including lands currently in or retired from agriculture. That there may be sacrifices based on *false* dilemmas means that it is crucial to explore how the justifications and terms of the sacrifice are constructed and negotiated. Who participates, shapes, and arbitrates matters of sacrifice, and under what kinds of power asymmetries?

Evidence from projects sited on public lands across the American Southwest supports many prior explanations for the social gap in renewable energy development. In many cases, it was not so much that individual USSE projects were industrializing desert landscapes but the threat of cumulative solar energy development across spatial scales. These proposals came into conflict with conservation priorities on public lands. The controversies around the early projects provided lessons learned to incorporate into more substantive planning processes that would be undertaken to avoid future conflicts.

Despite their problematic nature, there was very little formal opposition to the USSE projects fast-tracked during the ARRA period. Some researchers explain that this is because "when this sort of techno-optimism meets the desert, it cannot help to seek [*sic*] to transform it."[174] Arguments about how solar development can contribute to solving the climate crisis complicate conservationists' efforts to "protect nature for nature's sake."[175]

Social resistance to solar energy projects can be attributed to factors unique to this early moment in renewable energy deployment. Projects sited on more degraded or agricultural lands were far less controversial, as evident in the number of comments from environmental organizations on First Solar's Agua Caliente project, on farmland, compared to its Desert Sunlight project near Joshua Tree National Park. During an interview with a solar project opponent near Blythe, California, they pointed to an expanse of land across the highway, saying, "You see that one over there? We didn't fight that one. It was an old jojoba farm . . . retired some time ago."[176] Geographical and ecological context is central to understanding the degree of opposition, but also perhaps the legitimacy of the grievances.

Conservation groups viewed the BLM as an agency without a clear mission, highlighting a tension between patronage and administrative discretion and regulation.[177] The bureau is a legacy of the compression of two institutions—the General Land Office and the Grazing Service—into one land management organization in 1946. The natural resource agency seeks to pursue multiple missions simultaneously. Conservation of wildlife and ecosystems is one, but so is the development of energy and mineral resources, as well as recreation, hunting, and other multiple uses. Institutional drivers undermined land-use decision-making to promote stewardship of public lands.

Policies designed to foster investments in clean tech led to conflicts between USSE development and environmental organizations over natural resources in the American Southwest. There are significant challenges to simply "putting the market to work" for renewables because powerful industry actors can intervene in the process, steering the incentives away from more appropriate sites. Some projects were built with little opposition, while others were slowed by litigation or were literally reshaped by the threat by it. The approach to processing ROW applications lacked competition and was unable to guide developers toward more appropriate project sites. Early in the process it also clearly overwhelmed BLM staff, who were already facing cuts and other budgetary challenges and a high workload. The bureau was required to process and initiate environmental and cultural resource reviews for every ROW application, even when it was unclear whether a project was economically viable.

There were communities that invited projects, particularly those interested in local job creation and tax revenue. Local chambers of commerce enthusiastically endorsed projects in their public comments. Communities would reach out to developers as well. At a trade show during the 2011 Solar Power International conference in Los Angeles, California, among the many photovoltaic module manufacturers and input suppliers was a section of tables staffed by nonprofits and government agencies. The city of Needles, California, had a table, paid for by the Needles Public Utility Authority. The top of its display proclaimed to potential developers seeking land for solar projects, "We own the land! We own the water!" Some locals invited renewable energy development for job creation, tax revenues, and other activities spurred by construction. These rural regions also tend to have the highest unemployment rates in country.

Some view desert landscapes as wastelands and ideal sites for solar power. To others, renewable energy sprawl is needed to supply civilization's insatiable appetite. These controversies highlight the land-use and stewardship challenges associated with solar energy transitions. Why do locals oppose renewable energy developments? Policies to foster solar energy have largely favored investment banks and large energy firms, and, coupled with land managed by a federal agency with a history of serving industry, this ultimately shaped solar power plants' scale, environmental burdens, and impacts on cultural resources. However, in a world with numerous options for siting solar energy, the more pressing question is, why are projects heading to the most controversial places first? Institutional inertia helped widen the gap between social acceptance and institutional approval by making the siting process somewhat inflexible to various applicant-technology-project site combinations. With the right approach from the start, these trade-offs may not be necessary, allowing more harmonious integration of energy landscapes and the built environment.

The next chapter reviews an energy development framework that attempts to balance solar energy deployment with wilderness conservation. The early controversies of solar power projects point to the need for a comprehensive approach to land-use planning for solar power. Two land evaluation processes were initiated alongside the development of two major elements of public policy promoting solar energy development. The first was California's Desert Renewable Energy Conservation Plan, which was advocated by then Governor Schwarzenegger and signed into law alongside the state's RPS. A mostly parallel federal initiative called the Solar Energy Programmatic Environmental Impact Statement was also developed in response to the BLM renewable energy mandate and the opening of western lands to solar speculation. These planning processes aimed to minimize controversies like those surrounding the Imperial Valley, Sleeping Beauty, and Ivanpah projects, with greater expectations for consultations and collaboration, culminating in the Western Solar Plan.[178]

CHAPTER 6

The Western Solar Plan

Energy scholar Vaclav Smil anticipated the land-use challenges of solar energy deployment long before the controversies over utility-scale projects in the American Southwest. Impacts include direct wildlife mortality, disturbance of the soil surface and resulting dust emissions, road construction, and habitat fragmentation.[1] Yet, as finance was made available to build USSE projects on public lands, the policymaking community lacked sufficient information on the environmental and cultural impacts of these shovel-ready ARRA projects. Complicating these reviews was the fast-track status of over a dozen projects. The research literature was also not very useful. Instead of comparing less and more favorable sites or cataloguing species and forecasting impacts in specific regions, many published studies compared the impacts of solar energy to other technologies such as wind, coal, and natural gas.[2] These comparisons are perhaps useful for some other scale, or maybe some other time. But land-use planners rarely face decisions between coal, gas, and solar power plants. With the lack of information about where to site projects on public lands, it is no wonder that some USSE projects became so controversial. A science-based collaborative planning effort could minimize the ecological and community impacts of scaling up solar to terawatt levels.

The BLM approved sixty renewable energy projects, including thirty-six USSE projects, between 2008 and 2016; four were technically denied. Only nineteen of the USSE projects are operating or under construction. The others have been withdrawn, mostly due to financial considerations, but also due to public pressure. The staff resources needed by the BLM to review projects, and the conservation consequences of the decision-making, pointed to the necessity of a planning effort to minimize the impacts of USSE and reduce social resistance to deployment. A Solar Programmatic Environmental Impact Statement would be pursued to identify show-stopping issues, saving agency staff time and reducing the cost of mitigation and habitat-management plans for solar developers by finding the lowest-conflict sites. After a thorough review of the science of environmental impacts of USSE projects, this chapter reviews the governing framework—the Western Solar Plan—for solar energy development in the American West.

IMPACTS OF UTILITY-SCALE SOLAR POWER PLANTS

Land-Use Requirements

Numerous studies have examined USSE project land requirements and impacts on undisturbed arid lands. Smil estimates that the power density of USSE is 10 to 70 W/m^2, with an anticipated mean around 40 W/m^2.[3] Pasqualetti and Miller suggest that land-use requirements of solar energy facilities are in 5.79 acres/MW (42.7 W/m^2) to 12.36 acres/MW (20.0 W/m^2) depending on the technology and project-specific aspects (number of tracking axes, storage).[4] The land-use requirements—or "energy sprawl"—of USSE projects elsewhere are estimated at 15.3 km^2/TWh per year for CSP and 36 km^2/TWh per year with photovoltaics.[5] NREL finds that total land use averages 8.9 acres/MW.[6] These numbers translate into somewhere between 5 and 13 acres for every thousand homes powered.

Data were collected from development announcements and EISs to understand the land-use requirements for USSE plants operating, under construction, or under development. This research found an average power density of 35 W/m^2 across Southern California, remarkably close to what Smil anticipated in 1984.[7] Eighty percent of projects on private land have significantly greater land-use efficiency (35.8 W/m^2) than installations on public lands (25.4 W/m^2), both values well within the range predicted in the 1980s. The results for power density are similar for public (28.6 W/m^2) versus private lands (34.8 W/m^2). CSP facilities have the highest energy sprawl (7.6 acres/GWh per year), while USSE projects proposed on private lands (4.2 W/m^2) have the lowest. There were no differences by technology: CdTe thin-film photovoltaic farms were similar to crystalline silicon in energy sprawl values and power density (5.5 W/m^2 versus 5.9 W/m^2).

Land-Use Change and Ecosystem Impacts

The choice of where to site USSE projects can determine the severity of ecosystem impacts. Early research reported only minimal environmental impacts of installing and operating USSE, but only generically described ecosystem habitat as grassland, desert scrubland, or "true desert," limiting the relevance of these findings to planning efforts.[8] It found that the biodiversity of desert scrubland rivals forest ecosystems and argued that "true desert"—sand dunes lacking vegetation—lacks biodiversity.[9] However, a review of USSE projects with playas or sand dunes revealed the presence of endangered species. For example, a project proposed near the Big Dune area in western Nevada threatened the Giuliani's big dune scarab beetle (*Pseudocotalpa giulianii*) and three other beetles of conservation concern, a point emphasized by the environmental group Basin and Range Watch in its public comments and on its website.

A systematic understanding of the land types where projects are proposed and sited can reveal socio-ecological impacts. A study of California's deserts found that USSE projects most commonly disturb shrublands and scrublands (26%), and that less than 15% of facilities are sited on "compatible" lands.[10] A simulation of the deployment of 8.7 GW in the California desert by 2040 using a spatial risk analysis model found the western Mojave and Salton Sea areas to be the most compatible with USSE.[11] While the existing literature generally documents the land-use-change risks associated with solar development across the American West, it lacks detail on specific species and particular harms. There are also very few studies that incorporate public comments and interviews with conservation experts, whose voices often provide some of the critical information needed to make meaningful land-use decisions.

Wildlife Mortality

Direct wildlife mortality includes the death of individual animals from the scraping of land or vehicle movement that can occur during site preparation, construction, operation, and decommissioning. Direct impacts during operation include wildlife being killed on the roads built on-site or in collisions between birds and modules, fences, or other equipment. The issue of tortoises at Calico and the Ivanpah Valley was discussed in the prior chapter, where that population experienced the most translocations from the BrightSource and First Solar projects. The Moapa Solar Energy Center is believed to have similar numbers, according to biologists. But the Bureau of Indian Affairs does not require the disclosure of these data. Much is being learned about the behavior of translocated and resident tortoises from the tens of millions of dollars spent on tortoise research and mitigations. One research team is studying the space-use patterns and habitat use of tortoises that have been translocated and the receiving population using radio-telemetry data.[12] Finally, there is a great deal of confusion about translocation for conservation versus translocation for mitigation, and some in the scientific community have concluded that "mitigation translocations often represent a misguided conservation strategy."[13]

Among the charismatic mammals threatened by solar power development in the American Southwest are the Mojave and San Joaquin kit fox (*Vulpes macrotis mutica*), two related species that live in burrows on or near several sites receiving or considering ARRA support, stretching from California's Central Coast to the eastern Mojave Desert. Some early flashpoints came over the construction of SunPower's California Valley Ranch and First Solar's Topaz Solar Farm, where the San Joaquin kit fox was already losing habitat to housing developments. Biologists studying the site found eighteen distinct family groups on the site. Numerous studies were conducted, and both companies eventually agreed to construct artificial kit fox dens to help transition the kit fox out of the project area.[14]

The Genesis project has also impacted several resident populations of Mojave kit fox, another threatened species and the smallest canine in the world. To comply with the rules established prior to construction, the developer was required to evict the kit foxes, which it did by spraying coyote urine into their dens. Shortly after construction commenced, however, eight dead foxes were found on the site, killed by distemper, a disease previously never detected in the kit fox population.[15] Presumably, the attempt to drive away the animals played a role in their contracting the disease, though California Department of Fish and Game officials noted that domestic dogs or any wild carnivore could have transferred the disease. Officials sought to identify the species of virus to better understand the host animal. Dead animals continued to turn up to even after an electric fence was installed. According to an investigation by the California Energy Commission (CEC) in February 2011, an on-site biologist-monitor traced a radio-collared female kit fox to the toolbox of an on-site construction truck. The construction crew claimed that it could not find the key to the box, and the next day the biologists traced the signal to a nearby water tower, where they found the dead kit fox. State wildlife officials did not pursue legal recourse because there was not conclusive evidence of malfeasence, according CEC spokesperson Sandy Louey.[16] A necropsy later found that this kit fox too died of distemper.

Endangered species with fully protected status, such as the blunt-nosed leopard lizard (*Gambelia sila*), can lead to significant delays and project redesign. Much of this species' habitat has been lost to real estate development. One ecologist familiar with the species noted, "The blunt-nosed leopard lizard has very few habitat options left. The Central Coast communities of California have turned most of its habitat into housing."[17]

Many early USSE proposals were for CSP, raising concerns about bird mortality. Solar power towers create a heat flux that damages the feathers and skin of birds. USFWS officials stated on the public docket that solar power towers could be an ecological "mega-trap," luring insects and birds to their demise.[18] Bird mortality from solar power towers has been known since one was built near Daggett, California, in the 1980s, when one study found 70 bird fatalities involving 26 species over a forty-week data collection period.[19] It concluded that this represented a minimal impact on avian species overall. During the CEC proceedings to approve several projects, expert witnesses from conservation organizations illustrated how the science underlying the research undercounted bird mortality. USFWS biologists coined the term "streamers" to describe birds singed by solar flux at the Ivanpah site, which is particularly problematic when it is above the power tower receiver while the plant is in standby mode. Later research from Ivanpah raised the bird death totals upwards, with just under half of the deaths due to the heat flux.[20] One public letter, submitted by a USFWS chief biologist, asked that the CEC not approve any more solar power towers until data could be collected on the impacts of power towers on avian ecology. Unlike the challenges with tortoises, which can be avoided by siting projects on non-habitat, the solar power towers' impacts on birds may be unavoidable.[21]

More recent data suggest that most bird mortality at solar power plants is from collisions with photovoltaic arrays, heliostats, or fences.[22] While the issue of bird mortality was discussed in public comments, public attention was heightened when a peregrine falcon was killed at the Ivanpah facility. Some avian biologists have suggested that there is a "lake effect," or polarized light cues, from solar power plants, which attracts birds, particularly at night.[23] But other avian biologists note that there is no anatomical proof that birds have the receptors to see polarized light, suggesting that birds may be confounded or confused by the facility in some other way.

A wide variety of bird types have died at USSE plants.[24] Two endangered Yuma clapper rails (*Rallus longirostris yumanensis*), a population with only a thousand living individuals, were killed at the Desert Sunlight facility in Desert Center, California. At two solar power plants in the California desert (one photovoltaic farm and one parabolic-trough CSP), over 20 birds associated with aquatic habitat—yellow-headed blackbirds (*Xanthocephalus xanthocephalus*), great blue herons (*Ardea herodias*), eared grebes (*Podiceps nigricollis*), western grebes (*Aechmophorus occidentalis*), pied-billed grebes (*Podilymbus podiceps*), surf scoters (*Melanitta perspicillata*), red-breasted mergansers (*Mergus serrator*), buffleheads (*Bucephala albeola*), black-crowned night herons (*Nycticorax nycticorax*), double-crested cormorants (*Phalacrocorax auritus*), American coots (*Fulica americana*), and brown pelicans (*Pelecanus occidentalis*)—were found dead, apparently due to colliding with panels and mirrors, far from any sources of water.[25] Other species known to have avian-solar mortality include migratory birds such as the yellow warbler (*Setophaga coronate*), Vaux's swift (*Chaetura vauxi*), and loggerhead shrike (*Lanius ludovicianus*), and raptors such as the American kestrel (*Falco sparverius*), red-tailed hawk (*Buteo jamicensis*), golden eagle (*Aquila chrysaetos*), northern harrier (*Circus cyaneus*), and peregrine falcon (*Falco peregrinus*).

Some USSE sites have on-site ponds that may attract such birds. Polarized-light cues cause aquatic insects to lay their eggs on photovoltaic modules rather than in water, prompting some to argue for more research into how polarized light might affect insect, bat, and bird behavior near USSE installations, since water bodies are the only sources of polarized light in nature.[26] Many renewable energy advocates minimize the consequences of USSE bird mortality by comparing it to other sources such as cats, buildings, and automobiles, but this comparison seems incommensurate given that the impacts are cumulative, not trade-offs, and it does not distinguish between mortality of different types of birds. Mitigating USSE impacts on avian species will require greater scientific understanding of birds' perceptions and use of these facilities. A 2016 study of Southern California estimated that annually, existing USSE facilities kill between 16,200 and 59,400 birds.[27]

Two important laws protect avian species from development and require compliance from solar energy developers. These are the Migratory Bird Treaty Act and the Bald and Golden Eagle Protection Act. The Migratory Bird Treaty Act dates back to 1918 and makes it illegal to "pursue, hunt, take, capture [or] kill; attempt to take, capture or kill; [or] possess . . . any migratory bird . . . or any part, nest, or egg of any such bird."[28] The scope of the law includes activities such as power plant operations, and it is often applied to oil and gas companies, resulting in fines for violations. Intent or prior knowledge of the injury is not required when the statute applies strict liability.

Recent efforts have attempted to restrict the application to solar power plant and wind farm operators of this landmark bird protection treaty. What makes the current law difficult for solar developers is that there is no option for an "incidental take permit" for migratory birds. Hence, power plant construction hinges on the fact that solar power plant developers might be fined for "taking" birds protected under this treaty. Whether or not reduced liability shapes mitigation efforts is important because it is possible that if take permits are allowed, solar and wind developers will put less effort into mitigating impacts on birds.

Habitat Fragmentation and Loss

Habitat fragmentation can occur with USSE projects because the sites are usually devoid of vegetation and surrounded by fencing that restricts some wildlife movement. In the Panoche Valley, a solar power plant is under construction on private lands covering about half of the valley floor in habitat that is one of three in the state deemed critical for the recovery of the San Joaquin kit fox by the BLM, which manages tens of thousands of acres of surrounding public lands. On the Carrizo Plain, developers worked with an organization named Dogs for Conservation to identify where kit foxes moved through the proposed solar farm site, and built artificial dens and passes through the fencing as a mitigation.

Bighorn sheep and pronghorn antelope are two other large mammals that migrate across landscapes that could soon be occupied by solar power plants. Migration corridors are important in the climate change context, as it is expected that species ranges will shift in the foreseeable future. Putting large obstacles such as USSE sites in the path of migrating species could imperil them further by fragmenting their habitat. This could lead to genetic isolation of species and severe inbreeding depression. On the Calico Solar One project in the center of the Mojave desert, one ecologist noted, "Bighorn sheep use these ranges between the dry lakes to find water and shade, particularly during the summer months. With a typical project three or four miles on each side, we're going to see more decide to turn back, and possibly even more automobile encounters."[29] A species with protected status in several California coast range utility-scale solar projects is the giant kangaroo rat (*Dipodomys ingens*), which also has lost much of its habitat to real estate development and agriculture. The International Union for Conservation of Nature and Natural Resources lists the giant kangaroo rat as endangered because its range is small and fragmented, being restricted to small valleys and hills west of California's San Joaquin Valley.

A number of USSE projects and the notion of Solar Energy Zones raised concerns about bighorn sheep (*Ovis Canadensis*) movement and gene flow. Biologists Edward O. Wilson and Thomas Lovejoy and several bighorn sheep ecologists stood against a proposed USSE at a site called Soda Mountain, on a pass near Zzyzx, California. "We're all for solar projects," they wrote. "We need more of them. But not in this place."[30] Their op-ed followed a similar call earlier by two bighorn sheep ecologists, John D. Wehausen and Clinton W. Epps. Large developments can impede the movement of bighorn sheep across otherwise open basins to neighboring ranges. While the data collected for the EIS suggested infrequent bighorn sheep visitation and identified the interstate as the major barrier, the scientists argued that the site was an important restorable corridor for bighorns moving between the Mojave National Preserve and Death Valley National Park.[31]

Land-use change from USSE can damage cryptobiotic soil crusts, in which cyanobacteria, lichens, algae, mosses, and fungi help arid ecosystems fix carbon and nutrients and retain water. According to ecologist Eugene Odum, cryptobiotic crusts are the invisible component of biodiversity.[32] These collectives of organisms exist on the surface of arid soils. They act to stabilize the soil, fertilize it, and retain moisture. The development of solar power plants in undisturbed areas can lead to significant deterioration and destruction of cryptobiotic crusts when roads are built and land is heavily disturbed. Near Pisgah Crater in the central Mojave Desert, where the Calico project was proposed, there are extensive cryptobiotic crusts that are remarkably deep, possibly thousands of years old, and so delicate that even walking on them can damage them. Where they appear near hiking trails or other public use areas they are commonly roped off to prevent trampling. The time required to return arid systems to pre-disturbance conditions is estimated at two or three years for grasslands and decades in desert environments. Some research on land degradation in the Mojave suggests that the land can take between 50 and 300 years to return to pre-disturbance levels of biomass and plant cover.[33] Other studies find a longer period for full recovery of arid lands, though some ecosystem services, such as nutrient delivery and land cover, can begin to recover in on the order of 50 years.

USSE development can also impact the aeolian sand transport systems critical to the geomorphology of sand dune ecosystems. Ironically, aquatic ecosystems are put at risk, where land-use change causes prolonged drying of ephemeral water bodies, and disturbance of desert washes within the construction footprint of the facility can affect drainage and groundwater storage. Although deserts are dry by definition, they are also subject to flash flooding. In July 2012, a flash flood did several million dollars' worth of damage to the Genesis project and construction site. Water poured through the wash into the channel, damaging construction vehicles and many four-wheelers, and flooding some of the inverter rooms. Activists had warned that siting the project in a desert alluvial fan would make it vulnerable to flooding.

One ubiquitous but keystone plant species in the Mojave region is *Yucca schidigera*. Known by the common names Mojave yucca and Spanish dagger, the plant leaves behind hollows used by kit foxes and burrowing owls after the roots decay. Threats to plant habitat and risks to plant species from USSE are not well known, because very little of this region has been thoroughly documented. Pointing to the absence of Consortium of California Herbaria records for the desert regions in Southern California, one distinguished University of California research botanist pointed out that "roughly five to ten percent of the plant species in the Eastern Mojave have not yet been described. . . . How can we document the impacts when we don't know what's there?"[34] One project, later cancelled and now part of the Mojave Trails National Monument, threatened to disturb one of the few populations of threatened white-margined penstemon (*Penstemon albomarginatus*, sometimes called white-margined beardtongue), imperiling its presence in California. Other species, such as bighorn sheep and Joshua Trees, became focal points for other projects. The Mohave ground squirrel (*Spermophilus mohavensis*), another species that depends on habitat connectivity to prevent inbreeding depression, was one of the key species of concern for Ridgecrest and several other projects in the western Mojave Desert. Along California's Central Coast, species such as the blunt-nosed leopard lizard and San Joaquin kit fox dominated local conservationists' concerns. Along the U.S.–Mexico border the flat-tail horned lizard (*Phrynosoma mcallii*) triggered endangered-species concerns.

Road construction has key impacts, because where sites require new roads, there will be asphalt or gravel, which can be barriers to movement or sources of mortality for some wildlife and can lead to invasion of weed species such as Sahara mustard (*Brassica tournefortii*), cheat grass (*Bromus tectorum*), red brome, and buffelgrass, which can shift fire regimes. Permitting agencies may require weed management plans from solar developers. Critically, there are very few studies of the cumulative impacts of the multiple projects and extensive road infrastructure that serve USSE facilities. Design considerations should include how to accommodate wildlife movement or habitat at USSE sites.

Many projects that disturb wildlands seek mitigations by purchasing land or buying into a conservation easement bank. In California, a bill in the state legislature, authored by Alex Padilla, allowed solar developers seeking federal financing to mitigate by contributing to a fund used to purchase land.[35] One of the challenges with accepting mitigations as an added benefit of USSE development is that some areas purchased as mitigations are not under development pressure, so they are unlikely to be developed anyway. A mitigation used to offset the habitat loss from the Panoche Solar Power Plant was the purchase of 10,000 acres of the nearby Silver Creek Ranch. But it has very different habitat qualities (steep mountainous slopes versus flat, open fields), and the BLM reportedly was raising funds to purchase that land in the future anyway. This seems like a false mitigation. In other words, the lands purchased to mitigate impacts would have remained undeveloped and would have been in conservation anyway. But some mitigation purchases do stave off development, so this issue needs case-by-case treatment.

Freshwater Consumption

Significantly less water is used to produce, install, and operate solar devices and power plants than that needed to cool thermoelectric fossil and nuclear power plants. But since solar power plants are often sited in arid areas with high insolation and little rainfall, on-site water use is another important impact, especially where groundwater is used. Water use in USSE projects depends on the technology, with CSP with wet cooling using the most, followed by dry-cooled solar thermal, and then photovoltaics. Wet cooling for CSP is becoming less popular as dry cooling technologies are economically competitive. Lesser water needs for CSP include boiler blowdown (clearing the system of old water) and hydrostatic testing of the piping during the installation phase or during repairs.

Water use is also determined by the extent of land-use change that increases fugitive dust and particulate matter. The largest volume of water used across all types of USSE is for fugitive dust control, which is required when soil surfaces are intensively plowed or scraped. Making solar equipment uses water, from mining to chemical processing, but the biggest water use by volume at some photovoltaic farms is for dust control and cleaning during installation and operation. Utility-scale projects in the 230-to-550-megawatt range can require up to 1.5 billion liters (390 million gallons) of water for dust control during construction and another 26 million liters (6.9 million gallons) annually for panel washing during operation.[36] Water trucks are used for dust control at the Desert Sunlight Solar Farm project, where the annual amount of water used for dust control during the construction phase is permitted to use 1,557 acre-feet.[37]

Failing to control dust can lead to violations of the Clean Air Act and construction stoppage. Not all USSE projects require water for fugitive dust control, but it is common in the American Southwest. Some projects require water for fugitive dust control on roads for operations, not just during construction, though at much lower volumes. Cleaning heliostats or photovoltaic modules during operation uses the second-largest water volume for operations. Lesser quantities of water are needed for a workforce's potable drinking water; the number of construction workers is usually two orders of magnitude higher during construction. Several USSE projects also proposed demineralizing systems to remove impurities such as boron from the groundwater. Despite high variability in the amount of water it takes to produce a unit of electricity from solar power, the overall impacts of USSE projects on freshwater supplies are minimal compared to the region's availability.[38]

Falling groundwater levels could impact vegetation, as well as endangered species such as the desert pupfish (*Cyprinodon macularius*) and Moapa dace (*Moapa coriacea*), which depend on water bodies sustained by historical water table levels. Water overdrafts lower water tables and deplete fossil groundwater. Finding means to reduce on-site water use at USSE facilities will continue to be an issue across the American West, particularly in states with changing water governance regimes. California's Sustainable Groundwater Management Act could alter the availability and cost of water for USSE in that state.[39] Innovations in automated, waterless module and mirror washing are being explored which could reduce the second-largest source of water use at USSE sites.[40]

Fugitive Dust Emissions

Where projects cause heavy disturbance of underlying vegetation, they can become sources of fugitive dust emissions. The case study in fugitive emissions from a USSE project is the 280 MW Antelope Valley Solar Ranch. A series of windstorms in 2012 put the project's ARRA financing in jeopardy. Massive plumes of dust triggered a regulatory response from the U.S. EPA and regional air quality regulators, who together issued a ruling that the company should immediately stop construction and resolve the problem of airborne dust. Incidents on April 5 and 8 led to notices from the Antelope Valley Air Quality Management District (AVAQMD) of violations of federal ambient air quality standards for particulate matter.[41]

The series of windstorms also caused severe white-out conditions, which contributed to automobile pileups on roads near Lancaster, California, in the western Mojave Desert.[42] Dust emissions exceeded those permitted under the Clean Air Act. Federal projects are required to comply with California Environmental Quality Act (CEQA) and National Environmental Policy Act (NEPA) rules for regional air pollution control. Local resident Robert Kerekes, president of the Original Antelope Acres Town Council, described the cloud in the sky and its origins to the *Antelope Valley Press*: "Solar companies pulled out all the vegetation. It looks like a river, like a thick river. It's blowing like crazy. I never saw it that bad before."[43] The construction delays created a monetary concern about whether the power plant would be able to meet its ARRA construction deadlines.

AVAQMD operations manager Bret Banks told *Greentech Media* reporter Herman Trabish that there were "a myriad of things [First Solar] could have done that we didn't think they were doing to prevent the violations." Residents and policymakers remained upset about the situation. "We told them it is time to stop experimenting," added Norm Hickling, aide to L.A. County supervisor Michael Antonovich. "We need best practice standards."[44] First Solar describes experimenting with different materials to use on topsoil to prevent dust emissions, and is collaborating with AVAQMD and the Antelope Valley Resource Conservation District on identifying best practices for dust suppression.[45] At least one USSE developer is experimenting with polymer-based binding agents (polyvinyl acetate, sodium acrylate and acrylamide, and acrylic copolymers) dissolved in water and spread across the soil during construction and on roads to suppress dust.

Sources of fugitive dust emissions in the region are mainly agricultural fields. Off-road vehicle use also causes dust where riding can lead to erosion. Incomplete and unfinished housing developments are another important source of fugitive dust.

"We warned them not to grade and told them it was going to happen, and they ignored us," Humphreys told the Antelope Valley Town Council. "Now we live in Hell Valley."[46] Particulate matter and dust also increase exposures to soil pathogens. Valley fever is linked to dust containing spores of *Coccidioides immitis* in arid regions across the American West. But there are many sources of dust that contribute to valley fever, and the recent spate of cases in California has been attributed to agriculture.

CARBON DEBTS AND GREENHOUSE GAS RETURN ON INVESTMENT

Ecosystems fix carbon from the atmosphere into biomass and soils via photosynthesis. Disturbing soils and vegetation releases carbon back into the atmosphere. Topsoil loss and land-use change are already known as lost carbon sequestration and storage. Globally, large amounts of carbon are stored in arid regions like the Mojave Desert in caliche ($CaCO_3$), desert soils, and vegetation.[47] Desert ecosystems can take up to 100 grams of carbon per square meter per year from the atmosphere.[48] Higher rates have been found in the Mojave Desert, perhaps as high as a deciduous forest, suggesting that these landscapes may be carbon sinks. Land-use changes in deserts cause disturbances that release greenhouse gasses (GHGs). This can be as much as 50 grams of carbon per square meter per year from clearing desert vegetation and 150 grams of carbon per square meter per year from damaging caliche.[49] Land-use changes can result in lost or delayed sequestration potential, and research shows that the Mojave Desert increases rates of sequestration under increased CO_2 levels, so "old growth" desert ecosystems are important buffers against climate change.

The concept of carbon debt represents the GHG emissions associated with clearing land of vegetation for renewable energy production, as is applied mainly in studies of the impacts of biofuels.[50] Carbon debt must be paid down before renewable energy sources are considered to be reducing emissions. A review of the EISs for USSE projects fast-tracked or applying for federal grants revealed that only Chevron's Lucerne Valley quantified the GHGs from land-use change, though all the projects required grading of topsoil and vegetation removal on all or much of the project site for mounting equipment and roads. The relationship between carbon debts and GHG savings is important because simply adding two years of GHG debt to a USSE project due to clearing biomass can influence the GHG return on investment. Consider a USSE project with an expected 20-year lifetime that has 2 years of embodied GHGs to pay off from the steel, photovoltaic modules or heliostats, construction, copper wires, and so on, and another 2 years of land-use-change GHGs to pay off. The added debt could reduce the GHG return on investment by 50%, from ten ($20/2 = 10$) to five ($20/4 = 5$).

Albedo

Land-use changes for USSE projects can change the surface albedo (reflectivity), making a solar power plant act like a heat island, increasing local temperatures.[51] Several public comments from biologists noted that solar power plants could affect the thermoregulation of reptiles such as the desert tortoise.[52] In urban areas, because of changes in the thermal mass that absorbs solar radiation, photovoltaic and solar hot water panels can reduce the heat island effect.[53] Changes to albedo from USSE projects are site-dependent. Higher surface reflectivity can cause glare that impairs the vision of drivers and pilots. There is also evidence that the reflective glare from heliostats can have non-permanent ocular impacts on human vision.[54]

Microclimates are shaped by the temperatures, sunlight, and humidity in the first meter of air above the ground. These microclimates affect what kinds of organisms can live there. Solar power deployment will change the albedo and hence surface warming. Changes to surface roughness can affect windspeed, which can influence evaporation, temperature, dew point, and other meteorological variables. Where photovoltaics and heliostats cover significant areas, it will affect microclimates.

Occupational and Community Health and Safety

The occupational health and safety impacts of the solar industry have been repeatedly shown to be far smaller than other energy source when normalized on a per energy unit basis.[55] Worker exposure to spores that cause valley fever is one unique concern posed in the most arid regions of the West, including deserts and desert grasslands. The 550 MW Topaz Solar Farm and the 250 MW California Valley Solar Ranch, both just a mile north of Carrizo Plain National Monument, experienced a spate of occupational illnesses, as 28 construction workers were diagnosed with the often-debilitating illness. Both the developer and the construction company were cited multiple times by the California Occupational Safety and Health Administration for failing to reduce exposure to airborne dust.[56] One regulatory compliance worker notified the author of the use of hundreds of thousands of spray canisters of a zinc-galvanizing material (trade name High Performance Zinc Spray) that contains the neurotoxins ethyl benzene and xylene. A second USSE developer confirmed the use of a different galvanizing spray without those solvents.

Increase of particulate matter loads in some airsheds from dust emissions could pose environmental justice considerations where low-income communities reside nearby, particularly downwind of USSE projects. Careful planning and effective land stewardship are needed to ensure that USSE projects located near communities do not have environmental justice consequences. In general, proper screening for occupational hazard hotspots and cooperation with expert institutions such as the Centers for Disease Control and the Occupational Safety and Health Administration can help implement best practices to prevent occupational illness. Good-neighbor policies, too, can ensure that community health concerns are addressed.

One community safety and land-use-change risk is fire. USSE projects may present a risk of fire, but those risks still may be similar to those that would exist if the land remained vacant. A 71-acre fire at the ARRA-funded Antelope Valley Solar Ranch One was due to negligent driving of a truck and dry grass; fires could have similarly started from various other land uses. However, fires have occurred at early CSP plants, and the Ivanpah project caught fire in the summer of 2016 when the heliostats misdirected the solar flux at steam ducts and water pipes.[57] Solar farms and the associated transmission infrastructure could spark fires, but fire safety is reviewed in EISs and appears as a key design consideration for USSE projects in fire-prone regions. In numerous proposed USSE projects, questions were raised about the impact on community fire-fighting capacity and whether its presence demanded new equipment. Some USSE facilities have on-site fire-fighting equipment.

SOLAR FARMS VERSUS FAMILY FARMS IN PANOCHE VALLEY AND THE CARRIZO PLAIN

While much emphasis is put on alternatives to intact ecosystems, there are also groups that raise concerns about the loss of agricultural lands, particularly counties and state farm bureaus that regularly contend with a broader trend of land conversion to suburban sprawl. California continues to develop what was once prime farmland, and there are numerous pressures on the farmland that remains. Silicon Valley still bears many of the street names that once described the extensive agriculture and orchards in the Santa Clara Valley. These pressures have moved inland to the Central Valley, the most lucrative agricultural lands in the world. Groups such as the farm bureau in particular have voiced concerns about the loss of especially prime agricultural lands. In the Panoche Valley, where a utility-scale solar project was proposed, concern was even raised about the loss of 4,000 acres of class B farmland, which is probably best suited for ranching. One rancher interviewed in the local weekly newspaper said, "The land that they're proposing to build on would be the end of agriculture on those acres. . . . Done. Forever."[58] The following pages describe a few specific solar power plants where many of the above issues were raised. Figure 14 shows a sign voicing opposition to the use of prime farm land for solar development.

The Panoche Valley is a bucolic landscape deep in the Diablo Coastal Range of California, east of the rural community of Hollister, the county seat of San Benito County. The expanse is just out of reach for commuters to Silicon Valley, so it has been saved from the suburban sprawl that rolled over much of the areas closer to the Bay Area. To reach the area from the Bay Area requires a long traverse along narrow winding roads. The surrounding terrain of the Diablo Range gives the valley its distinct attribute, as it is the only flat area in an otherwise rugged region for fifty miles in any direction. The CEO of a USSE development company named Solargen, proposing a project in the Panoche Valley in San Luis Obispo County, called it "the valley God made to be a solar farm."[59] The Audubon Society list the same valley on its list of birding areas of global significance and filed numerous lawsuits aiming to stop the project. Solargen planned to build and install 1.8 million photovoltaic modules across 4,700 acres of what the California Department of Conservation rated as "prime farmland." The project received scrutiny first for numerous ecological and agricultural conflicts. But many also questioned the choice of technology: relatively inefficient amor-

FIGURE 14. "Solar farms versus family farms" is a theme where solar projects displace agriculture and ranching, such as in the Panoche Valley, California.

phous silicon thin-films manufactured by a company that had yet to ever sell a module.

The Santa Clara Valley and Fresno Audubon Societies organized a bird-sighting trip, followed by a meeting of other concerned local farmers, teachers, and business people who were alarmed at the prospect of industrial facilities transforming the rural character of the landscape. Unlike the western reaches of the Diablo Range, the Panoche Valley receives very little rain except between November and February. At less than ten inches per year, the region is considered desert grassland. Compare this with the Los Padres National Forest's Ventana Wilderness, east of Big Sur, which is only 70 miles to the west, but receives over 80 inches of rain annually. These conditions help make the area ecologically unique; it shares some qualities of the lush hills of coastal California, but also hosts many species also found in California's Mojave Desert.

While there are extensive protected areas in California, flat open space is disappearing. There is intense competition between agriculture and suburban housing, and the resulting high land prices give landowners strong motivation to sell their land to developers. For these reasons the BLM, which manages much of the surrounding Panoche Hills, has designated the valley floor as a core recovery area for the San Joaquin kit fox. California has three such areas, all of which are in the Coast Ranges. The San Joaquin kit fox population in its historic range in California's Central Valley was extirpated. A scientist at the meeting presented research on the genetic diversity of the three populations of kit fox that lived in the three core recovery areas, suggesting that the kit fox of the Panoche Valley is a genetically distinct subpopulation.

The land proposed for this solar farm had been in a Williamson Act grazing easement for thirty years. The Williamson Act is an important land conservation policy in California that offers landowners lower property tax rates in exchange for protecting their land from development. Williamson Act lands amounts to 15% of California's undeveloped lands. Several projects were proposed on Williamson Act lands. One memo about the Panoche Valley Solar Farm to a solar developer from a San Francisco legal office argued that solar farms are a "compatible use" with grazing under the Williamson Act. It argued that utility-scale projects could also support grazing, and the developers should not have to pay back taxes owed while under Williamson Act easement. In a 1985 front-page image in the *San Luis Obispo Tribune* of a solar power plant (built with ARCO solar modules in the 1980s), sheep grazed in the solar fields. However, with land at such a premium, USSE facilities seek to maximize the area for modules. Compatible use with grazing sheep or other smaller mammals will be geographically specific, but seems unlikely for utility-scale projects (except for occasional grazing for weed suppression). California developers were unable to convince Williamson Act officials that solar power is a compatible use, except where electricity is directly delivered to agricultural production, processing, or shipping.[60]

The project site contains class B soils, which are not the highest quality but still a productive medium for agriculture and very good for ranching. The reason the land was dedicated to ranching rather than farming had less to do with soil quality and more to do with the slightly higher elevation of the valley floor compared to the nearby San Joaquin Valley, arguably the world's most productive agriculture landscape. The cooler nights meant that crops ripened later, so farmers who tried their luck there found themselves selling into food markets when prices were lower due to excess supply; they missed out on the premiums received by growers whose crops could be harvested earlier. Hence, agriculture here had been restricted to animals.

The Panoche Valley case also illustrates concern about mitigations. While a mitigation bank would be later set up from which solar energy developers across the state could draw, Solargen proposed to buy an adjacent property. Silver Creek Ranch is a 10,000-acre parcel to the southeast of the Panoche Valley, with steep and rugged terrain and also legacy mercury contamination from the nearby New Idria Mercury Mine. This is a false mitigation because there is no real or immediate development pressure. More importantly, the BLM had long planned to purchase and preserve the parcel and add it to the regional public lands, including the Panoche Hills and the New Idria Mine. The project has been proposed numerous times, each time failing due to a lack of power purchase agreements and investors, and fear of lawsuits under the CEQA. In the latest case, the Audubon Society lost its CEQA suit and the project was approved.

The Carrizo Plain (also locally known as the Carissa Plains) is sometimes referred to as California's Serengeti. About 100 miles south of the Panoche Valley in the same coast range was another controversial solar project on agricultural land, pitting solar developers against local residents and conservationists. While the farms and vineyards of San Luis Obispo County were under pressure from real estate development, and the nearby San Joaquin Valley was transformed by industrial agriculture and oil development, the Carrizo Plain largely retained their rural character. It is an expansive landscape with a long wide valley of wildflowers, native grasslands, and vernal pools containing endangered longhorn fairy shrimp (*Branchinecta longiantenna*). The valley is home to the San Joaquin kit fox, American badger (*Taxidea taxus*), tule elk (*Cervus canadensis*), and pronghorn antelope (*Antilocapra americana*). Overhead fly birds such as the golden eagle (*Aquila chrysaetos*), mountain plover (*Charadrius montanus*), long-billed curlew (*Numenius americanus*), loggerhead shrike (*Lanius ludovicianus*), and California condor (*Gymnogyps californianus*). The iconic feature of this landscape is the long, exposed segment of the San Andreas Fault, the infamous strike-slip fault (clearly visible from the air) that is conveying the Pacific Plate northward toward the Aleutian Sea.

The Carrizo Plain and the largest town there, California Valley, in eastern San Luis Obispo County, are no strangers to solar technology. ARCO built the world's first photovoltaic farm there in the 1980s. The 5 MW power plant lasted less than ten years before being decommissioned in 1995 due to problems with the laminate on the modules, which caused dramatic declines in power output. The ARCO project also came under fire from conservation groups. But the project foot-print was far smaller, and the multi-use effort to graze sheep went over well with the local community. The new utility-scale solar projects would occupy 200 times as much land.

By 2006, there were three USSE proposals on the Carrizo Plain. Two of the three proposed projects were photovoltaic power plants. The first was SunPower's California Valley Solar Ranch, which would use photo-voltaic modules built by contract manufacturer Flextronics in Milpitas, California. SunPower is a long-standing resident of Silicon Valley and maintains its headquarters in San Jose, though much of its manufactur-ing capacity is in the Philippines. The second project was a photovoltaic farm proposed by OptiSolar with its amorphous silicon. The land would later be a part of the lands and strategic land rights acquired by First Solar in 2007. They allowed it to build several early large projects, like Sarnia (Ontario, Canada) and Blythe (Blythe, California). They also put First Solar in play for what would become its largest projects, in the Carrizo Plain (Topaz) and near Desert Center (Desert Sunlight). I had the opportunity to tour The Topaz site in 2009. Two of the First Solar employees who drove me around the site wore vests with OptiSolar's name and logo, an indication that they were part of a team that came with the purchase, which they confirmed. The other project proposed was a 177 MW solar thermal project using linear Fresnel reflectors, proposed by a startup named Ausra Energy. In 2009, soon after starting the permitting process, Ausra sold the project site to First Solar. These two parcels would become a single project known as the Topaz Solar Farm, part of which is shown in Figure 15.

The project drew criticism, including a scathing letter from the exec-utive director of the California Coastal Commission: "The argument that we must sacrifice fragile ecosystems for the common good (i.e., major impingement on the Carrizo to save the planet from climate change) is specious, relies on a false choice, and reflects a myopic view of the common good. Do we seriously believe a single coal-fired or nuclear plant will not be built or shut down if the Carrizo solar projects are constructed? Of course we must do our part to address climate change but not at the expense of an irreplaceable community jewel."

The development of the California Valley Solar Ranch and the Topaz Solar Farm revealed a new hazard to workers along the solar commod-ity chain: coccidioidomycosis. California's Department of Public Health, and the California Division of Occupational Safety and Health con-firmed the outbreaks of valley fever at both sites.[61] The spores that

FIGURE 15. The Topaz Solar Farm in San Luis Obispo County, California, uses CdTe thin-film photovoltaic modules.

grow on the tissues of the fungi Coccidioides cause valley fever. The fungi grow after rains, when the soil is wet. After the ground dries and hardens, the spores are released when the ground is disturbed by activities such as plowing. The disease is prevalent in California's Central Valley, where agricultural laborers are susceptible, and will become increasingly common in areas where solar power plants are located if similarly intensive construction practices such as scraping and plowing continue.

Somewhere in the public comments of almost every USSE project proposed in California during this research is a concern about valley fever. The claims often sounded hyperbolic, and even opponents of the projects dismissed the issue as a minor one. But soon after construction started, dozens of workers began to contract the illness. For communities, it is unclear whether those living downwind of solar farms are more likely to contract valley fever. People without access to healthcare and those with depressed immune systems are the most susceptible. This means that serious environmental justice issues might be present, because the populations that are most vulnerable to the illness are African Americans, Mexican-Americans, and Asian-Pacific Islanders. Fortunately, there were no fatal cases at the solar power plants on the

Carrizo Plain. The disease kills one out of every one hundred that contract it, but it often goes misdiagnosed as the common flu, so it also may be underreported. The Centers for Disease Control studied the 44 workers exposed to spores that caused valley fever.[62] They found that most workers did not wear simple protective equipment, like masks, that could have minimized exposure, suggesting that these occupational hazards can be addressed with proper training and best practices.

The California Valley Solar Ranch and Topaz Solar Farm projects were both considered for federal loan guarantees. Ultimately SunPower would receive a loan for the California Valley Solar Ranch, and First Solar sought other financing. As the Topaz project approached the deadline for the loan guarantee program, and on the eve of a visit by the DOE, the developers held an open house with the community. The DOE investment in the solar project was considered a federal action, and it was required to conduct its own EIS. The open house during the comment period was held on the main thoroughfare in California Valley and attended by nearly fifty people, mostly opponents of the project. Surrounding the room were easels with project maps and information about the species in the area. In the far corner of the room was a model of the photovoltaic farm and one of First Solar's signature black frameless CdTe photovoltaic modules.

At the public meetings and open houses, community members raised questions about how the solar farms would change California Valley and the neighboring Carrizo Plain National Monument, home to the iconic species of the region: the pronghorn antelope, kit fox, and California condor. One hold-out landowner was already surrounded by a sea of photovoltaic modules, "nine million, five hundred fifty-two, to be precise."[63] He complained that his family would be exposed to valley fever and that environmental groups were not supporting the cause of ecologically sensitive solar farm development. This homeowner shared a memorandum of understanding between the solar developers and the Sierra Club, Center for Biodiversity, Defenders of Wildlife, and NRDC, who, for $4 million, agreed not sue the project developers on these respective projects. This homeowner argued that the big environmental organizations had traded in their emphasis on wildlife recovery and making wildlife corridors to allow a project that would help reduce GHG emissions from electricity use in the state. Both projects would ultimately be built, and then sold to Warren Buffet's MidAmerican.

The visual impact of the valley is different now. Where before there was a valley lost to time, there is a landscape of 20 million photovoltaic modules. Journalist and environmental writer Chris Clarke probably said it best: "Hotter and more arid than the lands that surround it, it resembles nothing more than a desert valley that some determined giant had carved out of the western Mojave and towed halfway to Pismo Beach. It's a landscape that is unique in California, and it deserves better than being turned into another enterprise zone for the energy industry."[64]

Projects on Heavily Disturbed Lands

In the lower Sonoran desert between Yuma, California, and Phoenix, Arizona, the Agua Caliente Solar Project was built near the small town of Gila Bend, Arizona. The 290 MW solar project delivers electricity to PG&E territory in California, so it powers cities such as Oakland and San Francisco. The project is one of several built by First Solar and is owned by NRG, one of the largest power plant owner-operators in the U.S., and MidAmerican Renewables, a subsidiary of MidAmerican, owned by Warren Buffet.

Agua Caliente did not require public lands, instead acquiring land owned by Dole and several other agricultural producers. At the time of commissioning in 2014, this was considered the largest photovoltaic power plant in the world. Considering the benefits of displacing water-intensive agriculture in the hottest place in the Sonoran Desert alone yields substantial climate benefits by reducing water pumping. The project also did not lead to land disturbance, and predictably did not face significant opposition from environmental groups or local residents.

Hundreds of public comments and even some EISs made mention of land on an isolated stretch of the San Joaquin Valley called the Westlands Water District, which straddles Kings and Fresno Counties. Advocates of this property called it Westlands Solar Park and noted that it could support 5,000 MW of solar power. It also was crossed by the high-voltage transmission lines that serve as the backbone of the electricity grid in California, moving electricity up and down the state. The Westlands Solar Park consists of 30,000 acres of land no longer useful to agriculture, the soils long ago over-salted with selenium. The landowners were the Westlands Water District growers, who retained the water rights to the land. They owned 100,000 acres overall, having purchased it in the late 1990s for its water rights. The water comes from reservoirs in the Sierra Nevada foothills and delivered through canals to the San Luis Reservoir. The San Luis Reservoir is the site of California's giant "battery," a giant pumped hydroelectric storage system. John F.

Kennedy famously inaugurated construction of the earthen dam by setting off the first sticks of dynamite. Westlands would soon be turned into one of the Renewable Energy Transmission Initiative's new Competitive Renewable Energy Zones. The major challenge to siting projects in Westlands was that about two-thirds of the property was encumbered with Williamson Act contracts.

THE WESTERN SOLAR PLAN

On May 29, 2008, the BLM announced a Solar Programmatic Environmental Impact Statement (Solar PEIS). Dirk Kempthorne, secretary of the interior in the George W. Bush administration from 2006 until 2009, initiated Secretarial Order 3285A1 in 2009, saying, "We must use our own domestic energy resources as part of a balanced, rational and realistic national policy to secure a reliable supply of affordable energy for America's families and businesses. Expanded solar energy development is part of the solution, placing more control over energy supply in the hands of America."[65]

The Solar PEIS proposed 24 Solar Energy Zones (SEZs) covering 677,384 acres of the initial 22 million acres offered for development in 2005. SEZs identify lands more appropriate for siting utility-scale solar projects based on a number of physical and ecological criteria. The Solar PEIS aims to guide and streamline other environmental and cultural resource reviews, such as NEPA or CEQA, and ongoing processes at other natural resource agencies or at district BLM offices. SEZs can guide and incentivize lands for development that are more disturbed than intact ecosystems by making the NEPA/CEQA engagement less onerous. SEZs were described as previously disturbed or of low biological value, and aligned with transmission—meaning "solar ready" and unlikely to encounter resource conflicts.

The BLM undertook a process of identifying SEZs to prioritize for solar energy development within the 22 million acres available to solar energy developers. This would eventually be known as the Western Solar Plan. These lands needed various characteristics, including access to transmission and excellent insolation, and ideally would have minimal ecological and cultural resource conflicts. "With coordinated environmental studies, good land-use planning and zoning, and priority processing, we can accelerate responsible solar energy production," Secretary of the Interior Ken Salazar said in a press release.[66] SEZs would be the prioritized lands to sacrifice for solar energy. These SEZs would be governed in a way to incentivize the development of solar power across six western states.

Avoiding ecological and cultural resource conflicts was made a primary objective of the local BLM field offices tasked with participation in the Solar PEIS to identify the most appropriate lands for solar energy development. While marching orders would be handed down from Salazar to BLM director Bob Abbey, the local field offices would have an important role in identifying the many uses and attributes of these lands, because they are the ones who know the land and its people the best. The Solar PEIS process collected tens of thousands of public comments and held hundreds of hours of public meetings across dozens of cities.

The BLM is organized at the national and regional district levels. National BLM employees focus on the broad goals of the BLM's complicated agency mission. District BLM employees, on the other hand, are more like land managers. They deal with the everyday workings of the landscape, the people who use it, and the species that inhabit it. Hence, district staff were often less supportive of clean energy goals, given the way they were handed down from above. District-level priorities often contradicted the goals set at the national level. This was evident when comparing the enthusiasm for solar energy development in Washington, D.C., to the more muted support demonstrated by BLM district office employees. The first year of the Solar PEIS saw little progress, as the BLM lacked staff and collaborators in other agencies to make meaningful progress on SEZ development.

Orders for SEZs were handed down from the national to the district level. Interviews with BLM personnel suggest that the process of identifying polygons and boundaries for SEZs was complicated by the need to consider the views of multiple stakeholders. Land managers eliminated sites that could be problematic for various reasons, including grazing and habitat conflicts. District managers were asked to identify parcels of public land under their jurisdiction that would be suitable for renewable energy development. Some managers had several months of lead time, but others had to prepare maps of suitable lands in a very short time. One district manager simply assembled their staff and "broke out the maps, books, reports, GIS data, and tried to make the right decision given the circumstances, the lack of lead time."[67]

After substantial outreach, and expert and public feedback, including 80,000 public comments, 19 SEZs were finalized in 2013, representing up to 27 GW of power potential across the American Southwest that would be available by competitive auction, another key change in the solar program. Several of the initially proposed SEZs were eliminated and several others subsequently reduced in size. Two SEZs in California, Iron Mountain and Pisgah, were eliminated after being heavily targeted in public comments due to the high-quality habitat. Other SEZs were modified around the margins as more information was collected to determine resource conflicts. This process illustrates how public participation and collaboration from locals and environmental organizations can shape the direction of energy transitions. Later several new SEZs would be added.

From the onset, several conservation organizations were concerned about a lack of focus on energy efficiency, conservation, and rooftop solar, and felt that the process assumed that public lands are required for increased solar deployment. The most detailed documentation of impacts was carried out by Basin and Range Watch, a public lands and wilderness advocacy organization in the region. It reported on the various species and histories of the sites. These groups rejected the idea of SEZs and the use of public lands altogether. The public participation process that informed the Solar PEIS literally shaped the SEZs, as the final proposed polygons are different in shape and several SEZs were eliminated through public comment and further evaluation. The PEIS was finalized in 2013 and came into force for projects proposed after 2014, ushering in a new regime of USSE development and public lands governance in California. While on the surface the Solar PEIS appears limited in that it allows USSE development both within and outside SEZs, the BLM would have authority to deny projects before entering into the full NEPA (or CEQA, where state jurisdiction or a joint action was needed in California).

The PEIS identified SEZs across six western states. Initially, three SEZs were proposed in California when the policy was originally released for public comment in 2009. After successive efforts to edit the boundaries and reconsider the SEZs, the final Western Solar Plan was released in 2012. Of the 24 proposed SEZs, several were removed after further consideration, while several more were added. Of the four original SEZs announced in 2009 for California, Pisgah and Iron Mountain were removed from the final plan after scientists and numerous public comments identified them as inappropriate. More data were collected on other parcels, identifying where to remove and expand SEZs. The BLM even received input on where to propose new SEZs, so the final plan included a new Western Mojave Desert SEZ, an area that is more developed than the eastern Mojave. The Western Mojave Desert also had species of conservation concern present on the sites, including the Mohave ground squirrel (*Xerospermophilus mohavensis*), golden eagle, and burrowing owl.

The SEZs proposed by the BLM had several distinguishing characteristics, including a minimum size of 2,500 acres, a slope of less than 2%, proximity to existing transmission corridors, at least 6.5 kWh/m² per day of insolation, and a lack of impacts on special-status lands such as the U.S. Fish and Wildlife Service's critical habitat for sensitive species, or areas of critical environmental concern.

The Western Solar Plan identifies 285,000 acres of SEZs deemed appropriate for development and removes nearly two million acres of exclusion zones from development. Complicating the plan was the BLM's decision to classify 19 million acres as "variance zones." These were places where companies could propose projects outside of SEZs, opening the door to repeating the mistakes of Ivanpah. In essence, the Western Solar Plan made the right-of-way process simpler in SEZs, but companies could propose projects outside SEZs so long as they prepared an EIS. In September 2014, the BLM ruled on its first variance zone application, the controversial Silurian Valley Solar Project. The project was deemed incompatible with the newly minted Western Solar Plan, setting a high bar for future projects. This meant that the project would not advance through the NEPA process. But in early 2015, a second application, for the Soda Mountain Solar Project, was approved. A letter in the *Los Angeles Times* from E. O. Wilson and Thomas Lovejoy popularized concerns about the solar project. Through public pressure, the buying entity, Los Angeles Department of Water and Power, decided it would not purchase the electricity, given the controversy.

A few other planning processes were going on at the BLM as well, including Chocolate Mountains in California and one by the Arizona BLM. The latter is called the Arizona Restoration Design Energy Project, which prepared an EIS to review the most suitable land for renewable energy, with a close focus on previously developed or heavily disturbed sites. The approach was generally praised for its work, and several SEZs were removed as a result of their EIS.

THE DESERT RENEWABLE ENERGY CONSERVATION PLAN

The DRECP was authorized by the legislature at the same time that California passed its RPS to help site projects in the California Desert

Conservation Area. It was anticipated that the mandated markets for renewable energy would attract solar and wind investors to the desert regions because of the wind and solar resource quality, cheap private lands, and the policies recently put into play at the BLM. Congress established the California Desert Conservation Area in 1976, ushering in a new management regime and offering protection to 25 million acres. As a result, public lands in the California deserts have higher protections than public lands in other states, across Imperial, Inyo, Kern, Los Angeles, Riverside, San Bernardino, and San Diego Counties. The purpose of the DRECP—a collaboration between the California Energy Commission, California Department of Fish and Game, USFWS, and the state and national BLM offices—is to help the BLM and other land managers inform and plan for the rapid growth of renewable energy in California deserts where there are important conservation considerations. The DRECP would help developers navigate the NEPA and CEQA processes by identifying public lands with the fewest conflicts and was officially given staff and resources after a memorandum of understanding was signed between California governor Arnold Schwarzenegger and interior sectary Ken Salazar.

One key feature of the DRECP is an adaptive management framework to deal with the potential conflicts associated with solar development and wildlife. Adaptive management is an approach developed in fisheries and wildlife management by C. S. Holling and Carl J. Walters.[68] It refers to a decision-making process that is flexible enough to account for uncertainties and makes use of the best available science. Any plans must be adaptable to new information and circumstances. The DRECP would build data sets of species occurrence and connectivity issues for species such as tortoises and bighorn sheep. The DRECP identified teams of researchers working on relevant research that could be used in adaptive frameworks, including work done at the University of California, Davis, UC Santa Barbara, and the U.S. Geological Survey.

Other stakeholders in the process took the view that the DRECP would make for a more secure investment environment. Southern California Edison, serving much of Southern California, characterized it as providing "the regulatory framework necessary to support investment in renewable energy resources and related transmission, while ensuring effective protection and conservation of the state's wildlife, plants, and natural communities."[69] The utility was recognizing that the scientific process would result in rules and regulations that would make it less likely for opponents to block local, state, and federal permits without clear and justifiable reasons. The argument is that investors find these kinds of projects risky because some of the agencies involved have authority to stop the project beyond just the BLM, including state and federal wildlife agencies and the U.S. Army Corps of Engineers. Nevertheless, the solar industry strongly opposed the DRECP, saying that it cut off future opportunities to site where the greatest value is. In February 2018, the Trump administration announced plans to repeal the DRECP.

The Silurian Valley sits at the bottom of a landscape ringed by national parks and reserves in San Bernardino County, California, just north of Baker. Death Valley, the Salt Creek Hills, and the Mojave Preserve are all within a few minutes' drive. A Spanish company, Iberdrola Renewables, proposed both a wind farm and a 200 MW photovoltaic solar project on over 7,000 acres. The project drew significant opposition because it was being proposed after developers had a sense of what the Western Solar Plan offered for development, and after the formal adoption of the plan by the BLM. There were several habitat and cultural concerns, including some historic trails, dating back to Native American and Spanish times, that would be disrupted by the project. Among the public comments were letters from a handful of federal agencies. In November 2014, the BLM rejected the project on the grounds that it was incompatible with the objectives of the Western Solar Plan.[70]

The Soda Mountain Solar Project is just west of Baker, California, on 4,179 acres of BLM land proposed for development by engineering firm Bechtel. Soda Mountain lies above Soda Dry Lake, an oasis and unique desert ecological feature, where the California State University system has a research outpost called Zzyzx ("ziz-zex"). The site was a former mining claim turned resort. The site developer believed the minerals in the seasonal lake promoted health and operated until the BLM shut it down in the 1960s. Shortly after the draft EIS was published for Soda Mountain, two researchers—John Wehausen and Clinton W. Epps—published an op-ed in the *Daily Bulletin* suggesting that the project threatened to take away a potential connectivity corridor critical to the survival of desert bighorn sheep in the context of gene flow.[71] They noted the threatened status of bighorn sheep and the long history of human interactions with the species, dating back to the times they were depicted in Native American drawings in the area. Soda Mountain eventually was approved by the BLM and planned to deliver electricity to Los Angeles, but was stopped before construction after a groundswell of negative publicity led the mayor to cancel the power purchase agreement with the city-owned utility.

In September 2016, the DRECP was finalized in alignment with the Western Solar Plan, identifying where future USSE projects would be sited across 388,000 acres in the California Desert Conservation Area. A provision added between the draft EIS and the final EIS allowed unallocated lands—another 400,000 acres—to also be available for USSE. Developers choosing these unallocated lands would not receive the same streamlining for USSE permitting. In all, the DRECP allows solar development across 847,000 acres and conserves 10.8 million acres of California desert habitat by making it off limits to solar development.[72] The DRECP extended strict conservation protections to the proposed but denied Silurian Valley solar site.

A LAND ETHIC FOR SUSTAINABLE SOLAR ENERGY TRANSITIONS?

Finding appropriate places to site solar power on public lands poses an intractable problem. In the Ivanpah Valley and numerous other public lands, biodiversity-conservation concerns clashed with a particular plan to decarbonize electricity. Advocates of utility-scale solar power plants stressed that some nature will have to be sacrificed to save human civilization from climate change. To maximize power generation, solar power plants were needed in the areas with the greatest solar resources. These were framed as environmental trade-offs necessary to help society achieve GHG reduction.

Most research on USSE focuses on questions about land-use change. However, the most significant challenges inherent in USSE projects may have to do with birds. The "streamers" seen near Ivanpah's and Crescent Dunes' solar power towers and the prevalence of collisions with photovoltaic panels and heliostats suggest this may be a challenging problem. Solar power towers pose immitigable impacts because of the impact of solar flux not just on birds living on the site, but on any birds that pass through it. Hidden Hills was a project proposed by Bright-Source on 3,277 acres of partly public, partly private property. The company would build the same type of solar power tower project as Ivanpah, but a third taller. California Fish and Game and the USFWS raised concerns about the impacts of the Hidden Hills Project on birds. Early in the EIS process for Hidden Hills there were dissenting wildlife officials. One went so far as to write a public letter during the open comment period asking the BLM to put a moratorium on solar power towers until further evidence could be collected on avian impacts, especially for threatened and endangered species such as golden eagles, falcons, and owls. And the dissident agency scientists were far more numerous than the lone official who risked his job by posting the comment. Many agency staff even within the BLM voiced such concerns in public comments and in the public meetings.

What does a sustainable land ethic for solar energy development look like? We need attention to designs that minimize how USSE projects interrupt the interconnectivity of ecosystems and wilderness in the western deserts. Several power plants, Ivanpah for example, for all its other ecological controversies, did not scrape the entire project site and instead left topsoil and plants intact (Figure 16), except where there were roads. So reducing the degree of land transformation is one way to improve the ecosystem compatibility of solar projects. For example, prohibiting scrapers as a practice can ensure that topsoil will be protected, which could mean less water use for dust control and better relationships with neighbors. Going further, constructing USSE sites with ecosystem services in mind with help push the frontiers of sustainability practices. For example, integrating hedgerows or honeybee forage onto sites can help enhance agro-ecosystem services, while ecological restoration can help native pollinators. In 2018, a study suggested there are over 860,000 acres of agricultural lands near USSE sites that could benefit from increased pollinator services.[73]

Agricultural lands can also make appropriate sites for solar energy development.[74] For example, there are many agricultural lands in the Colorado River Valley that will have to be retired due to the overdrafting of water from the river in California. The Westlands Solar Park too—where the soil is not useful to agriculture because of selenium contamination—would qualify as a responsible development site because it lacks the sensitive habitat issues found on more controversial sites. These "land-sparing" opportunities for solar development could provide synergistic outcomes.[75]

Through the RE-Powering America's Land Initiative, the EPA identified over 11,000 sites with nearly five million acres of brownfields—industrial lands not readily available for use in housing or other real estate developments—landfills, transfer stations, and other severely degraded lands that make excellent opportunities for solar development in the U.S. The RE in RE-Powering stands for renewable energy. These sites are also generally closer to their loads, and so do not incur the line losses associated with moving electricity over great distances.

FIGURE 16. BrightSource's Ivanpah project left much of the plant and topsoil intact, unlike most utility-scale projects in the American West.

The DRECP's scientific advisory panel recommendations emphasize that to maintain compatibility between USSE and conservation, program development must use a "no regrets" approach to siting projects. In other words, proceed with caution where there is the potential to compromise species or ecosystems or the possibility of assaulting the rights and cultures of Native Americans. This requires a participatory, collaborative, science-based planning approach. The land-use impacts of distributed photovoltaics are essentially zero, and where they are actually on land causes less fragmentation. There also may be benefits to siting closer to load or making consumers more aware of their electricity sources, so better incentives and emphasis on distributed generation in grid design can provide win-win solutions. Photovoltaic canopies can provide shade for parked cars, lowering vulnerable populations' exposure to heat stress and reducing the heat island effect. Photovoltaics are the only electricity source that can be built over human developments. "Floatovoltaics" are being installed at water treatment plants, on reservoirs, at the near shore, and atop of other water bodies. The world's largest floatovoltaic power plant, a 200 MW floating solar farm built over an aquaculture operation, opened in China in 2017. Across India they are putting photovoltaics over irrigation canals. Both of these strategies reduce water evaporation, which in some places also means saved energy. In California, 2–3% of the state's electricity is used to convey and pump water.

There are no simple explanations for the social gap in USSE. In cases across the American West, projects sited on public lands attract a broad set of environmental organizations interested in conservation. The bulk of public comments and numerous analyses and stakeholder processes have concluded that much of California's solar electricity could be sited on already disturbed lands or integrated into the built environment. Basin and Range Watch argues that "the most effective way to conserve the California desert, in the context of renewable energy development, is to not make the California desert the focal point of solar energy development in the state."[76] Should public lands continue to be developed for solar energy? The issues of habitat loss and fragmentation and cultural resources will continue to surface even with the Western Solar Plan in place. The problems will likely get more challenging as cumulative impacts come under consideration.[77] At the time of writing, the Trump administration is considering ending both the DRECP and the Western Solar Plan. Reversing these policies could set back progress toward mitigating cultural resource, ecosystem, and wildlife impacts by opening up millions of acres of public lands to development again. Building effective institutions to guide responsible land use will help places around the world develop solar power expeditiously, with fewer land use and resource conflicts.

CHAPTER 7

Breakthrough Technologies and Solar Trade Wars

We should be making some higher risk loans. These would be much more innovative, might be more likely to fail, but could create bigger changes in the long run.

—U.S. energy secretary Steven Chu, March 2009[1]

CHASING BLACK SWANS AND BREAKTHROUGH TECHNOLOGIES

One long-standing debate in energy transitions is whether innovations are needed to deploy low-carbon solutions, or if existing technologies will suffice. Carl Pope, the former chairman of the Sierra Club, once debated "skeptical environmentalist" Bjorn Lomborg about investments in clean technology innovation.[2] Lomborg argued that scarce economic resources should be directed toward research and development (R&D) to develop breakthrough technologies that could supplant existing conventional electricity sources. Lomborg preferred that investments go toward research on better, more efficient low-carbon technologies, rather than deploying existing ones. Pope argued that while R&D and technological improvement are clearly important, key innovations occur as manufacturing processes are scaled up, supply chains are developed, and manufacturers' operations mature. Hence, Pope argued for investments in existing photovoltaic technologies. Princeton University professors Stephen Pacala and Robert Socolow agree that low-carbon energy deployment can be done with technologies available today, and that investments in R&D, while important, are not as important as other energy and climate policies.[3]

Today this debate manifests in the space between the "wind, water, sunlight" energy strategies camp, led by Stanford professor Mark Jacobson, who argue that existing renewable technologies could be deployed to meet energy demand. Groups like the Breakthrough Institute and experts at think tanks like the Council on Foreign Relations emphasize the need for next-generation renewable, storage, and nuclear technologies. More concretely, they argue that pursuing deployment of existing renewables could lead to a lock-in that makes decarbonization impossible due to costs.[4] Do technologies evolve by way of disruptive or breakthrough advances—or in a more incremental way? Does deployment of certain technologies lock in or lock out other, more innovative technologies? All of these questions are difficult to answer and are at the core of several debates in the low-carbon-energy space.

These two positions typify an important split in the energy innovation space, and they map onto investments in thin-film innovations versus crystalline silicon photovoltaic deployment. There were stark differences in investment patterns between China and the United States. While the U.S. invested heavily in breakthrough innovations in thin-films, crystalline silicon photovoltaic manufacturing capacity and output grew rapidly in China, Taiwan, South Korea, the Philippines, and Malaysia. The split tended to reflect the interests of market analysts and venture capitalists on one side, versus semiconductor-industry mainstays on the other.[5] The former saw flows of capital as evidence of the technology's potential, while the latter were concerned about the technical challenge of scaling up the application of semiconductor compounds uniformly as thin films, and pursued more conventional technology.

Breakthrough technologies are those capable of transforming economies and particular ways of doing things in everyday life and work. Sometimes called disruptive technologies, they sometimes displace incumbent technologies, leaving them obsolete. For electric power generation devices like photovoltaics and solar power towers, attaining the status of a breakthrough technology would mean displacing today's conventional electricity sources, such as coal and natural gas, reaching what economists call grid parity—the point at which renewable sources are able to compete with commercially available ones. A disruptive technology would allow solar-powered electricity to have deeper market penetration and higher rates of adoption. Breakthrough technologies are critical to economic growth, because they usually translate into greater productivity, and they are engines of individual wealth generation. Breakthrough technologies are the Holy Grails of the venture capital community because of the possibility of capturing economic rents through patents.

The Section 1705 DOE loan guarantee program made explicit the goal of generating breakthrough green technologies. When the DOE solicited project applications for the loan program there was a great deal of enthusiasm about thin-film photovoltaics among technological futurists, engineers, and speculators in the clean-tech space. Thin-films were viewed enthusiastically as having game-changing, disruptive technological potential. After thirty years of consistent cost declines, crystalline silicon prices in 2007 and 2008 began to rise as the industry experienced a polysilicon shortage. Thin-films' lower material and energy requirements seemed to promise cost savings critical to driving down the cost of solar power. These attributes stirred interest from the venture capital community and clean-tech policymakers. The most publicly known of these was the controversial $535 million loan to thin-film CIGS (copper indium gallium diselenide) manufacturer Solyndra to build a manufacturing facility in Fremont, California. These investments in clean-tech innovation were framed around familiar themes of geopolitical advantage, green jobs, and economic recovery. Jonathan Silver, executive director of the DOE loan guarantee program, said, "Deploying innovative clean energy technologies will have an enormous impact on our global economic competitiveness, energy security and the environment, as well as on our continued economic recovery. Equally as important, deploying commercial technologies will help the country regain control of its energy future in the near term, reduce oil consumption and strengthen our domestic supply chain."[6]

The clean-tech sectors include technologies that reduce GHG emissions and promote more sustainable uses of energy, such as batteries, electric vehicles, solar energy technologies, wind, geothermal, smart grid applications, and energy-efficiency devices. Clean tech is one of the most rapidly growing global industrial sectors, with the United Nations estimating $243 billion annually in investments.[7] In 2011, the photovoltaic sector alone was valued at $80 billion, despite still providing only 1% of global electricity supply. Its value was over $161 billion by 2018 as it surpassed 2%.[8]

Economists have proposed numerous models for how the innovation process generates breakthrough technologies. Early scholarship in science and technology studies focused on the extent to which innovations come out of investments in basic scientific R&D. As a measure of success, they would look at nation-states and compare the numbers of innovations compared to the national investments. But this lacked meaningful granularity. Economists' theories of innovation combine factors from firm-level engagements—R&D, marketing, and management strength—to the macro-economy, attributing success to government incentives to trade protections.

Black swan theory is an idea attributed to a book by Nassim Nicholas Taleb.[9] The premise is that high-profile innovation "events" are often unanticipated and rare. Many of these chance occurrences are undirected outliers that defy human expectations. Black swan theory suggests that centering the evolution of science and technology around human agency and intentionality is partly misguided. Investors and engineers work to make innovations happen, but in reality, the innovations that stick do so for numerous reasons beyond the work and effort of innovators and entrepreneurs. The reasons might be behavioral, cultural, or simply attributable to serendipity. Investors are looking for black swans when they evaluate investment opportunities. They are rare, and sometimes unplanned, but they have large impacts on economy and society when they occur.

Venture capital investments in clean tech topped $2 billion by 2011, with the bulk of this spent in the solar energy sector, particularly photovoltaic module manufacturing.[10] Though only a fraction of overall investment in the solar energy sector, venture capital in high tech is an important engine of innovation, where more "private serendipity"—time and space to explore "adjacent possibilities" and create new things—is permitted and cultivated.[11] This views startup companies in the solar energy sector as candidates for the next eBay, Google, Adobe, or Apple that will emerge with a breakthrough, game-changing technology, finally allowing solar to reach the elusive grid parity. The role of venture capital from Sand Hill Road in cultivating the creative spaces to develop innovative new technologies is an important causal trope in Silicon Valley mythology of how innovation works.

The DOE's framework for innovation sees the problem as the lack of availability of capital for risky investments. Low-cost loans through the DOE allied the interests and cultures of venture capital with the state, as the program was meant to serve technologies that were "innovative"— a strictly defined litmus test (which was ultimately unevenly applied to loan grantees, as explained below). The program intended to take innovative companies and their technologies from pilot to commercial-scale production, giving them an infusion of capital to help them through the "valley of death," a popular heuristic for one of the stages of a successful innovation.[12]

Critics would later argue that through this program the government was playing venture capitalist, and picking winners and losers. The program became a lightning rod for the Obama administration's plan to invest in clean energy and green jobs. On March 7, 2011, U.S. congressional legislators from the Republican Party, most notably California Congressman Daryl Issa, opened an investigation of the decision-making process for the DOE loan guarantee program. This was months before Solyndra would collapse, and was mainly driven by concerns about the Ivanpah Solar Electric Generation Station project, which was making news because of delays related to exceeding an incidental take permit for desert tortoises. Meanwhile, in Washington, D.C., the loan controversies fanned the flames of the polarized times in Congress. The Obama administration was accused of currying favors for friends and campaign donors, and intervening in the innovation process, which critics argued was best left to the market. Moving beyond what might be best described as cronyism arguments, however, there are more straightforward explanations for why some investments failed to become commercial successes.

CLEAN-TECH DEVELOPMENTALISM AND CONSTRUCTING THE INNOVATION PROCESS

The breakthrough-technologies narrative of innovation directed U.S. investments toward thin-film solar technologies. Innovation takes on a peculiar mythos in policy conversations about transforming energy systems toward renewables. Investments in breakthrough technologies are tied to the expectation that some investments will pay off in real-world scenarios. Particular ontologies of innovation and disruptive technology impute agency to venture capital, assuming that simply pairing capital and technology alone can be harnessed as a force to drive energy transitions. Venture capital enterprises fail far more often then they succeed, but when they succeed they can deliver a large payoff. The assumption underlying the loan guarantee program as a mechanism to drive innovation is that if you make enough investments, eventually one will pay off.

This idea that bringing venture capital into contact with disruptive technologies would hasten innovation shaped the particular investment strategies of the U.S. ARRA investments in clean-tech were seen as first steps toward building an industrial foundation for low-carbon innovation. Public policy reflected how venture capital sought out game-changing technologies, and applied that mode of thinking to government investments in pre-commercial technologies. Thin-film investments were justified by their potential to be cheaper, but also because they could be protected by intellectual property rights regimes. This, rather than the brute force of economies of scale and mass production, was a strategy the U.S. could use to compete. The only real chance the U.S. had to keep pace with China in the clean-tech race was believed to be through technologies that could be protected by intellectual property rights schemes.[13]

This public policy approach in the DOE loan program is a form of clean-tech developmentalism. "Developmentalism" generally refers to the belief that progress depends on the expansion of human consumer society and the unleashing of market forces.[14] Geographers have urged a deeper investigation into "the complex of institutions, discourses, and practices" that constitute green developmentalism, which "reflects efforts by relatively far-sighted capitalist actors to overcome barriers to accumulation."[15] The DOE, energy experts, and ultimately Congress identified barriers to success in the solar industry during their expansion of the loan guarantee program to include "innovative" technologies. The barriers included a lack of available capital just at the moment that companies took on considerable debt to build out factories. Numerous thin-film companies, for example, were developing proof-of-concept ideas into pilot facilities. But many lacked access to capital markets to enable meaningful investments in new manufacturing facilities that could deliver at the scale needed to drive down costs.

An understanding of green developmentalism is necessary to understand the logic and rationale emerging from energy policies embedded in neoliberal modes of institution-building and governance. Neoliberalism can broadly be described as a political rationality justifying particular modes of governance that deliver social forms such as privatization, deregulation of environmental and social protections, free markets, capital mobility, and commodification. It is *liberal* in the sense that it underscores the notion of personal freedom.[16] It is *neo* in that it interprets "neoclassical" economic principles strictly, emphasizing an opposition to state intervention and centralized state planning. While neoliberalism's ideological roots are in the writing of Freidman, Hayek, and the Mont Pelerin Society, accounts of its ascent to mainstream public policy start with the Reagan and Thatcher governments of the early 1980s.[17] On the heels of Hayek's (1974) and Friedman's (1976) Nobel Prizes in economics, which gave legitimacy to the project, this period saw neoliberal governance in practice: state antagonism toward labor unions, the retreat of environmental protections, new spaces for accumulation, structural adjustment, and a purge of all Keynesian influences from public policy. This logic was embedded in the political rationalities for the laissez-faire industrial policy pursued by the U.S., in which startup companies were awarded loans to build facilities, but had no real supports or protections from the state to ensure their success.

Sociologist David Hess uses the notion of the "green developmentalist state" in his exploration of green jobs and green energy, but in a slightly different vein, one that echoes the Keynesian sentiment that there is a role for the state in fostering and facilitating innovation.[18] New York Times contributor, Princeton professor, and Keynesian economist Paul Krugman takes the view that state-led investments in jobs would better facilitate economic recovery.[19] This is opposed to the widely held view that the burden of debt from such investments will stall private-sector investments. ARRA investments are an instance of state-led green developmentalism, or more specifically related to the solar investments, a form of *clean-tech developmentalism*; one guided by a strong sense of how to deliver green and clean technologies, and one benefitting private interests rather than Roosevelt's New Deal public works. At the same time, the financial risks associated with investments in pre-commercial innovations are shifted from private companies to the public. The culture of high risk / high reward made investments in thin-films seem obvious from the perspectives of the institutions and firms that helped facilitate their development.

A very different vision of innovation emerged across the Pacific, as described earlier. China saw innovations in the supply chains and business model driving economies of scale, with crystalline silicon photovoltaics as the future, and hedged on that solar energy sub-sector. By 2012, before the many loan disbursements to the thin-film companies SoloPower and Abound were distributed to build manufacturing operations, it was already apparent that the focus on crystalline silicon was a more successful public policy investment strategy. The overall proportion of thin-films in the market dropped from a high of 12.5% in 2011 to less than 5% today.[20] Only two major thin-film manufacturers, out of about a hundred at the pilot stage a decade prior, were commercially successful at scale by 2017 (those two were First Solar and Solar Frontier).

Thin-film technologies, given the advances in equipment for depositing thin films on substrates, were seen as a perfect fit for the kinds of strategic investments the U.S. government could make to give them the edge in the solar power space. Hence, thin-films became the obvious technology to support because of available international intellectual property protections. The DOE loan program would attempt to balance both perspectives to some degree, but breakthrough technologies that could be protected by intellectual property regimes would be the dominant emphasis in the program.

The loan program is evidence for the deployment argument, as its purpose is to show to capital that renewables at this scale are a worthy investment. "We are trying to identify potentially transformative technologies, which can grow to scale and do important things for the country, but also demonstrate to private capital markets that these projects are indeed viable," said Jonathan Silver in a 2010 interview.[21]

The loan program emerged as the primary means to channel investments toward innovative but financially risky energy technologies. The program was heavily criticized because of the controversy over the Solyndra bankruptcy in the fall of 2011, as well as others that eventually met the same fate (notably Beacon and Abound Solar, and the mothballing of a power plant constructed by SoloPower). The bankruptcies left U.S. taxpayers on the hook for the balance of payments on over a billion dollars in loans.

Innovation is risky business. That is why there is a patent system and other intellectual property laws that reward inventors and their investors for taking the time and capital to advance technologies that have no guarantee of coming to fruition. Investment is risky business too. That is why loans are more expensive as it becomes less likely they will be repaid. But investing in innovation can yield technologies. National investments in high-tech innovations through R&D support the development of more competitive technologies. Throughout much of the history of solar energy, the technology has been dependent on government support for basic research and new commercial enterprises. The rewards from public investments as technologies go commercial are new sources of tax revenues and the other intangible benefits from technological advancement and the associated jobs. In some cases the national labs that develop the technologies obtain patents, which also generate revenue.

The major criticism of the loan guarantee program is that the risk is socialized, which reflects broader trends in the high-stakes financialized global economy, where the most speculative ventures have private benefits and social risks. The DOE loan guarantee program socialized risks by being the guarantor of any loans that go into default. For some of the projects, the program subordinated the interests of the government to the benefit of private investors. For example, in the debt structure offered to Solyndra, private investors would recoup their losses before the government.

The context for understanding the anatomy of the DOE loan guarantee program is the situation investment capital and the clean-tech space found themselves in after the collapse of Lehman Brothers and Bear Stearns. There were significant cash flows toward clean tech running up to the financial crisis, with nearly 150 venture capital–backed startups and over 100 thin-film companies. Immediately after the crash, the first in a series of financial bubbles would crush several companies under debt burdens they were unable to support. With capital markets frozen, it was difficult for many of the emerging clean-tech companies to survive the valley of death—the period from concept through profitability that many companies do not survive. Numerous companies once heralded as the future of solar energy found themselves auctioning off equipment and real estate. OptiSolar is the best-known early failure, as are thin-film manufacturers such as Advent Solar (owned by Applied Materials), Ready Solar (acquired by SunEdison), and Applied Solar (owned by David Gelbaum's Quercus Trust). The importance of providing stable, public sources of investments in new manufacturing facilities was summarized by Brett Prior, a senior analyst with Greentech Media: "With the newer technology, banks are not comfortable lending, so the idea is for the government to step in."[22]

President Obama appointed Jonathan Silver, a former venture capitalist, who promised to run the program like a "shadow bank," making risky bets on technologies that no investor—institutional, private equity, or venture capital—would make.[23] The DOE loan guarantee program for solar energy technologies has focused on two primary places in the value chain. "Innovative" manufacturers using thin-film technologies were awarded loan guarantees to help them cross the valley of death. The valley of death is the period between when a company takes on debt to scale up manufacturing and when it becomes profitable through the sale of products. DOE program literature and graphics called the valley of death the "pre-IPO gap." IPO stands for initial public offering, which is when private companies raise public finance by offering stock to the public. The idea is that most often some company would go public, by selling shares on Wall Street, to raise money to bridge this valley, where debt burdens are high and revenues are low. Most solar companies were unable to do this without an infusion of capital because the market still treated these investments as having too much risk.

Silver saw tremendous potential to help the U.S. innovate its way out of the economic crisis and geopolitical clean technology race through disruptive technologies. In an interview with *Ethanol Producer Magazine,* Silver noted that the goal of the program was not to pick winners and losers in clean energy, as critics would contend. "Think of us as a shadow bank," he said. "We are active in energy sectors where private capital markets have not yet become meaningfully involved, and when they do, we exit."[24]

What made projects like Solyndra, which did not require large swaths of land to build a manufacturing facility, "shovel ready" was that manufacturing facilities did not require the lengthy environmental reviews because they were usually in areas zoned for industry and relatively small compared to utility-scale solar power plant facilities. All three manufacturing facilities built with ARRA support had formal environmental assessments with findings of no significant impact.

SOLYNDRA, SAND HILL ROAD, AND THE SOLAR BUBBLE

From its inception in 2005 as Gronet Technologies, founded by Chris Gronet, Solyndra was the poster child for the solar innovation revolution that would be fueled by venture capital. Gronet was a product of Stanford University's engineering school. The first innovative technology he produced (with James Gibbons, a former dean of Stanford's School of Engineering) was patented by G-Squared Technology, which was eventually bought by semiconductor titan Applied Materials in 1991.[25] Assimilation into larger firms is the fate of many startups. Gronet stayed with Applied Materials until 2002. After a short time out of industry, he joined U.S. Venture Partners in Menlo Park, California, and started spending more time at the National Renewable Energy Laboratory, in Golden, Colorado, where several researchers were working on CIGS and other thin-films.

Gronet put together emerging semiconductor technologies with a novel design invented by Ratson Morad.[26] Gronet made solar cells with a p–n junction made of a CIGS absorber layer and a cadmium sulfide (CdS) buffer layer. Crucially, he changed the form factor of the photovoltaic modules. Instead of a flat profile, which takes rays most efficiently from directly overhead, Gronet and his team made the modules as a parallel series of round glass tubes, so the sun would be directly facing the rounded edge of the module as it moved across the sky. The tubes could also absorb diffuse and reflected light.

Solyndra was a rising star in the venture capital community, which backed the startup with a pilot-scale factory prominently located in Fremont, California, along the major interstate connecting Oakland to San José. Solyndra first applied to the loan guarantee program in late 2006, as one of sixteen companies invited to submit a full application to the DOE (143 companies applied in the first round). In the lead-up to the loan it was receiving a good deal of hype in the blogosphere and was caught up in the CIGS fever that ran through Silicon Valley from 2005 to 2011. By 2007, clean-tech bloggers began to write about Gronet's company, which by then had changed its name to Solyndra.

In December 2008, the George W. Bush administration's DOE said the company was "not ready for prime time." Ratings agencies such as Fitch and Moody's had arrived at a B+ grade, which investors classify as below investment quality, or "junk." But in March 2009 it received a conditional loan guarantee commitment for $535 million from the DOE and the Federal Financing Bank.[27] The pre-commercial status of the various companies in the loan applicant pool made most of the ratings across the board rather poor. Solyndra's was actually the best rating of all of the loans made, even those to utility-scale solar power plants. And the interest rate of 1.025% offered to Solyndra would be the lowest offered across the entire portfolio.

The site of the pilot facility would later become important, as the loan was used to build a factory on an adjacent lot. Nearby, the New United Motor Manufacturing Inc. plant—a long-standing joint venture between General Motors and Toyota—had recently closed, leaving behind a very skilled workforce of about 1,000 that would take up new green jobs at the Solyndra factory (and the nearby Tesla factory).

CIGS technology attracted investments because it had higher theoretical efficiency limits than amorphous silicon and cadmium telluride (CdTe) thin-films, which were also attracting venture, multinational, and government capital. *Time* magazine named Nanosolar's CIGS one of best inventions of 2008.[28] *MIT Technology Review* and the *Wall Street Journal* called Solyndra's CIGS modules one of the most innovative technologies in clean tech at the time.[29] MiaSolé was attracting capital from Silicon Valley investment titans. Of the various technologies, CIGS thin-film companies were the most numerous in Silicon Valley. All sought to achieve commercially successful high-efficiency thin-film photovoltaics to catch up with industry leader First Solar, the largest photovoltaic manufacturer in the world by 2010. Praise from the widely respected technology and financial press helped move

Solyndra to the front of the green jobs line for ARRA support to build a factory.

Solyndra was the first recipient of a loan guarantee from the DOE, from a program mothballed since the 1980s. The $535 million loan was for a new fabrication facility known as Fab 2. Their private investors included a who's who of the venture capital and clean-tech investment circles, including a number of firms with iconic Sand Hill Road addresses. This road in Menlo Park, California, near Stanford University, has become a metonym for venture capital, much as K Street (in Washington, D.C.) is associated with lobbying, and Wall Street represents financial services. Sand Hill Road is considered the most expensive U.S. street address for office space. Among Solyndra's investors were USVP, CMEA Ventures, Rockport Capital, Redpoint Ventures, Argonaut Ventures Private Equity, Madrone Capital Partners, Masdar, Artis, and the Virgin Green Fund. These were some of the biggest investors in the clean-tech venture and angel investment community. Most venture capital is from New York City, San Francisco, or Silicon Valley, so these companies were well positioned geographically to seek out these sources of high-risk investment capital.

Solyndra's innovative design positioned it as the Apple of photovoltaic modules. The modules had an elegant, thoughtful design; the look of the device itself helped attract customers. The parallel CIGS tubes would absorb direct sunlight at all times of the day as the sun moved across the sky, plus light reflected from the rooftop. Solyndra's modules were also efficient for thin-films, approaching 13%. This was already higher than the CdTe modules used at the time in solar farms built by First Solar, but lower than crystalline silicon. Solyndra's modular design was another key innovation, as it would be easier and safer to install than bulky flat-plate photovoltaics. Videos made by the company showed how easily workers could move up ladders and across the roof while installing frames and then each tube. The cylindrical shape of the tubes also prevented most of the buildup of dirt, snow, and other substances that can accumulate on the surface of conventional flat photovoltaic modules. The form factor also had better aerodynamics, which would keep wind from blowing the photovoltaic modules off rooftops, and reduced the weight that rooftops would have to support.

Enthusiasm around Solyndra soared when it was offered the loan, with a business plan ostensibly scrutinized by DOE experts. The loan offering conveyed a sense of approval and technological vetting that perhaps it did not deserve, at least not from the DOE approval process.

FIGURE 17. Solyndra's innovative CIGS thin-film photovoltaic form factor on display at Solar Power International, Los Angeles, in 2010.

The Solyndra booth was always the most crowded on the floor of several solar energy events from 2009 to 2011, such as the Solar Power International, InterSolar North America, and PV America trade shows and conferences. Figure 17 is a view of Solyndra's modules from the floor of Solar Power International in Los Angeles, California, in 2009. Onlookers asked lots of questions and closely inspected the tubes, which looked nothing like any other modules at the trade show. But the promise shifted from novelty to legitimate cutting-edge technology with the legitimation of the technology by the DOE.

Solyndra's modules appeared to have generated enough buzz to warrant closer consideration for a loan, but also helping Solyndra to the front of the line for a loan guarantee was Oklahoma oil billionaire George Kaiser. Kaiser was a "bundler"—a super fundraiser—for the 2008 Obama campaign, raising millions of dollars for the successful presidential bid. His Kaiser Family Foundation's investment arm, Argonaut Ventures, had the majority $342 million dollar stake in Solyndra. Kaiser made sixteen trips to the White House to meet with Obama and aides during the loan process, according to a later congressional investigation led by Energy and Commerce Committee chairman Fred Upton (R-Michigan) and Oversight and Investigations Subcommittee chairman Cliff Stearns (R-Florida).[30] The two congressmen were vocal critics of Solyndra, despite Upton having made efforts to bring a loan guarantee to amorphous silicon thin-film manufacturer United Solar Ovonic, based in Auburn Hills, Michigan. Stearns took a harder line on the program, demanding that the president turn over his Blackberry and threatening to subpoena White House chief of staff Rahm Emanuel.

Madrone Capital Partners is another Sand Hill Road venture capital firm, also with ties to a powerful political family. Robson Walton, the oldest son of Sam Walton, founder of Walmart, owned roughly 11% of Solyndra, but also invested in competing technologies such as those made by companies serving the oil and gas industries. Madrone and Argonaut were also donors to Democratic campaigns. Their involvement came to symbolize crony capitalism and green pork: promises of government investments in exchange for campaign contributions.

Solyndra immediately became the symbol for a new turn in cleantech innovation policy. DOE secretary Steven Chu and vice president Joe Biden spoke to the company leadership via satellite and were slated to attend a ribbon-cutting ceremony before the deal was finalized. When Obama attended the completed Solyndra factory in May 2010, he stood alongside former governor Schwarzenegger and proclaimed, "The true engine of economic growth will always be companies like Solyndra."[31] The president even toured the facility with Gronet and other members of the manufacturing team, which produced some of the best images of the inside of the factory to date.

A month after President Obama's visit, Solyndra canceled a planned IPO. Goldman Sachs had planned to take Solyndra public to raise money on Wall Street, but the company had a cash burn rate of $10 million per week.[32] More disconcerting, Solyndra shuttered its pilot factory, several properties away in an industrial park in Fremont, California. The company had previously announced that it intended to operate both facilities to maximize production.

The situation changed even more dramatically only a few months later. In August 2011, Solyndra asked the DOE for an addition $5.5 million to cover invoices and paychecks due to suppliers and workers. Wall Street analyst Lazard was hired by the DOE to explore refinancing options to see whether the troubled solar firm could be rescued. The agency ultimately decided against granting Solyndra an additional loan. On hearing this, Solyndra locked out 1,100 employees and filed for

Chapter 11 bankruptcy. The news made headlines immediately, with conservative media outlets calling the project President Obama's boondoggle. The loan program head, Jonathan Silver, notified Solyndra that it was in default, having spent $528 million of the loan. The effects extended beyond the company and its workers. Companies such as Xyratex, a publicly traded Silicon Valley firm that made equipment for Solyndra, were left holding unpaid invoices.

Within a week there were numerous lawsuits against Solyndra by former employees. The FBI and the DOE's inspector general raided the company's headquarters and the homes of its president and CEO. Congressional investigations followed, including one in which Solyndra executives Brian Harrison and Bill Stover pled the fifth, in front of the House Energy and Commerce Committee, on September 23, 2011. During one hearing, Representative Bilbray (R-California) asked Energy Secretary Chu what kinds of solar technology he thought would make good investments. Bilbray said that DOE experts had told him that thin-film companies were more likely to go out of business because they were experimental, and he proceeded to illustrate this point by listing recent bankruptcies of thin-film companies across the U.S. and Europe.

Republican lawmakers had opened an investigation into Solyndra in March 2011, well before the troubled photovoltaic manufacturer went bankrupt. For the next several months, the FBI subpoenaed documents and emails between the administration and Solyndra, particularly those from the White House Office of Management and Budget and analysts at the DOE who seemed critical of this and some other loans. Even at this early stage, the administration came under fire for "picking winners and losers" and putting taxpayer monies at risk. Such a framing was in some ways disingenuous, because both houses of Congress knew of the risks of the DOE loan guarantee program: they had appropriated $10 billion to set aside to pay the debts of failed companies. Lawmakers had already planned for some failures, and the failure rate turned out to be far smaller than they anticipated and regularly experienced in defense contracting.

At the core of the congressional controversy was why the DOE had agreed to subordinate taxpayer money in the loan repayment structuring. The loan terms were negotiated so that the investors would recoup losses before the federal government, so Argonaut, Madrone, and others received portions of the debt back from the sale of real estate, equipment, and other things auctioned off from the shuttered factory. Key questions asked by lawmakers included: Did the DOE exercise good judgment in subordinating debt owed to the government to private investors? Was it even legal to do so? The DOE was accused of exercising poor judgment throughout the process, though it should be pointed out that career DOE employees continued to have internal reservations regarding the projects, according to the trove of emails uncovered in the investigations by Congress, the DOE, and the Government Accountability Office.[33]

Solyndra's goal was to make CIGS cheap enough to compete with thin-film technology leader First Solar, but the manufacturing costs stayed too high. Several trips by the author to an e-waste processing facility in 2009 and 2010 in nearby San Jose suggested that Solyndra might be plagued by low manufacturing yields. They were also putting thin-films on a cylindrical tube instead of cheaper flat glass. This form factor had the benefits described above, but made it difficult to take advantage of price drivers elsewhere in the value chain. When several different companies source the same materials and inputs, they benefit from economies of scale. As the photovoltaic industry boomed, costs came down for manufacturing equipment, glass, backsheet, and other components. But Solyndra's manufacturing process required more expensive, specially ordered equipment to handle the tubes. The innovative form factor that had generated enthusiasm across Silicon Valley and inside the Beltway also made it impossible for Solyndra to recoup its debts by selling off equipment to other manufacturers, because no other company used such equipment. Figure 18 shows Solyndra's state-of-the-art $733 million manufacturing facility in Fremont, California, with a FOR SALE banner facing busy Interstate 880. It would be eventually be purchased by data storage manufacturer Seagate in 2013.

All of these factors contributed to the ultimate problem for Solyndra, which was the cost of making its modules. The DOE predicted that the 300,000-square-foot Fab 2 would produce 7 GW worth of photovoltaic modules over its life, at a rate of 100 MW annually. However, only 500,000 modules were ever produced, or about 50 MW. Solyndra's costs came down to about $3 per watt for its thin-film CIGS, according to Securities and Exchange Commission filings, but some analysts suggest that the actual cost was nearly $6 per watt.[34] At the time, industry leader First Solar was making thin-film CdTe modules for $0.73 per watt, and photovoltaic modules made from crystalline silicon imported from China were widely available for under $1 per watt by 2011.[35]

On October 29, 2011, the White House ordered an investigation into all of the loans given out by the DOE since 2009. President Obama appointed former Troubled Asset Relief Program overseer and former

FIGURE 18. The Solyndra factory, along Interstate 880 in Fremont, California.

Treasury official Herb Allison to oversee the process, particularly in the context of future loan monitoring and management. One question raised by Allison is whether the DOE asked for a stake in Solyndra for an additional loan. Asking for a stake could suggest that DOE believed that the firm would succeed.

Republican members of the House of Representatives proposed a new law they called the American Taxpayer and Western Area Power Administration (WAPA) Customer Protection Act of 2011.[36] Under ARRA, the WAPA was given authority to borrow money to lend to developers building new transmission lines. The bill would repeal the WAPA's borrowing authority, blocking it from offering any more loan guarantees. Republican Doc Hastings, chairman of the House Natural Resources Committee, wrote a letter to Secretary Chu, expressing concern about a provision of the WAPA program: "If, at the end of the useful life of a project, there is a remaining balance owed to the Treasury under this section, the balance shall be forgiven."[37] The emails also revealed that the former CEO fired two months before the bankruptcy negotiated a $456,000 severance, though those funds were never taken. In November 2011, the "Solyndra rule" was proposed in Congress by

Alabama senator Jeff Sessions. The rule was that the federal government could not raise taxes without eliminating wasteful spending. Solyndra has come to represent wasteful Washington spending influenced by business allies instead of the breakthrough technologies and innovations it promised.

In the fall of 2011, the House Energy and Commerce Committee opened its own investigation into several of the loans, including Solyndra's. The committee subpoenaed all emails from the White House that referred to Solyndra and the loan program. The emails provide one window into the early energy policy contemplations of the administration and the DOE. Dan Carol, the research director and an energy and environmental analyst at the Congressional Budget Office, suggested that the relationships between donors to President Obama's election campaign and DOE deals would come back to haunt the administration.[38] The emails cover areas such as the influence of campaign contributions on the loan restructuring; the involvement of the White House in the decision to grant to loan; and the White House role in subordinating the government interest in the loan restructuring. Around this time, Bloomberg reported that Solyndra spent excessively and unnecessarily on items such as robotic equipment that whistled Disney songs, spa-like showers, and glass-walled conference rooms.[39]

But despite all of the political connections, possible incompetence, and faulty assumptions, most parties turned to the Pacific to place blame for the failure of Solyndra and other DOE investments. "China is cheating!" Ron Wyden would later charge on the launch of a trade suit against China's illegal subsidies to its crystalline silicon photovoltaic sector. Argonaut, the Kaiser Family Foundation, the DOE, Obama, and others would describe Solyndra's failure as if it was simply caught off guard by a natural calamity. Secretary Chu would surmise, "This company and several others got caught in a very, very bad tsunami."[40] China and the U.S. were pursuing different innovation trajectories. The U.S. pursued breakthrough technologies, with big risks and rewards, intellectual property protections, and high-tech innovations. This strategy put many chips on one bet, and by the time it was clear these large-scale investments were unable to produce much return in terms of technological evolution, it was too late to go down the other path. China took the older crystalline technology, mass-produced it, and quickly drove out U.S. competitors.

The photovoltaics industry has always been volatile and tumultuous. Many in the industry refer to these waves of highs and lows as the "solar coaster." The top names in the high-tech and energy sectors have come and gone throughout the industry's history; ARCO, General Electric, British Petroleum, IBM, Exxon, Kodak, Boeing, Mobil, Westinghouse, Shell, and Chevron all have owned or invested in photovoltaic manufacturers over the years. While numerous multinationals hold solar energy companies, the composition of the industry today remains a largely obscure list of both private and public companies. Yingli Solar is one of the few recognized brands, in part because of its sponsorship of FIFA soccer matches. The major players in photovoltaics are not household names: Jinko Solar, First Solar, Trina, SunPower (now owned by French multi-national oil major Total), Canadian Solar, and so on. However, the supply chain for the sector sounds like the Fortune 500, with Dow, Dupont, Wacker, Mitsubishi, and Saint Gobain (a major flat glass producer) leading the pack.

Manufacturers are subject to the whims of price fluctuations for several key inputs made by a handful of companies. These price pressures include falling costs of products produced by competitors (many benefiting from generous subsidies) that compete in an economy where prices are ultimately set by investor-owned utilities (for homeowners and businesses) and the price of conventional energy generation. The influence of supply chain costs can be seen in the series of price spikes that have affected the industry at various times over the past decade for inputs such as polysilicon, indium, and tellurium. Indium prices rose for CIGS because of more widespread consumer purchases of devices with flat-panel displays, from televisions to smartphones. Prices surged for tellurium, as one thin-film manufacturer consumed more than 40% of the global supply.[41]

While many thin-film manufacturers were tinkering with pilot production and raising venture funds, the polysilicon shortage hit, with tenfold spot price rises. A persistent tail of doubled prices would haunt crystalline silicon manufacturers for several years, while billions of dollars of dedicated photovoltaic polysilicon production came online, largely in China but also in new facilities in the U.S., Europe, Qatar, and Korea.

In 2012, venture capital investments in solar were only 50% of the year before.[42] Scores of the heralded thin-film companies had gone bankrupt or left the industry. Even some multinationals, such as BP, were divesting. China's successful scaling-up of its domestic photovoltaic manufacturing sector led to significant oversupply, and the liquidation of this oversupply was driving prices down once again, further reducing the profitability prospects for troubled solar energy companies.

The falling prices of photovoltaics let the technology take a significant market share of utility-scale solar projects from CSP facilities. Many utility-scale power plant projects that were initially proposed as Stirling engines, power towers, or parabolic troughs in 2005 had switched to photovoltaic solar farms just a few years later, particularly late in the decade, when Chinese imports drove down prices. In 2006, when the BLM opened 22.5 million acres of land to solar development, more than 50% of right-of-way applications were for CSP, but by 2009, many projects had already switched to cheaper crystalline silicon or thin-film photovoltaics.

In numerous congressional hearings on the topic, energy secretary Steven Chu pointed to the rapid ascendance of China and the subsequent crash in prices for photovoltaics.[43] He said that no one expected the collapse of photovoltaics prices across the sector so soon and that the dominance of imports imperiled U.S. investments in thin-film technologies. Others pointed out that the market was softening in Europe, as German, Spanish, and Italian feed-in-tariffs, which were very favorable for consumers of photovoltaics, began to come off the books starting around 2011. The administration would argue that a solar bubble instigated by China's overproduction undermined the portfolio of U.S. investments in thin-film manufacturing. This would be echoed by solar technology researchers and advocates, like Ken Zweibel, the director of the George Washington University Solar Institute in Washington, who commented, "The artificially low prices resulting from Chinese overproduction have nearly destroyed a second generation of photovoltaic technologies based on thin film. This has been a huge setback for the U.S. competitive position."[44] Zweibel had founded CdTe manufacturer Primestar Solar, which was acquired by General Electric. The company announced it was delaying investment in a new factory by eighteen months after it was clear that the market was in a severe decline.

Others have pointed out that thin-films' failure could be tied to the global natural gas rush of 2006 through today, much as natural gas prices sank the solar energy industry in the 1980s. Natural gas sets the clearing price for electricity markets, and falling natural gas prices lower the value of power purchase agreements between utilities and solar developers. Utilities are likely to seek severance of contracts that fail to meet deadlines or benchmarks. So the risk of these projects is also linked to the natural gas boom from hydraulic fracturing and horizontal slant drilling.

The mainstream media pinned the failure of thin-film investments like Solyndra on crony capitalism. This was particularly true for conservative media outlets like Fox News and the *Drudge Report*. This version of the story puts the blame for the poor investments squarely on President Obama's administration and the DOE.[45] Donors to the Democratic Party, particularly bundlers for the 2008 Obama presidential campaign, supposedly received special treatment in the DOE loan guarantee program application process. Conservative think tanks like the Heritage Foundation also argued that the root problem was crony capitalism. In their minds, "market discipline" is the solution, and this was another instance of the government interfering in markets, only this time with real impacts on taxpayers. One congressional hearing began with a map of China under the caption "President Obama's job program."

There is a more nuanced explanation for the failure of innovation policies in the clean-tech space with ARRA. Policymakers with roots and experience in venture capital went to Washington with this venture capital mindset, and made certain kinds of assumptions about how disruptive technologies would displace conventional ones in the energy space. They were willing to let the public accept financial risks because of the potential upsides of breakthrough, game-changing technologies. Such "inevitable mistakes" are typical of venture capitalism, which tends to see "failing faster" as a virtue. Only a rare fledgling becomes a black swan.

Some critics criticized the program for offering numerous moral hazards, because the parties taking the risk had very little skin in the game. ARRA made cash grants available from Treasury for up to 30% of a project's costs, so in some cases, between the loan guarantees and the grants, some developers could have skin in the game of less than 10% of the total cost. For example, a 2014 headline noted that BrightSource used a Treasury 1603 grant of over $500 million to pay off a portion of the $1.6 billion it borrowed to build the Ivanpah solar project.

While much funding was available for pre-commercial technologies, there were a few projects that received loan guarantees for otherwise commercial technologies. These projects ostensibly took away from investments in breakthrough technologies. Antelope Valley Solar Ranch One is a 230 MW CdTe photovoltaic farm on 2,100 acres of private lands in the Western Mojave Desert near Lancaster, California. The project is near the Antelope Valley California Poppy Reserve, about 100 miles north of Los Angeles, and lies across the border between Kern and Los Angeles Counties. The $1.36 billion project was built by First Solar with a $646 million DOE loan guarantee and is currently owned by the energy conglomerate Exelon Corporation. The Western Mojave area has long been viewed as a possible future desert metropolis. This is true especially near the desert town of California City, locally known as the planned sister city for Los Angeles. That vision, too, was halted, and today families and workers at nearby Edwards Air Force Base largely occupy the city. The Antelope Valley received many public comments on environmental impact assessments, including the Solar Programmatic Environmental Impact Assessment, that suggest it would make an excellent place for responsible solar development. Proponents of development here pointed to the numerous flat expanses of abandoned agricultural land.

NextLight Renewable Power first proposed the site for a solar farm in 2006. NextLight was a project developer and proposed to use crystalline silicon photovoltaic modules for Antelope Valley Solar Ranch One. For a separate plant, Agua Caliente, it proposed to use amorphous silicon photovoltaics. NextLight received an invitation from the DOE to the due diligence stage of its program for both projects in 2008. After a series of failures in its business plan, NextLight sold off its lands and its place in the queue at the loan guarantee program to the highest bidder. First Solar acquired these assets from NextLight in 2010.

When First Solar began the process of financing its four major projects, it planned to submit all four to the DOE loan guarantee program. Three were ultimately supported with $4.5 billion in loan guarantees and Treasury 1603 grants. Antelope Valley Solar Ranch One is an interesting case in understanding the flexibility the DOE had to define an innovation, which was a key element of the loan guarantee program. First Solar had two projects that used very loose interpretations of innovation. Antelope Valley's single-axis tracker was innovative only because First Solar's thin-films had never been put on trackers to make the modules follow the sun. Neither the trackers nor the modules alone were considered innovative technologies; but the combination of both allowed it to qualify for the innovative category. Likewise, First Solar's Agua Caliente Solar farm used an inverter technology that had been used in Germany for a decade, but never in the U.S.

A Congressional investigation would later find that the "innovativeness" requirement might not have been met by this technology. For a technology to be considered innovative under the loan program it could not already be a commercially successful product. NextLight planned to use innovative, pre-commercial technologies. When the project was sold to First Solar, a highly successful manufacturer of a widely used and mature technology, it was forced to be creative in defining the innovations in its projects. The investigation found that the project only added very minor features. Email exchanges between staff at the DOE noted that First Solar's CdTe technologies were "commercially proven" and "deployed since 2001."[46]

The DOE ultimately classified the projects as innovative. To meet the technical eligibility criteria for innovation, Antelope Valley Solar Ranch One employed these innovative single-axis trackers for its power plants, as opposed to the fixed-axis trackers used in other First Solar projects. While it may be considered new, some onlookers claimed that putting commercially widespread CdTe photovoltaic modules on trackers that have been around for decades is not driving innovation as the DOE loan guarantee program was intended to. The critique extended to Aqua Caliente as well, because it used an inverter technology that had no impact on energy production and had been used in commercial projects in Germany, Spain, and Italy since 2010, as found in the House of Representatives' investigation.[47] It is not clear that this definition is consistent with the spirit of the law because of how scalable or game-changing these solutions are. On the other hand, these were well-functioning, successful projects, which helped show off the technological capabilities of solar power technologies offered by a leading U.S.-based firm, First Solar.

First Solar's pipeline of proposed projects helped drive up its stock price in 2009, to over $300 per share. It was building utility-scale power plants using its own CdTe technology through subsidiary LLCs. Access to the loan program was seen as critical to First Solar's success in difficult financial times. A spokesperson said, "The DOE's loan guarantee program provided an important source of liquidity to help provide debt financing during a difficult time in the financial markets."[48] However, after buying these projects, it became clear that First Solar's thin-film modules did not fit the definition of "innovation" required by the DOE because its modules were already commercially available and in fact quite successful. First Solar had long before successfully crossed the valley of death and thus risked being kicked out of the loan program queue.

However, First Solar planned to build a photovoltaic manufacturing facility in Mesa, Arizona, to garner political support for its projects. Seeing the opportunity for more manufacturing jobs, the DOE allowed a broader definition of innovation and financed $4 billion worth of CdTe thin-film power plants using First Solar modules. To qualify for the loan, First Solar needed to prove that its project fit the definition of an innovation. For the DOE 1705 loan program, an innovative technology is one that is pre-commercial. For most companies, this means operating at pilot scale, but not yet commercial scale.

Public records obtained through a House of Representatives investigation by the Committee on Oversight and Government Reform suggested that there were internal efforts at DOE to recast First Solar projects as innovative. Many of the dissenting opinions came from agency staff. "Be clear this is not an innovation," wrote Dong K. Kim, DOE's director of the loan program's technical division. Specifically, Kim's email read, "Someone keeps changing [Antelope Valley Solar Ranch] Technical slides to include single axis trackers as an innovation. Be clear that this not an innovation. The record will show that we did not grade this as an innovative during intake review. It will not stand up to scrutiny if compared with CVSR [California Valley Solar Ranch] trackers. Whoever continues to make this change needs to understand that Technical does not support the 20 percent of the CVSR field with trackers as an innovative component."[49]

There was also a "one technology per sponsor rule," according to the House of Representatives investigation. But emails from loan program head Jonathan Silver showed that the agency would treat First Solar's three applications as a single package. In exchange, First Solar promised to build a $300 million manufacturing facility to supply the projects in Mesa, Arizona, in an area where a car manufacturing facility was mothballed several years earlier. First Solar eventually canceled plans for this factory that it had used as leverage and provided the bulk of its photovoltaic modules from manufacturing operations in Malaysia.

The DOE's Loan Programs Office provided an important financing "bridge" at a time when the U.S. private debt markets had little or no experience financing first-of-their-kind utility-scale solar projects, and the capital markets remained constrained in the wake of the global financial crisis. Another email illustrates the pressure on the program to finance projects: "If First Solar's project applications are not approved, or if they're delayed beyond September 30 [2010], we believe it could jeopardize our ability to close financing (both debt and equity), jeopardize construction of 1,620 megawatts of solar capacity and, frankly, undermine the rationale for a new manufacturing center in Arizona."[50]

In an email to other DOE officials in June 2011, Matthew Winters, senior adviser for loan programs at the DOE, wrote: "We have often talked about how the 3 FSLR [First Solar] projects were are [sic] considering will support the building of a manufacturing facility in Arizona.

Can one you [*sic*] please quickly draft a 1–2 sentence blurb that states exactly how this is the case, and give the location, size, and expected construction date of the mfg facility? This will go into a document for the White House that describes the manufacturing impact of the projects in our pipeline."[51] But the projects would never materialize, as even the thin-film market would be affected by the larger trend of falling prices, making it wrong to presume that a U.S.-based facility would be viable.

Abound Solar received a loan guarantee to produce a thin-film module very similar to the CdTe one made by First Solar. Interest in the Colorado State University spin-off came from optimism that Abound Solar could also achieve the significant cost reductions achieved by First Solar. General Electric also had just acquired CdTe startup PrimeStar Solar, and Q-Cells was promoting its thin-film spin-off, Calyxo, so excitement around this semiconductor type was high, no doubt partly owing to the success of First Solar. The Longmont, Colorado–based startup would borrow $400 million from the Federal Financing Bank with the goal of constructing a factory in Tipton, Indiana, on the site of a defunct Chrysler factory. The company planned to increase the output of its CdTe modules from 45 watts to 90 watts and raised $260 in venture capital from Invus Group, Bohemian Companies, DCM, GLG Partners, Technology Partners, BP Alternative Energy, and West Hill Investors. Bohemian Companies' head, Pat Stryker, is a billionaire and contributor to the Obama 2008 presidential campaign, although congressional investigations did not focus on these connections, in part because Abound also received favorable tax treatment ($12 million in tax credits) from Indiana's Republican governor, Mitch Daniels.

Abound had only drawn down $70 million of the loan when it abruptly halted production in 2012. Soon after, the company filed for bankruptcy, after only selling roughly 65 MW worth of modules over two years. Over half of these modules were shipped to India with the support of an Export-Import Bank export subsidy. Experts attributed the decline of Abound to lower efficiency and a high defect rate in manufacturing, which made its modules far more expensive than competitors'. The sublimation process used by Abound Solar posed challenges to company scientists and engineers, who were unable to produce modules at a cost and performance to compete with the vapor deposition process First Solar used. But while Abound Solar struggled to keep manufacturing costs down, its real challenge was lack of a market.

CdTe photovoltaic modules are best suited for utility-scale projects because they are less efficient than crystalline photovoltaics, although that is changing. Numerous utilities operating under RPS goals were signing power purchase agreements with other photovoltaic developers. But there were very few utilities or developers in discussions to use Abound Solar's modules. The failure to coordinate production from this facility with some of the loans for power plants is another example of the challenges of a laissez-faire approach to innovation policy.

Many countries subsidize renewable energy, and in the context of innovation in the solar energy sector, the U.S. happened to direct public resources to thin-film technologies. Thin-film technologies were left out of the scope of the trade conflict. For thin-film advocates, China's rapid growth is an easy explanation for the failure of many startup thin-film manufacturers. Zweibel's PrimeStar Solar was a startup thin-film manufacturer that was acquired by General Electric in 2007. The Colorado-based company had plans to build a manufacturing plant near Denver to employ 350 people in well-paying factory jobs. When photovoltaic prices began to plummet with the rapid ascent of Chinese manufacturers, General Electric sold PrimeStar's intellectual property to First Solar for 1.72 million shares, valued at $82 million in 2012.[52] First Solar was believed to have hit an efficiency wall that PrimeStar's technology help overcome to set new efficiency records for CdTe thin-films.[53]

Similar sentiments about China's influence on U.S. investments in innovation were echoed by Senator Wyden: "Failure to address China's practices will undercut U.S. innovation. It will also make it more difficult for the United States to act against China's cheating in other areas on everything from the manipulation of its currency to its export restraints on resources such as rare earth minerals."[54] China became the scapegoat for failures to deliver to market the breakthrough technologies needed to make the U.S. the global leader in photovoltaics.

RPSs, mandatory quotas, and domestic content requirements (DCRs) are green industrial policy tools and incentives, but global trade politics can add an unexpected layer of cost and complexity in photovoltaics. Ultimately, the solar power industry still depends on favorable tax equity schemes and renewable energy portfolio standards for electric utilities to pull them onto the market. In a *New York Times* story, a representative of Sunzone identified in the story as Mr. Zhao observed that, "Who wins this clean energy race really depends on how much support the government gives."[55]

THE ASCENT OF CHINA AND TAIWAN IN THE PHOTOVOLTAIC INDUSTRY

American solar operations should be rapidly expanding to keep pace with the skyrocketing demand for these products . . . but that is not what has been happening. There seems to be one primary explanation for this. China is cheating.

—Oregon Senator Ron Wyden[56]

The emergence of China and Taiwan in the photovoltaics industry is arguably the most significant story in renewable energy over the past decade. The industry was historically dominated by the United States, Japan, and Germany, but they were all overtaken by the rapid expansion in China and Taiwan. In 2006, when China's Renewable Energy Law went into effect, the United States imported less than $50 million in solar cells and modules from China, which overall maintained a 12% share of a $20.3 billion market—about $2 billion.[57] By 2010, China accounted for 33% of a $76.1 billion market—or about $25 billion.[58]

By 2017, the photovoltaic industry had climbed over $200 billion, with China and Taiwan occupying over 64% of global market share in cells and 59% in modules—hoisting the combination of China and Taiwan's market size over $120 billion. Ninety-five percent of global photovoltaic manufacturing capacity that year was in Asia (in Malaysia, the Philippines, India, Singapore, Japan, and South Korea).[59] By 2020, the industry is expected to be worth $350 billion. The growth in solar bodes well for China's carbon intensity as it begins to substitute solar for coal for electricity. However, more than half of the photovoltaic modules made in China and Taiwan are still exported to the U.S. and Europe as of 2018. This rapid ascent would have several implications for the DOE's thin-film investments as well as the possibilities for green industrial policy.

China's rapid rise to the top of the photovoltaic industry sent shockwaves through industries in the U.S., Europe, and Japan. From 2011 through 2013 and part of 2014, manufacturing growth in China led to a global oversupply and glut of photovoltaics. The scale of the manufacturing capacity in China rapidly came to dwarf the size of factories that drove the early deployment in Europe. U.S.-based manufacturers could not compete with this scale of production. Companies based in China and Taiwan began to liquidate this accumulated oversupply below cost at the peak of the crisis.[60] It would not be until late 2014 that prices returned to normal and production better matched capacity. In the meantime, the same companies dominating market share were losing money, including Yingli Solar, LDK, and Suntech—some of the largest manufacturers in the world. Some of these manufacturers benefited from infusions of capital from state-owned banks and private entrepreneurs. This made it difficult for U.S. photovoltaic manufacturers to compete, leading lawmakers to assert that Chinese manufacturers were selling far below the cost of production, violating free trade rules.[61]

As crystalline silicon photovoltaics began to flood the U.S. market from Taiwan and China, U.S. manufacturers began to close their facilities. BP cut manufacturing in Maryland, where it had made photovoltaics since the 1990s, laid off a significant portion of its workers in Spain, and soon after, left the industry entirely. A once-promising startup, Evergreen Solar, with a unique wafer-manufacturing approach, moved its manufacturing from Devens, Massachusetts, to China, eliminating 300 U.S. jobs. The factory had received $100 million from the Massachusetts governor to expand the facility it was now leaving, which would create negative political fallout for his next re-election campaign. Table 11 lists sixteen U.S. photovoltaics manufacturers that went bankrupt in the period of oversupply in 2011–2013. Other major global players from Germany, such as Q-Cells and Solon, also shuttered facilities during this period. Soon the crystalline silicon photovoltaics industry was the arena for a full-on solar trade war between China and the U.S. and between China and the E.U., with tariffs added to the cost of modules and cells to protect domestic industries against cheap imports.

For some parts of the U.S. domestic solar industry, the growth of imports from Asia was good news. Consumers were seeing the costs of photovoltaic installations fall to all-time lows. As projects became more bankable and less risky from a financial perspective, banks were willing to loan at lower interest rates, helping drive down the cost of these "no-money-down" alternative-financing schemes. This policy innovation helped many customers ultimately pay less for solar electricity than they were paying their utility and put a little money down on a solar lease. Cheaper photovoltaics were critical to installers' profit margins from leasing programs. To many players in this installation and financing part of the industry, cheaper photovoltaics were precisely what were needed.

Staff from Oregon senator Ron Wyden's office described China's approach to growing the industry a "grab for green jobs."[62] Oregon had several manufacturers, including the largest U.S. domestic manufacturer, SolarWorld, and SoloPower, an ARRA-supported thin-film manufacturer the state had attracted away from Silicon Valley to Portland. The

TABLE 11

Company	Status	Online	Closed	State	Products
Abound Solar	Closed	2009	2012	CO	Module
Evergreen Solar	Closed	2008	2011	MA	Wafers
Helios USA	Closed	2010	2013	WI	Modules
MEMC Southwest	Closed	1995	2011	TX	Ingots
Nanosolar	Closed	2009	2013	CA	Modules
MX Solar	Closed	2010	2012	NJ	Modules
SolarWorld Americas	Closed	2007	2011	CA	Modules
Solon America	Closed	2008	2011	AZ	Modules
Solar Power Industries	Closed	2003	2011	PA	Cells, modules
Solyndra	Closed	2010	2011	CA	Modules
SpectraWatt	Closed	2009	2011	NY	Cells
BP Solar	Closed	1998	2012	MD	Cells, modules
Energy Conversion Devices	Closed	2003	2011	MI	Cells, modules
Suntech	Closed	2010	2013	AZ	Modules
Sharp Solar	Closed	2003	2014	TN	Modules
Sanyo	Closed	2003	2012	CA	Wafers

SOURCE: Michaela D. Platzer, "U.S. Solar Photovoltaic Manufacturing: Industry Trends, Global Competition, Federal Support," U.S. Congressional Research Service, Washington, DC, 2015.

lawmaker questioned whether China was benefiting from U.S. tax equity subsidization of photovoltaics. The policy was intended to spur photovoltaic manufacturing, but that growth was not occurring in the domestic market. Instead of growing the U.S. photovoltaic manufacturing base, dozens of manufacturers went bankrupt or offshored production to China, South Korea, Taiwan, the Philippines, or Malaysia.

A major reason Chinese manufacturers dominate the photovoltaic sector is the effective use of contract manufacturing and original equipment manufacturers in electronics and semiconductor industries more broadly.[63] Multinational computer manufacturers like Apple, which worked with original equipment manufacturer Foxconn to make its flagship products, notably popularized these organizational forms. In 2010, a widely reported flurry of suicides at Foxconn dormitories, where migrant Chinese workers stayed while employed at the facilities that made iPhones, prompted many to question whether Apple understood the social impacts and environmental practices of its suppliers.[64] New innovative organizational forms may emerge as solar energy transitions unfold. Foxconn announced in 2014 that it was getting into photovoltaic manufacturing, making modules for SunEdison, and later acquired long-time manufacturer Sharp's solar assets.[65]

Another important element driving the growth of photovoltaic manufacturing in China is that numerous companies are state-owned enterprises or funded by state-owned banks. The legacy of the hybrid hypercapitalist yet staunchly communist government has produced a number of contradictory institutions. Forty percent of China's gross domestic product is from Chinese state-owned enterprises.[66] These organizations get preferential access to loans, favorable interest rates from Chinese state-owned banks, and favorable tax treatment, atop any other benefits conferred by local officials or policy. Several major Chinese photovoltaic manufacturers are owned outright or partially by state-owned enterprises. One example was LDK, a vertically integrated polysilicon-to-module manufacturer. They received direct state-bank support to build one of the world's largest polysilicon refineries in 2008, and by 2011 were one of the top ten manufacturers in the world based on volume of modules sold.[67] The company never turned a profit, however, and it declared bankruptcy in 2014, blaming overcapacity in both the cell and module markets for crystalline silicon photovoltaics.[68]

Soon some of the companies accused of dumping would have import duties imposed. Important context for the solar trade war that would soon unfold requires a better understanding of the broader context of U.S.–China relations. Two major geopolitical and economic themes are foremost. The first is the U.S.–China trade deficit, which was about $278 billion in 2012, when the Department of Commerce filed the trade case.[69] A 2012 Economic Policy Institute report suggested that 2.7 million U.S. jobs had been lost to various subsidized Chinese exports from 2001 to 2011.[70] Whereas China previously competed well in labor-intensive production, success in high-tech industries illustrated how competitive the country could be in capital-intensive goods.[71]

While private companies generally enjoy very lucrative relations across the Pacific Rim, some companies traded assets to expand operations in China. One partnership between First Solar and China's state-owned Power International New Energy Holdings proposed to build the largest solar power plant in the world—2 GW—in Ordos, Mongolia, and bring the manufacturing to China.[72] The deal, which never materialized, would have included the development and sale of the world's largest photovoltaic power plant and the transfer of intellectual property explicitly to China, property partially developed through the National Renewable Energy Laboratory's Thin-Film Photovoltaic Partnership.[73] Several congressional hearings in 2011 covered China's Indigenous Innovation Production Accreditation Program, which began in

2009 and compels foreign companies to enter into partnerships over intellectual property in exchange for market access.[74] Intellectual property protection remains an important issue, and some even accused China of stealing intellectual property through reverse engineering of equipment purchased from the U.S. and Germany or hacking.[75]

Overlaying the solar trade politics were preexisting tensions between the U.S. and China over currency exchange rates between the renminbi and the dollar. The U.S. believes that China undervalues the renminbi and that currency manipulation is an illegal subsidy to their exporters. Some estimate that China spends billions of dollars per day intervening in currency markets.[76] The U.S. has not kept its view on this issue secret, often lambasting China in the press. Obama's treasury secretary, Timothy Geithner, noted in 2010, "We are concerned about the depth and breadth of the measures they have taken," later adding, "we will be aggressive on the trade front in terms of fighting anything that is clearly discriminatory."[77] This vehemence is tempered by China's purchase of U.S. Treasury bonds, which underwrites the purchasing power of U.S. consumers, benefiting the U.S. economy enormously.

China's move into crystalline photovoltaics occurred rapidly, and numerous stakeholders and analysts said that the DOE and Treasury had failed to anticipate it. The ambitious state-organized plan aimed to turn the same technology invented at Bell Labs in the 1950s into a high-volume commodity. Foreseeing opportunities to ride the clean-tech wave of investments in the U.S. and Europe, China's 11th five-year plan (2006–2010) included incentives for photovoltaic manufacturing in addition to local government enticements such as zoning changes, lower property taxes, and cheaper water and electric utilities to attract clean technology firms. Soon after, several major Chinese photovoltaic and polysilicon manufacturers traded on the New York Stock Exchange and made bankable, high-quality crystalline silicon photovoltaics rivaling the quality of modules made in Germany, the U.S., and Japan.

China's first wave of investments hit the polysilicon industry in the areas of Sichuan, Wuhan, and Xuzhou. The timing was fortuitous, as the photovoltaic industry was only beginning to source dedicated supplies of polysilicon, historically relying on discards from other semiconductor industries that required higher-purity silicon. A global polysilicon shortage shook the industry in 2007 as demand outstripped supply. The shortage forced manufacturers to consider alterative supplies, including scrap polysilicon and even recycled photovoltaic modules. Several manufacturers incorporated scrap polysilicon into their feedstock. Germany's SolarWorld experimented with recycled polysilicon feedstock as a hedge against price volatility.[78]

The polysilicon manufacturers that dominated the industry at the time (Wacker, Hemlock, REC, MEMC, OCI, GT Advanced Technologies, and Tokuyama) undertook production expansion, but there were also many new market entrants. In 2008 alone, at the height of the polysilicon shortage, 133 polysilicon plants were built in China.[79] A quarter of these projects were in Sichuan Province. By 2009, these plants were producing 86,000 tons of polysilicon per year.[80] These supplies almost immediately satiated demand, and within months of their coming online, polysilicon prices began to come down again. It would not be long before the industry was in a glut.

One of the first reactions to the polysilicon glut was a prohibition of scrap polysilicon imports. To protect its infant domestic polysilicon industry, China instituted a ban on scrap polysilicon in 2009.[81] China's Environmental Protection Ministry claimed that the polysilicon scrap was tainted with "heavy chemicals" that made the materials too hazardous for handling. This essentially eliminated an alternative supply of polysilicon during a time when it was most needed, angering scrap traders and brokers.

By 2012, less than a decade after the initial wave of investment, 80% of these new polysilicon manufacturers in China were out of business.[82] The companies that lasted through the difficult times included GCL-Poly Energy Holdings, China's state-owned and largest polysilicon manufacturer. Other companies were able to weather the difficult economic conditions because they held diverse portfolios of production or vertically integrated product offerings. Yongxiang Solar, for example, was able to make money from PVC and cement production, while producing polysilicon at a loss. Despite the emphasis on developing a domestic polysilicon industry, China still relied heavily on U.S. imports of polysilicon. In 2011, the U.S. exported $873 million in polysilicon to China.[83]

In the subsequent five-year plan (2011–2015), China shifted emphasis toward photovoltaic module manufacturers and supported domestic photovoltaic consumption. China's energy policy called for 21 GW of solar power capacity to be installed by 2015.[84] Some demand would be created domestically. The Golden Sun program provided subsidies to cover 50–70% of the costs of 1.5 GW of utility-scale solar projects. The State Grid Corporation of China, the nation's largest state-owned utility, agreed in 2012 to allow solar power plants smaller than 6 MW to connect to the electricity grid, opening up important new markets. By the end of the five-year plan, China was the largest manufacturer *and* consumer of photovoltaic modules. Coupled with strong government subsidies and incentives, contract manufacturing has helped rocket China to the global center of photovoltaic production with a market share exceeding 84% as of 2017.[85] Crystalline silicon photovoltaics have become a commodity, as in bulk, standardized goods, where parts are more standardized and interchangeable.

SOLAR TRADE WAR

The World Trade Organization (WTO) was established as an extension of the General Agreement on Tariffs and Trade (GATT), part of the original Bretton Woods system of monetary and fiscal policy in the postwar era, alongside the World Bank (first called the Bank for Reconstruction) and the International Monetary Fund. As an official institution for managing the affairs of the GATT, the WTO came into being in 1995. The rules established and enforced (or not) by the WTO aim to prevent countries from subsidizing export sectors that could put competing industries out of business. The reason countries apply tariffs to imported products is to protect domestic industries. Imagine a scenario where a country decides to subsidize a grain crop, with the intention of exporting it, to put another country's farmers out of business. That could present severe food security threats to that nation. Historically, international trade has operated under a regime of protectionism. Global free trade agreements seek to manage the opening of markets and removal of tariffs as a barrier to trade.

The major objectives of the WTO include removing tariffs in global markets, as well as removing non-tariff barriers to trade. The latter has become a sticking point with environmentalists, because some contested non-tariff barriers include environmental regulations. The WTO aims to prevent countries from blocking imports without justification. In such cases, WTO tribunals often side with exporting countries, arguing that imports can only be blocked if there is a difference in the product itself, not the means of production. Environmental regulations can stand where justified by science or economics, but nations assume larger burdens to prove that their laws are not discriminatory.

WTO members agree that it is illegal to subsidize exports.[86] Grants, low-interest loans, and other means to support producers for domestic consumption do not violate trade rules. The way the WTO enforces free trade is that members must declare all national, state, and local subsidies. When member states complain of illegal non-tariff barriers to trade, a dispute settlement tribunal convenes in Geneva, Switzerland. The panel deliberates behind closed doors and is not required to produce any evidence to support its decision.

The mechanisms for filing a complaint with the WTO in the U.S. draw on the 1974 Trade Act, passed initially to aid the GATT process. A company or industry association files a Section 301 petition with the U.S. International Trade Commission and Department of Commerce when it believes that it has suffered injury from a trade policy. The agencies convene an expert panel and conduct an independent investigation into whether the claims about damages can be verified. If damages are found, the U.S. Trade Representative files a complaint before the WTO dispute settlement tribunal. If informal negotiations are not possible, the settlement board will convene in Geneva and issue a ruling. If the tribunal agrees with the assertion of damages, the country filing the petition is allowed to apply a tariff to imports equivalent to the costs of the subsidization.

The Section 201 petition under the same trade act allows a company to petition for global protection from all imports. Rather than targeting a particular country with tariffs, the tax is paid by all sellers into the market. The Section 201 petition also allows a price floor to protect domestic industries. The Section 201 issue would emerge in 2017 when SolarWorld and Suniva petitioned the Trump administration for immediate protective tariffs. In February 2018, the administration ruled that all crystalline silicon photovoltaic imports (with a handful of exemptions) would have a 30% tariff imposed. The ruling set off a number of responses as South Korea weighed tariffs on U.S. products and Canadian solar manufacturers sued the Trump administration for imposing tariffs outside the process required by the North American Free Trade Agreement (an agreement the administration would like to renegotiate).

Several notable conflicts between the U.S. and China at the WTO since China joined in 2001 have involved the renewable energy industry. A major source of contention was China's support for its domestic photovoltaic and wind turbine manufacturing industries with a generous grant program and access to cheap capital. The Chinese government gave three solar manufacturers and one wind turbine manufacturer $23 billion in credit from the China Development Bank from April through September 2010. The United Steelworkers filed a Section 301 complaint with the Department of Commerce on behalf of employees in the U.S. domestic wind industry. In the 5,800-page document, it asked that the U.S. Trade Representative investigate China's trade practices in the renewables sector. The Commerce Department sided with the Steel-workers and filed a complaint with the WTO. But a tribunal was never convened because China decided to abandon the grant program.

By 2011, the implications of China's practices were much more apparent. Renewables industries were borrowing from the Treasury and the DOE loan guarantee program, and it was becoming clear that these investments would be challenged to compete with the low prices of photovoltaics from China. When a reporter asked U.S. President Obama if the U.S. would take action to protect companies like SolarWorld, he replied, "We have seen a lot of questionable competitive practices coming out of China when it comes to the clean energy space."[87] Oregon legislators lobbied for President Obama to enact a domestic sourcing requirement to protect to the U.S. solar industries, though the Solar Energy Industries Association (SEIA) strongly argued against tariffs.

Ironically, until 2008, before becoming the dominant player in the solar space, China had an 80% domestic content requirement for solar, to protect its fledgling domestic industry from global competition. It had a 70% domestic requirement for wind. Domestic sourcing requirements are allowed under WTO rules so long as those industries are not exporting to other countries.

Another area of tension between China and its international trade partners at the time involved a controversy over the uncommon elements tungsten and molybdenum, the latter of which is used in some thin-film technologies. In 2010, China imposed restrictions on rare metal exports, which affected the supply chains of several renewable energy and clean technologies, including the permanent magnets used in wind turbines, photovoltaic semiconductors, and materials for fuel cells and electric car batteries.[88] The DOE suggests that some of these are critical materials not only for clean and renewable energy technologies, but also for national defense.[89]

The first real shot across the bow in the solar trade war was taken October 19, 2011, when a group of seven U.S. solar manufacturers founded the Coalition for American Solar Manufacturing (CASM) and filed petitions with the U.S. International Trade Commission and Department of Commerce. The Section 301 petitions sought relief from injury caused by Chinese imports of crystalline silicon photovoltaics.[90] The complaint included reference to over two hundred different subsidies from 2006 through 2011. The petition singled out $30 billion in loans from the state-owned Chinese Development Bank to JA Solar, LDK, and Yingli Solar.

The coalition argued that China was violating two aspects of free trade. First, the anti-dumping petition sought duties to offset Chinese dumping. The WTO defines dumping as "the practice of selling goods in the U.S. at less than home market price or cost of production. Dumping is prohibited by the WTO agreements and by U.S. law, if it results in material injury to a competing industry."[91] SolarWorld's leadership went on record as many public venues repeated the claims. "Artificially low-priced solar products from China are crippling the domestic industry," Gordon Brinser, president of SolarWorld, told the press.[92] He told a conference on American manufacturing, "Since 2010, employees of at least 12 U.S. solar manufacturing companies—in Arizona, California, Florida, Maryland, Massachusetts, New Jersey, New York, Pennsylvania, Texas and Wisconsin—have become road kill along China's five-year planning superhighway."[93]

The second petition was a countervailing duty petition, alleging that China was illegally subsidizing its crystalline silicon solar industry. Subsidies are defined as financial assistance from the government to benefit the production, manufacture, or exportation of goods. Article 3 of the WTO Agreement on Subsidies and Countervailing Measures provides guidance for interpreting whether or not a program is a subsidy. The allegations in the petition described hundreds of subsidy mechanisms for Chinese manufacturers.

Several instances of heavily discounted land, power, and water provided by local governments were also documented. Local governments are willing to pay for electricity and water, as well as the interest on loans, to attract companies to their region. In Changsha, Hunan Province, one of the epicenters of clean-tech and renewable energy development in China, Hunan Sunzone Optoelectronic received valuable real estate from the municipal government.[94] The local government of Zhuzhou, a city near Changsha, was even more generous. In an interview with the *New York Times,* He Jianbo, deputy director of Zhuzhou's high-tech zone, said, "For really good projects, we can give them the land for free. . . . This land subsidy is not available to traditional industries, only high-tech industries."[95] The petition also alleged that Chinese manufacturers were receiving discounted polysilicon and aluminum inputs, which are necessary for photovoltaic production.

The CASM alleged that Chinese manufacturers also received multi-billion-dollar preferential loans and credit from state-owned banks. Government tax policies provided exemptions, incentives, and rebates. It was claimed that China supported its exports with grants and subsidized insurance. The Export-Import Bank of China offered preferential rates for exports as opposed to modules that remained in China. Finally, fitting with the larger narrative about Chinese trade policy in general, the complaint argued that China was undervaluing its currency to benefit its exports.

Photovoltaic modules made in China are 18–30% cheaper than U.S.-made modules, according to a report published by the Kearny Alliance.[96] The rapid scaling-up of production—some companies were adding 1 GW of capacity in a single year—led to oversupply, forcing companies to clear inventory below cost. Other key factors affecting photovoltaic prices included declining prices for key inputs (most importantly polysilicon, but also glass) and competition from lower-priced thin-film products. Waning overseas government incentives and demand subsidies were also prime motivators for China to scale up as rapidly as it did so as to take advantage of tax equity policies and cash grants in the U.S. Chinese manufacturers also made extensive efficiency improvements, driven by learning and innovation.

The China Development Bank made approximately $40 billion available in loans to manufacturers. The challenge with state-owned banks is that they are not eager to acknowledge bad loans and take large write-offs, preferring to lend more money to enable the repayment of previous loans, a practice that resembles a Ponzi scheme. Many Chinese manufacturers offer vendor financing (a 60-day window for payment), which is attractive to wholesalers and installers who may not have cash on hand. This gives them an advantage over other manufacturers.

Despite the political organization CASM having a dozen or so members, SolarWorld was largely leading the solar trade war charge alone. Some support came from Ron Wyden, the Democratic senator from Oregon, where the company turned polysilicon into high-quality photovoltaics at a factory in Hillsboro, a suburb of Portland. At 550 MW of capacity, this was the largest crystalline silicon factory in the U.S. from 2009 through 2018.[97]

Many U.S. companies in other segments of the solar supply chain that benefited from the cheap imports for their business model criticized SolarWorld and its CEO, Frank Asbeck, and president, Gordon Brisner. This position coalesced into a political organization headed by SunEdison founder Jigar Shah, the Coalition for Affordable Solar Energy (CASE). In blog posts and other pubic venues, Shah specifically called on SolarWorld to drop the trade petition before it "destabilized" the global solar industry and led to "inadvertently spiraling trade wars."[98] According to the CASE website, "Global competition is making afford- able solar energy a reality in America and around the world. Solar- World's action to block or dramatically curtail solar cell imports from China places that goal at risk. . . . Protectionism harms the future of solar energy in America and negatively impacts consumers, ratepayers, and over 100,000 American solar jobs. The coalition is committed to growing a domestic solar industry, promoting innovation, and making solar an affordable option for all Americans."[99]

Danny Kennedy, a former Greenpeace activist and the founder of installer and solar lease pioneer Sungevity, embraced the move to Chi- nese manufacturers. "There's not a lot of coverage of the fact that China and others are doing with an American invention—the solar panel— what we should be doing with the inevitable power platform of the 21st century—investing in it. Instead, there's a lot of angst about whether the U.S. can catch up with regimes we don't even admire in a declining, dirty industry that showed its greatest promise in the 19th century."[100] Kennedy was also quoted as saying, "This industry doesn't need mis- sionaries anymore, we need mercenaries."[101] In other words, the need is no longer to convince people of the benefits but to go out and deliver a more economically competitive product. Sungevity was a key innovator in the retail business model space, responsible for over 100,000 instal- lations before declaring bankruptcy in January 2017.

SEIA at first refused to take sides, but eventually aligned with the global solar industry interests over U.S. manufacturers and against the tariffs. "Solar energy is a global industry, as reflected by our 1,000 member companies from around the world that serve residential, com- mercial property and utility customers in the U.S.," said long-time SEIA president Rhone Resch. "We believe global competition benefits con- sumers. SEIA is focused on expanding the U.S. solar energy market and strongly supports open and fair global competition regulated by a strong, enforceable, rules-based international trade system."[102] SEIA pitched the need for U.S. subsidies in order to keep up with China.

> Regardless of the outcome of this petition, it is clear that U.S. domestic clean energy policies need to be strengthened. We have fallen behind European and Asian nations and need to develop strong and stable clean energy policies that stimulate the U.S. solar market, address financing challenges, expand domestic manufacturing, grow domestic jobs and increase clean energy exports. For the U.S. to meet its economic, environmental and energy security

goals, we must adopt a national comprehensive energy policy. Specifically, U.S. policy-makers should extend the Treasury [cash grant] to support the private financing of solar projects, replenish funds borrowed from the DOE loan-guarantee program and improve application processing, and extend tax credits for investment in solar manufacturing.[103]

Arno Harris, CEO of the utility-scale solar project developer Recurrent Energy, described the tariff intervention as contrary to the interests of photovoltaic customers. Canadian Solar, a crystalline silicon manufacturer with operations in China that supplied Recurrent Energy projects, would soon purchase his company. "This is not in the interest of American consumer," he said. "The best thing the industry can do is drive down the cost of solar. We don't want to see this kind of development."[104]

In March 2012, the Department of Commerce and International Trade Commission announced a preliminary determination in the countervailing duty investigation. Different companies would have different tariff rates. Suntech Power received a preliminary countervailing duty of 2.9%; Trina Solar, 4.73%; and all other Chinese producers, 3.61%. In May, the Department of Commerce announced preliminary anti-dumping duties on Chinese manufacturers of crystalline silicon photovoltaic cells. Suntech Power (31.22%) and Trina Solar (31.14%) were assessed additional preliminary tariffs. Fifty-nine other manufacturers, including JA Solar, Yingli, Hanwha SolarOne, Canadian Solar, LDK Solar, and Jiawei Solar China, were assessed a tariff of 31.18%. All other Chinese producers received a preliminary tariff of 249.96%. These added costs of course were passed on to purchasers of photovoltaic modules, such as homeowners or electricity ratepayers.

The final ruling was issued in November 2012. The Department of Commerce and International Trade Commission found that China "enabled subject importers to gain market share at the expense of the domestic industry." "In sum, the significant and growing volume of low-priced subject imports from China competed directly with the domestic like product, was sold in the same channels of distribution to the same segments of the U.S. market, and undersold the domestic like product at significant margins, causing domestic producers to lose revenue and market share and leading to significant depression and suppression of the domestic industry's prices."[105]

In April 2013, the Department of Commerce suggested that Chinese crystalline silicon photovoltaic module manufacturers were underreporting imports: "The data suggest that some importers may either be improperly declaring merchandise as not subject to the AD/CVD [anti-dumping / countervailing duty] orders, or may be understating the value of the imported merchandise declared as subject to the relevant orders."[106]

To avoid the tariffs, many Chinese manufacturers were routing modules and cell assemblies through Taiwan or South Korea. Others were announcing or breaking ground on new U.S. factories, such as one proposed by Suntech in Arizona. The CASM ultimately filed an appeal to fix the loophole, and by 2014, modules from Taiwan received similar treatment to those from China.

The U.S. was not the only economy to take issue with the emergence of Chinese solar. Several European countries pushed the European Union to investigate as well, and tariffs were placed on Chinese crystalline silicon photovoltaic imports there. In 2015, the scope expanded to include Malaysia and Taiwan, two countries also displaying strong solar industry growth.

While low-cost modules from China challenged U.S. crystalline silicon photovoltaic manufacturers, the U.S. continued to be a major producer of polysilicon, with manufacturing sites in Michigan, Texas, Tennessee, and Washington, and smaller ones in Alabama, Idaho, Ohio, and Montana. China's solar industry relies significantly on U.S. polysilicon exports for its crystalline photovoltaic manufacturing, spending $873 million in 2011, on the eve of the solar trade war.[107]

In 2012, China announced its own investigation, into solar-grade polysilicon dumping by the U.S. and South Korea that injured China's nascent domestic industry.[108] While the U.S. claimed that China was subsidizing its domestic industry, major U.S. polysilicon manufacturers were receiving subsidies themselves. Hemlock Semiconductor received $169 million in state support for a plant in Tennessee in 2010, which would close by 2015.[109] Renewable Energy Corporation received a $155 million tax credit for its Moses Lake, Washington, polysilicon facilities, which ceased production in 2016. The Chinese manufacturers used these instances to point out that other parts of the photovoltaic supply chain were receiving state or local government support too.

China's tariffs on polysilicon imports would negatively affect the U.S. industry. These retaliatory measures were exactly what Danny Kennedy and Jigar Shah feared. In an industry that spends hundreds of millions of dollars on R&D, looking for ways to make solar cheaper, tariffs were placed overnight on polysilicon. These costs would be passed on to consumers.

In November 2012, China filed suit with the WTO against subsidies to European manufacturers. In Europe, the same processes were at work. The key difference is that German and Spanish photovoltaic module manufacturers were already leaving the country as the market came closer to development. In anticipation of looming tariffs, major Chinese manufacturers began to flood U.S. ports and affiliated warehouses with millions of photovoltaic modules to bring them to market before tariffs would apply.

In July 2013, China imposed a 53.3% preliminary anti-dumping tariff on Hemlock's polysilicon, adding 6.5% a few months later.[110] Not all polysilicon producers were affected, as Wacker Chemie negotiated a price settlement with China. In December 2014, Hemlock Semiconductor, owned in majority by Dow Corning, abandoned a polysilicon facility in Clarksville, Tennessee.[111] The plant had produced 10,000 metric tons of polysilicon annually to complement the company's production in Michigan, where it has been making polysilicon for 53 years.[112] Renewable Energy Corporation's Moses Lake polysilicon facility, too, ceased operations after many years. The U.S. continues to be a top polysilicon producer, but it no longer is the leading producer.

DCRs—requiring that specified proportions of a product be made domestically—are at the center of several other global trade disputes. In Canada, the Ontario Electricity Act mandated that electric utilities in Ontario implement a generous feed-in tariff to pay customers who installed photovoltaics for all the solar-powered electricity they delivered to the grid, regardless of how much they consumed, so long as the modules met the domestic sourcing requirements. For systems to be eligible, 60% of the module had to be made within the province. The goal of DCRs is to grow a domestic industry to create jobs or local manufacturing capacity. But Japan and the European Union successfully challenged DCRs at the WTO, arguing that they unfairly discriminate against free trade.[113] In 2014, the DCR was removed for large projects and lowered for smaller projects in Canada. Later, Canada retaliated by launching its own investigation into silicon metal dumped into its domestic market by manufacturers in China.

Another major conflict over DCRs is in India. With its abundant sunshine, India is an enormous market for photovoltaics. India aims to have 100 GW of photovoltaics by 2022.[114] Building domestic photovoltaic manufacturing capacity is critical to India's ambitious solar energy initiative, the Jawaharlal Nehru National Solar Mission. With such a large portion of Indians in energy poverty, and experiencing grid congestion and overgeneration that causes blackouts, India implemented a domestic sourcing requirement in 2008 to support the growth of its own industry linked to its subsidy program. To be eligible for the subsidy, the Ministry of New and Renewable Energy required that a portion of the modules be made domestically. The policy ratcheted up India's domestic production by starting with the requirement that modules only needed to be assembled in India, until 2011, when solar cells were also required to be domestically sourced.

Companies like First Solar became outspoken about India's policy, despite themselves receiving Export-Import Bank loan subsides from the U.S. federal treasury to support the sale of thin-films in India. Ohio congressman Sherrod Brown notified the Department of Commerce of the potential trade conflict in 2013. Soon after, U.S. Trade Representative Ron Kirk pursued formal consultations with the WTO regarding India's DCR.[115] After several years of deliberations, the U.S. successfully challenged India's policy at the WTO. India also remains in trade conflict with China, Taiwan, and Malaysia; there is an active complaint as of 2018 about these countries dumping modules in India's market.[116]

Within a few years, most of the photovoltaic modules entering the U.S. were able to skirt the trade laws by relocating to countries outside the scope of the Section 301 petition. The petitioners turned to another tariff tool, Section 201 of the Trade Act, which offers protections to industries from all imports. The Trump administration, under pressure from SolarWorld and Suniva, pursued a tariff, ultimately set at 30% for all imports. Most solar analysts and advocates suggested that this would put tens of thousands of U.S. jobs installing photovoltaics in the balance, leading to a decline in the growth rate of industry installations. But within three months Jinko Solar announced plans for 400 MW of solar cell and module production in Florida; SunPower bought an underutilized 550 MW factory in Hillsboro, Oregon, from the petitioner SolarWorld; and First Solar announced the construction of the largest solar factory ever in the United States. While the net impact on jobs may still be negative if there are fewer installations, the reshoring of photovoltaic manufacturing to the U.S. in early 2018 took analysts by surprise.

INDUSTRIAL POLICY FOR SOLAR

[The U.S.] can't compete with China to make solar panels and wind turbines.

—U.S. Representative of Florida, Cliff Stearns[117]

Science and technology studies scholar David Hess argues that the "U.S.'s long pursuit of ongoing trade liberalization and laissez-faire approach to industrial policy no longer matches its declining position when faced with a highly competitive global economy and the aggressive trade and industrial policies of rising economic powers."[118] In other words, the problem is not countries and companies skirting fair rules of play in free trade, but instead originates in the institutions that facilitate innovation in industry and research centers. The U.S. approach of offering loans to a handful of thin-film startup manufacturers and utility-scale power plant developers is an example of how laissez-faire innovation policy can be undermined by competing actors that have strong industrial policies that provide resources to meet specific goals and objectives.

What is needed, argue Hess and others, is an industrial policy that can help nurture green economies, as opposed to this quasi-hybrid form of free-market governance where green industries are given cash to be innovative and then left to fend for themselves in the market. Scholars Adrian Smith and Rob Raven argue that technological success in energy markets depends on policy interventions that create "protective space" to protect newly emerging technologies.[119] For example, connecting thin-film manufacturing investments to the renewable portfolio standard program could have been an alternative approach, though one that goes against the free market norms at the WTO. It will be long debated why most thin-film manufacturers failed in 2009–2012. Some will argue that the technology was truly undermined by Chinese imports, while others will suggest that the manufacturers were unable to overcome cost, quality, and manufacturing yield issues when scaling up from the bench to pilot, or pilot to commercial production. Furthermore, First Solar, the thin-film company that succeeded where 95% of other thin-film manufacturers failed, continues to be the most successful of all photovoltaic manufacturers, from revenues to market cap, further complicating narratives about the fate of thin-film technology.

A narrow interpretation of innovation and the need for preordained "disruptive technologies" set U.S. solar manufacturing back as thin-films faltered and the relatively small crystalline silicon players could not compete without significantly scaling up or cutting costs. The approach was led by energy secretary Steven Chu, whose claim to fame, aside from a Nobel Prize in physics, was being the head of the Lawrence Berkeley Laboratory, where he helped secure a $500 million grant from BP for the University of California, Berkeley, to examine second- and third-generation biofuel technologies. Chu strongly believed that transformative technologies would emerge with proper investments in R&D.

However, his new role at the DOE would be slightly different. Instead of directing a basic research lab with some of the best young scientists in the world, his team's task was to identify technologies that they believed would be commercially viable. For this more complicated question the DOE would build a team more familiar with the world of venture capital. But even this team was no match for the merciless global markets, the brute force of Chinese commodity manufacturing, and the fact that the technology never performed as engineers expected it to.

The loan guarantee program may be a policy worth reforming. Numerous economists and policymakers argue that loan guarantees can be critical, given how financial markets view investments in renewables. Numerous people interviewed for this research claimed that many of the larger projects would not have been built without the loan guarantees, because no projects at that scale had ever been built. There are no bonanza investments with renewables; they basically yield what they yield on a predictable schedule. This predictability is one reason solar companies developed the innovative policy to allow the formation of a company or venture between several companies that owns power plant assets that generate electricity and produce predictable cash flows (called a "yield co"). This enables the securitization of solar assets and pooling of resources across multiple companies to build projects. Any loan program reintroduction would need to explore lessons why they failed to deliver innovative, game-changing technologies as promised, or explore whether commercial technologies should be eligible instead of just pre-commercial ones.

Clean-tech developmentalism—pursuing policies to drive clean energy through market solutions—without more holistic and comprehensive climate and energy planning, informed by and coordinated with industrial policy, may undermine efforts to develop low-carbon energy technologies. The Solyndra investigations, and later some poor investments in nuclear power, have left little public appetite for loan guarantees. Despite opponents of the loan program at times describing it as a form of socialism, the laissez-faire innovation policy of the DOE loan guarantee program is not a true industrial policy. Industrial policies aim to annihilate wealth and facilitate the creative destruction and transformation of energy commodities. Karl Marx used the phrase "creative destruction" to describe how wealth was annihilated in technological transformations such as we are experiencing in energy today—he argued this was inherent in the condition of capitalism.[120] Schumpeter borrowed the term to theorize how innovation in technological development in democratic capitalism evolves.[121] Marxist interpretations of innovation see technological change as driven by the aim of reproducing existing power relations or colonizing new aspects of exchange relations. As solar energy becomes an increasingly large portion of global energy supplies, there will be new photovoltaic technologies that replace the old. But the form and type of the technology is not preordained. There are promising thin-film and organic/polymer solar cells on the horizon, but also combinations of amorphous silicon, perovskite, and crystalline silicon exceeding 24% efficiency as of 2018. The crystalline silicon workhorse from Bell Labs has surprised us again and again, and with 30-year warranties on the market, it's difficult to imagine that technology waning anytime soon. The new area of innovation in photovoltaics is integration with electric vehicles and energy storage systems.

Perhaps energy should be understood differently from other technologies and commodities, because its demand is derived. Consumers do not necessarily want solar power; they simply want electricity. Electrons from photovoltaics cannot be made more attractive; they simply must be cheaper. While innovations in other consumer products rely on the allure of consumer demand, commodities that have a derived demand can only compete on price. There are of course customers who are willing to pay more for energy from photovoltaics, but most solar customers aim to keep their bill the same or lower it.

The ARRA investments in thin-films show that what constituted an "innovation" was interpreted very unevenly across the DOE solar investment portfolio. Technologies that were *not* considered innovative by the DOE were still eligible for the DOE's 1703 program. But the coveted 1705 program, built and financed out of ARRA, is where the expensive interest was better subsidized, and more attractive to investors. Some DOE staff had a high bar for what set a technology apart as innovative, while others were grasping at straws for something new to say about a technology that made it deserving of a loan. Investments in innovation were supposed to foster significant progress in technological development. Instead, some technologies were defined as innovative despite being commercially available for many years. It is not clear that this fostered much evolution of U.S. technological development, as it did not yield new patents, new processes, or new firms.

One major criticism of the loan program is that the projects supported were at a scale "too big to fail." Moving from pilot to commercial-scale production entails countless assumptions about chemical processing, construction costs, markets, and policy that make it challenging to know *a priori* what the prices will be as technologies are deployed. Coupled with competition for some key ingredients (driving up prices for indium, for example), this makes cost projections challenging. Sometimes the proposed "cheap" or "inexpensive" technology turns out to be a bit more expensive to make in the end.

The effort to chase down black swans can often yield no results, and has the opportunity cost of turning a blind eye to existing commercially viable technologies and firms. The low likelihood of investments panning out makes these kinds of energy policies more difficult for the public to accept. Solar energy deployment trends demonstrate that Chinese manufacturers are dramatically driving down the price of solar through technical innovations, metallization pastes, anti-reflective coatings, business model innovations like contract manufacturing in some operations, and economies of scale and experience. These are not iPhones or flat-panel displays. Solar electricity competes against electricity that is already entrenched in the economy and flowing to consumers.

The market will challenge pursuit of "breakthrough technologies" lacking integration of innovation policy with an industrial policy. Startups like Solyndra and other solar companies will fail—this is a venture capital program, one in ten do fail—until they are better protected by industrial policies. A number of renewable energy policies could come to represent watersheds in industrial policy designed to encourage solar energy, but linked to domestic sourcing requirements or as qualifiers for investment tax credits. Countries like Canada, China, India, and Germany that are committed to seeing their own Green New Deal investments bear fruit have already done so, whether it be tied to carbon taxes, renewable portfolio standards, loan guarantees, or feed-in-tariffs. A policy could have used government procurement to purchase Solyndra modules, presuming they were of the quality claimed, to help buy down the early high costs. But this requires a level of market intervention that falls outside of the norms of what is typically in play.

U.S. innovation policy under ARRA and the DOE loan guarantee program was challenged by three considerations. First, the loan program was based on a narrow conception of what constitutes innovation. The definition of innovation was limited to pre-commercial technologies, when even commercial technologies are constantly innovating. Second, program administrators assumed that "disruptive technologies" had agency and would survive on their own in the market. The underlying assumption was that the innovation process alone, aided by finance and crossing the valley of death, would foster socio-technical transitions away from conventional energy sources. Third, the loan program failed to foresee how creative destruction works in the solar energy space. The focus on material innovations—thin-films and new materials—overshadowed the cost reductions that are possible with supply chain innovations and economies of scale.

This interpretation, if accurate, is important because there was a significant backlash against several failed ARRA investments, leading policymakers to eschew investing in clean tech. The tens of billions of dollars once available for clean-tech deployment have been reduced to tens of millions for R&D. This was exacerbated by cost overruns at two nuclear power plants in the U.S. The government may have proved that it is not good at playing venture capitalist, but these failures do not mean that government cannot execute industrial policy. The cases of thin-film investments from ARRA and the growth of China's domestic photovoltaic industry suggest that without industrial policy, more global solar companies would be underwater, not just Solyndra.

In the end, the very things that shifted manufacturing to China, and Asia more broadly, the low costs of production and the flexible workforce, have been eroded by the higher costs of importing photovoltaics from China due to tariffs. The retaliatory tactics used across the Atlantic and Pacific Oceans significantly narrowed the margins of U.S.-based installers of photovoltaics. In 2015, Germany, the largest photovoltaic market, saw prices increase for the first time in almost a decade. While the European Union ended its trade war with China in 2018, the U.S. remained locked in a trade war with China. As policies are designed to facilitate sustainable and just solar energy transitions swiftly, technological and institutional innovations will be needed to ensure that reduced costs do not undermine environmental and worker health and safety.

CHAPTER 8

Solar Power and a Just Transition

I'd put my money on the sun and solar energy. What a source
of power! I hope we don't have to wait til oil and coal run
out before we tackle that. I wish I had more years left!

—Thomas Edison[1]

There is little question that solar power will play an essential role in the
future energy supplies for human civilization. But there are still opportunities to shape how solar deployment occurs and how the social transition comes, with maximum opportunities to improve sustainability and
avoid environmental injustice. This book argues that there are many
remaining challenges to delivering on the promises of green and clean
energy. Green technologies do not always have the intended consequences, thus the importance of designing institutions and structures of
governance to protect the most vulnerable and prevent the least desired
outcomes. A constellation of political, economic, cultural, and material
forces shape socio-ecological materialities, and how and where impacts
ultimately touch down. One goal of this work is to evaluate how power
asymmetries—hence the use of solar *power* in the title—drive how the
impacts of solar are distributed, and who benefits and loses.

What are the institutions and practices that can guide a vision for just
and sustainable solar energy deployment? Green innovations in solar
energy manufacturing are being pursued in research laboratories in universities, national labs, and multinational corporations. Researchers at
Rohm & Haas Electronic Materials, a subsidiary of Dow Chemical,
have identified substitutes for the hydrofluoric acid used in solar cell
manufacture. One good candidate is sodium hydroxide (NaOH), which
has been explored as a substitute for over fifteen years.[2]

Although NaOH is itself caustic, it is easier to treat and dispose of than hydrofluoric acid and is less risky for workers. It is also easier to treat wastewater containing NaOH. Fluorinated and chlorinated chemicals are another suite of inputs that researchers are seeking to green. Researchers at the National Renewable Energy Laboratory, in Golden, Colorado, have made polysilicon from ethanol instead of chlorine-based chemicals, avoiding the creation of silicon tetrachloride waste associated with the Siemens and modified-Siemens processes.[3]

Cadmium compounds pose environmental, worker, and community risks up the thin-film supply chain toward smelters and zinc mining activities.[4] The best way to avoid exposing workers and the environment to toxic cadmium is to minimize the amount used, or eliminate use cadmium altogether. Already, two major CIGS photovoltaic manufacturers—Avancis and Solar Frontier—are using zinc sulfide, a relatively benign material, instead of cadmium sulfide. Researchers at the University of Bristol, the University of Bath, Stanford University, the University of California, Berkeley, and many other academic and government laboratories are trying to develop thin-film photovoltaics that do not require toxic elements like cadmium or rare elements like tellurium. First Solar, meanwhile, has been steadily reducing the amount of cadmium used in its photovoltaics, while at the same time setting new efficiency records for CdTe solar cells. First Solar's effective and responsible management of cadmium compounds lends credence to the claim that companies can continue to use waste products like cadmium in production processes without negative consequences for workers or the environment. The question is whether this is exemplary leadership, and whether companies that were less informed or capitalized would also manage these compounds responsibly. This issue will continue to be raised as the industry continues to use lead in crystalline silicon and turns toward perovskite solar cells, which also can use lead compounds.

Recent innovations in solar energy are marked by strong tensions with environmental justice. By opening the black box of solar energy, it becomes possible to anticipate and address environmental injustice. In the case of solar energy transitions, groups such as the Silicon Valley Toxics Coalition (SVTC) and Solar Done Right have emerged as critical industry watchdogs along two segments of the commodity chain. Both groups support solar energy, but argue that manufacturing and siting can be done more sensibly to meet the collective expectations of decarbonization, sustainability, and social justice.

Take SVTC's campaign for extended producer responsibility in the U.S. photovoltaics industry. Extended producer responsibility requires companies to develop end-of-life management schemes to collect and recycle or responsibly dispose of their products. Photovoltaics sold in Europe already fall under such a scheme under the WEEE Directive, but the U.S. industry has not (even though the same companies operate in both regions). The Solar Energy Industries Association, a U.S. national trade organization, has discussed it in committee since 2009 and began listing preferred recycling vendors on its website in 2017. SVTC's goal is to raise awareness of photovoltaic recycling issues through its annual Solar Scorecard, where extended producer responsibility is prominently featured. The aim is to get consumers to ask for brands with end-of-life management plans. Since consumers have a significant influence in the market, companies might find this pressure compelling. SVTC has also worked to enroll investors who control over $2 trillion in funds held by socially responsible investment companies. Photovoltaics companies need to cultivate a clean and green image to maintain the confidence of socially responsible investors. News of pollution from photovoltaics factories can affect stock prices if investors dump their shares. This opens the industry to societal criticism, unlike many other energy-industry sectors, perhaps affording social movements, nongovernmental organizations, and communities greater influence in shaping solar energy's innovation trajectory.

Western Watersheds, Solar Done Right, and Basin and Range Watch are monitoring the rush of renewables projects, and have closely tracked the progress of utility-scale solar energy projects on public lands. These groups oppose biodiversity loss and cultural damages from utility-scale solar projects and instead promote policies to encourage the deployment of more distributed generation on rooftops and in degraded landscapes. Larger environmental groups have more recently voiced similar concerns, including the Sierra Club, the Center for Biological Diversity, and the Nature Conservancy. Distributed photovoltaic power could obviate the concern that large-scale solar energy farms disproportionately impact marginalized groups and other species. Solar Done Right advocates a different energy paradigm, one where public lands are not paved over to power the air conditioners of Los Angeles, and energy development is driven by broad participation and stakeholder engagement. Utility-scale solar projects can lose up to 20% of the power they generate to AC/DC conversion, transmission, and distribution.[5] Advocates of local, distributed solar power also find these arguments about efficiency and appropriate technology compelling. Some studies suggest that net metering—the policy in the U.S. that allows rooftop solar customers to spin their electricity meter backwards when delivering power—has many quantifiable but hidden benefits for ratepayers, including avoided fuel use, avoided transmission and distribution losses, and displacing some of the grid's most expensive electricity.[6]

Solar energy can be compatible with environmental justice, particularly because unlike many other energy options, it does not involve combustion pollution. The electrons generated from photovoltaic modules displace the need to generate electricity from other sources. By installing photovoltaic devices, a consumer can displace greenhouse gas emissions, and mitigate some climate change effects on human health, food supplies, and water availability, which can be unequally distributed across populations. It may displace peak power plant generation, which often burns oil and natural gas in and around fenceline communities. Environmental justice analysis must be ultimately considered in the wider context of displaced fossil fuel energy production, because there may be implications for climate justice. Nowhere in this book has it been argued that the use of photovoltaics should be eschewed, only that appropriate attention should be given to the full range of environmental justice implications so that, as Van Jones said earlier, "the green tide lifts all boats."

From the options foreseeable today, it appears that future energy supplies will be harnessed from the sun. The question is, how quickly will this solar energy transition happen? What forms of social organization will drive the composition of these future energy supplies? There are multiple strategies to pursue socially just and sustainable outcomes, but much depends on the willingness of solar energy advocates to open black boxes and see what is inside. Too often the consequences of climate change and the urgent need for climate action silence criticisms of solar energy. There are many good reasons, if not obligations, to ensure that solar energy commodity chains evolve in a just and sustainable way. This book has explored the how and why of potential challenges to solar energy transitions by focusing on real-world projects financed through major public investments.

One main feature of the rise of attention to sustainability is the effort to certify and verify production practices based on social and environmental criteria. Socially responsible investors, social and environmental organizations, and others began to call for companies to disclose more about their manufacturing organizations, supply chains, and the like.

Since then there has been a proliferation of standards used to evaluate the social and environmental dimensions of products and organizations. The Global Reporting Initiative is a standard developed through a partnership between a number of organizations, including the United Nations Global Compact. The framework offers a broad range of reporting criteria and metrics to capture the environmental performance of companies. As more solar power companies report on these metrics it will become more evident which regions and manufacturers produce the greenest solar-powered electricity. SVTC's Solar Scorecard borrows from the reporting and disclosure framework developed by the Global Reporting Initiative.

Researchers at the National Photovoltaics Environmental Research Center, at Brookhaven National Laboratory in Upton, New York, have published many studies of the possible environmental hazards of photovoltaics. More recently, formal environmental performance ratings for the solar industry have started to emerge. Basel Action Network and SVTC collaborated on a stakeholder process to develop best practices for photovoltaic recycling. In 2016, the Green Electronics Council, the National Standards Foundation International, and SVTC have embarked on a plan to add photovoltaic modules to the EPEAT registry (a whitelist of products that can be purchased by government institutions that are tasked with buying the greenest electronics) through a sustainability leadership standard.

The Solar Energy Industries Association has proposed new industry guidelines, in "Solar Industry Environment & Social Responsibility Commitment," aimed at preventing occupational injury and illness, preventing pollution, and reducing the natural resources used in production. The document urges companies to ask suppliers to report on manufacturing practices and any chemical and greenhouse-gas emissions. These are important trends for the solar industry to adopt early in solar energy transitions, as early learning about sustainability issues will put the industry on surer footing moving forward.

The two notions of sustainability and justice have long come into conflict, and while they clearly have interests in common, there are some areas where one set of issues has turned a blind eye to the other. *Just sustainability* attempts to bridge these paradigms by meeting the "need to ensure a better quality of life for all, now and into the future, in a just and equitable manner, while living within the limits of supporting ecosystems."[7]

The transition toward low-carbon energy futures must be careful not to homogenize the renewable energies technologies that will supplant the old ones. All technologies and commodity complexes should *not* be treated equally, nor should the technologies we use to measure their performance. It matters what is inside the black box, the conditions of production, and how lives and communities are affected by the commodity's life cycle. Most decarbonization researchers agree that photovoltaics will play a prominent role in lessening energy externalities. Yet, advocates of solar energy will have to ensure a commitment to principles that encourage clean production and chemical stewardship.

As human civilization moves to tackle the multiple challenges to Earth's life support systems and the Anthropocene's push on planetary boundaries, it will have to rearrange energy resources and geographies. The transition toward solar energy will reduce the overall impacts of energy production, but it will have material implications for people living near sites of production, generation, and disposal. A just transition toward solar energy could pursue the following principles.

Manufacturing processes based on inherently safer design principles can help eliminate worker, health, and safety risks. For example, the redesign of processes that use the pyrophoric (capable of spontaneous combustion) gas silane has led to zero accidents in crystalline silicon manufacturing. Most injuries and deaths related to silane handling occur in amorphous silicon manufacturing, a technology very few continue to explore today. The history of semiconductor and electronics manufacturing suggests that environmental justice and occupational health issues could be problematic.[8] Key performance indicators can help benchmark progress toward making production processes more in line with green jobs ideals and the discourse of the "circular economy."[9]

The elimination of heavy metals in semiconductors of thin films would align with the goals of green chemistry and green design.[10] Cadmium and lead are candidates for replacement, though cadmium's use is more complicated, and some studies suggest that replacements for cadmium could contribute more cadmium pollution from a life-cycle perspective.[11] Green chemistry and alternatives assessment are tools for assessing how to replace the most toxic chemicals used in production with more environmentally benign ones. Recent advances in chemical engineering have explored replacing the cadmium chloride used in CdTe manufacturing with magnesium-based salts.[12] Similarly, there are multiple research initiatives dating back decades aimed at eliminating chlorine-based chemicals from crystalline silicon photovoltaics.[13]

The long-term disposal and resource-availability questions can both be solved through a robust take-back and recycling policy based on the principle of extended producer responsibility. Photovoltaic modules degrade in power output over time until they are no longer capable of generating sufficient electricity or no longer meet the needs of the customer (homeowner, power plant operator, etc.). Companies work diligently in research and development to minimize this degradation, and many modules can operate for many decades without an appreciable output decline. However, ultimately modules need to be replaced, presenting end-of-life waste issues which should arise in the next 15–20 years, as the past decade has seen a significant volume of deployment and photovoltaic modules are warrantied for 20–25 years. Crystalline silicon photovoltaics contain a number of valuable materials with secondary recycling markets, such as copper, aluminum, and glass. Several grams of precious silver are also used in each crystalline silicon photovoltaic module. Some materials can be recycled into high-value glass or downcycled into lower-value secondary glass products. Recycled silicon wafers also have value, since a significant energy investment is required to make them (and some of that energy can be saved by recycling them). Tellurium, indium, gallium, and ruthenium are among the rarest of the metals used in thin-film technologies, as they are among the rarest elements in the Earth's crust.

Wildlife and land-use issues will be the most critical facing solar energy geographies in some places. The scientific advisors to the Desert Renewable Energy Conservation Plan emphasize siting with "no regrets" to minimize ecological disturbance from solar farms.[14] Because of the unique situation in the United States, where a large portion of solar projects are sited on public lands, there are particularly vexing siting issues, because these tend to be habitats in relatively if not completely undisturbed ecosystems and open spaces. Among areas that should be avoided are Bureau of Land Management lands with special designations, as well as sites under consideration for conservation of threatened, endangered, and sensitive species habitat, and critical cores and linkages for wildlife habitat; citizen-proposed wilderness areas; and other lands with wilderness characteristics. Areas outside the U.S. will also face similar challenges, particular near the world's deserts, although the quality of habitat on public lands may be somewhat unique in triggering these controversies. Germany also relied on public lands for utility-scale solar deployment, but these were former military lands, so less controversial regarding wildlife impacts. There is some evidence of similar land-use conflicts in Morocco, though with more emphasis on the dispossession of local and indigenous communities.[15]

Shifting to degraded farmlands may lessen some of the ecosystem impacts related to loss of habitat, although some agricultural lands are still used and frequented by wildlife, and some risks, such as bird collisions, will still be present. There are also concerns about the loss of prime farmland, which has been an important issue particularly in the U.S. and Europe due to suburbanization. In developing countries, the loss of agricultural lands could pose food security or livelihood risks to poor farmers.

The challenges to making solar energy more sustainable are exacerbated by the fact that there are few advocates for sustainable solar, since much of the advocacy is still faced on promoting solar in light of the greater environmental and social harms of fossil fuels. In other words, most sustainability and renewable energy advocates see solar energy as sustainable on its surface, without questioning the environmental impacts of production or questioning whether more sustainable alternatives are out there. Yet, the issues described here suggest that there are important sustainability and environmental justice considerations that the solar industry and society will have to plan for to minimize impacts, including reducing chemical use, eliminating toxics, participating in extended producer responsibility and recycling programs, and siting solar power plants with no regrets. Solar energy could be a dominant energy as early as mid-century, but this will require rapid scaling-up of the technology. The development of sustainability leadership standards, green procurement, and eco-labels could go a long way toward enabling and rewarding companies with greener designs, better sustainability initiatives, and stronger worker health and safety practices.[16]

Human civilization has always depended on solar energy, because the sun powers photosynthesis, food webs, and the water and weather cycles. Warnings about the scarcity, security, and geopolitical concerns related to fossil and fissile nuclear fuels, and later concerns about air, water, and climate issues, led to the rise of solar power as a priority research topic, policy objective, and object of investment. The multiple environmental benefits of solar energy led many to emphasize solar energy as the vanguard of clean, low-carbon, and renewable energy transitions. But solar comes in different forms and enrolls different landscapes and geographies, suggesting that the way solar power is integrated into decarbonization strategies will have implications for some communities and even social organization. Social planning for energy transitions aims to direct society more toward the utopian solar narratives, and not the dystopian pictures painted by a small number of accidents in a relatively young industry that magnified environmental and worker injustices, similar to stories from electronics. Anticipating the socio-ecological challenges that will accompany solar energy transitions offers an opportunity to understand and govern the planning challenges associated with a just transition to solar power.

Notes

INTRODUCTION

1. Fraunhofer Institute for Solar Energy, *Photovoltaics Report,* Freiberg, Germany, November 17, 2016.

2. U.S. Energy Information Agency, *Annual Energy Outlook,* Washington, DC, 2001.

3. Fraunhofer Institute for Solar Energy, *Photovoltaics Report,* 5.

4. Sammy Roth, "World's Largest Solar Plant Opens in California Desert," *USA Today,* February 10, 2015.

5. California Independent System Operator, "California ISO Solar Production Soars to New Record" (press release), July 14, 2016.

6. Tom Kenning, "Solar Accounts for 1% of Global Electricity, How Long Will the Next 1% Take?" *PV Tech,* June 29, 2015.

7. Varun Sivaram, *Taming the Sun: Innovations to Harness Solar Energy and Power the Planet* (Cambridge: MIT Press, 2018).

8. John Perkins, *Changing Energy: The Transition to a Sustainable Future* (Oakland: University of California Press, 2018).

9. U.S. Department of Energy, "SunShot Vision Study," February 2012, Office of Energy Efficiency and Renewable Energy, https://www1.eere.energy.gov/solar/pdfs/47927.pdf.

10. Peter Newell and Dustin Mulvaney, "The Political Economy of the 'Just Transition'," *Geographical Journal* 179, no. 2 (2013): 132–40.

11. Intergovernmental Panel on Climate Change, *Climate Change 2013: The Physical Science Basis* (5th Assessment Report), Geneva, 2014.

12. Christophe Lecuyer and David Brock, *Makers of the Microchip: A Documentary History of Fairchild Semiconductor* (Cambridge, MA: MIT Press, 2010).

13. Ted Smith, David Allan Sonnenfeld, and David N. Pellow, *Challenging the Chip: Labor Rights and Environmental Justice in the Global Electronics Industry* (Philadelphia, PA: Temple University Press, 2006).

14. Vaclav Smil, "On Energy and Land: Switching from Fossil Fuels to Renewable Energy Will Change Our Patterns of Land Use," *American Scientist* 72, no. 1 (1984): 15–21.

15. Dustin Mulvaney, "Identifying the Roots of Green Civil War over Utility-Scale Solar Energy Projects on Public Lands across the American Southwest," *Journal of Land Use Science* 12, no. 6 (2017): 493–515.

16. Jeffrey E. Lovich and Joshua R. Ennen, "Wildlife Conservation and Solar Energy Development in the Desert Southwest, United States," *BioScience* 61, no. 12 (2011): 982–92.

17. Robert I. McDonald, Joseph Fargione, Joe Kiesecker, William M. Miller, and Jimmie Powell, "Energy Sprawl or Energy Efficiency: Climate Policy Impacts on Natural Habitat for the United States of America," *PLoS One* 4, no. 8 (2009): e6802–13.

18. Karen Eugenie Rignall, "Solar Power, State Power, and the Politics of Energy Transition in pre-Saharan Morocco," *Environment and Planning A* 48, no. 3 (2016): 540–57.

19. Kara A. Moore-O'Leary, Rebecca R. Hernandez, Dave S. Johnston, Scott R. Abella, Karen E. Tanner, Amanda C. Swanson, Jason Kreitler, and Jeffrey E. Lovich, "Sustainability of Utility-Scale Solar Energy: Critical Ecological Concepts," *Frontiers in Ecology and the Environment* 15, no. 7 (2017): 385–94.

20. Rebecca R. Hernandez, Madison K. Hoffacker, Michelle L. Murphy-Mariscal, Grace C. Wu, and Michael F. Allen, "Solar Energy Development Impacts on Land Cover Change and Protected Areas," *Proceedings of the National Academy of Sciences* 112, no. 44 (2015): 13579–84.

21. Benjamin K. Sovacool, Matthew Burke, Lucy Baker, Chaitanya Kumar Kotikalapudi, and Holle Wlokas, "New Frontiers and Conceptual Frameworks for Energy Justice," *Energy Policy* 105, no. 6 (2017): 677–91.

22. Newell and Mulvaney, "Political Economy," 140.

23. Kirsten Jenkins, Darren McCauley, Raphael Heffron, Hannes Stephan, and Robert Rehner, "Energy Justice: A Conceptual Review," *Energy Research & Social Science* 11 (2016): 174–82.

24. Karen Bickerstaff, Gordon Walker, and Harriet Bulkeley, *Energy Justice in a Changing Climate: Social Equity and Low-Carbon Energy* (London: Zed, 2013).

25. Paul Robbins, *Political Ecology: A Critical Introduction* (Oxford: Blackwell, 2011).

26. Clark Miller and Jennifer Richter, "Social Planning for Energy Transitions," *Current Sustainable/Renewable Energy Reports* 1, no. 3 (2014): 77–84.

27. Vasilis Fthenakis and Hyung Chul Kim, "Photovoltaics: Life-Cycle Analyses," *Solar Energy* 85, no. 8 (2011): 1609–28.

28. Apple for example made headlines when its iPads and iPhones were linked to deaths at migrant worker dormitories operated by their contractor Foxconn. Jenny Chan and Ngai Pun, "Suicide as Protest for the New Genera-

tion of Chinese Migrant Workers: Foxconn, Global Capital, and the State," *Asia-Pacific Journal* 8, no. 37 (2010): 2–10.

29. The formal name was the National Energy Policy Development Group. Eric Dannenmaier, "Executive Exclusion and the Cloistering of the Cheney Energy Task Force," *New York University Environmental Law Journal* 16 (2008): 329–80.

30. The U.S. Department of Interior manages 245 million acres of public land and 700 million acres of subsurface mineral rights. Department of the Interior, "Draft Solar Programmatic Environmental Impact Statement," May 29, 2008.

31. Dustin Mulvaney, "Are Green Jobs Just Jobs? Cadmium Narratives in the Life Cycle of Photovoltaics," *Geoforum* 54 (2014): 178–86.

32. Paul D. Moskowitz, Vasilis M. Fthenakis, L. D. Hamilton, and J. C. Lee, "Public Health Issues in Photovoltaic Energy Systems: An Overview of Concerns," *Solar Cells* 19, no. 3 (1987): 287–99.

33. Silicon Valley Toxics Coalition, *A Just and Sustainable Solar Energy Industry*, San Jose, CA, 2009.

34. Ken Zweibel, "The Impact of Tellurium Supply on Cadmium Telluride Photovoltaics," *Science* 328, no. 5979 (2010): 699–701.

35. Debra Kahn, "U.S. Clean Tech Outpaced by China—Chu," *Energy & Environment News*, March 9, 2010.

36. Chris Martin, "China Flooded U.S. with Solar Panels before Trump's Tariffs," *Bloomberg*, February 14, 2018, https://www.bloomberg.com/news/articles/2018-02-16/china-flooded-u-s-with-solar-panels-before-trump-s-tariffs.

37. Vincent Castranova and Val Vallyathan, "Silicosis and Coal Workers' Pneumoconiosis," *Environmental Health Perspectives* 108, no. 4 (2000): 675–84.

38. Adel Valiullin and Valentin Tarabarin, "Archimedes' Burning Mirrors: Myth or Reality?" *History of Mechanism and Machine Science* 11 (2010): 387–96.

39. Alfred W. Crosby, *Children of the Sun* (New York: Norton, 2006).

40. Willoughby Smith, "Effect of Light on Selenium during the Passage of an Electric Current," *Nature* 7, no. 303 (1873): 303.

41. *New York Times*, "American Inventor Uses Egypt's Sun for Power; Appliance Concentrates the Heat Rays and Produces Steam, Which Can Be Used to Drive Irrigation Pumps in Hot Climates," July 2, 1916.

42. Energy Information Agency, "Annual Energy Outlook 2017," U.S. Department of Energy, 2017.

43. John Perlin, "Silicon Solar Cell Turns 50," NREL/BR-520-33947, National Renewable Energy Laboratory, Golden, CO, 2004.

44. John Perlin, *From Space to Earth: The Story of Solar Electricity* (Cambridge: Harvard University Press, 1999).

45. Fraunhofer Institute for Solar Energy, *Photovoltaics Report*.

46. Jenq-Renn Chen, Hsiao-Yun Tsai, Shang-Kay Chen, Huan-Ren Pan, Shuai-Ching Hu, Chun-Cheng Shen, Chia-Ming Kuan, Yu-Chen Lee, and Chih-Chin Wu, "Analysis of a Silane Explosion in a Photovoltaic Fabrication Plant," *Process Safety Progress* 25, no. 3 (2006): 237–44.

47. Nichola Groom, "Solar Company Evergreen Files for Bankruptcy," *Reuters,* August 15, 2011.

48. Neil Thompson and Jennifer Ballen, "First Solar," Massachusetts Institute of Technology Sloan School of Management, 2017.

49. Chris Mooney, "Why Big Solar and Environmentalists are Clashing over the California Desert," *Washington Post,* August 15, 2016.

CHAPTER 1. SOLAR POWER

1. Quoted in Steven Mufson, "The Green Machine." *Washington Post,* December 9, 2008. http://www.washingtonpost.com/wp-dyn/content/article /2008/12/08/AR2008120803569.html.

2. Ted Nordhaus and Michael Shellenberger, "The Death of Environmentalism: Global Warming Politics in a Post-Environmental World," Breakthrough Institute, 2004, http://www.thebreakthrough.org/images/Death_of_Environmentalism.pdf.

3. Ted Nordhaus and Michael Shellenberger, "Green Jobs for Janitors," *New Republic,* October 6, 2010.

4. Rachel Pinderhughes, "Green Collar Jobs: Workforce Opportunities in the Growing Green Economy," *Race, Poverty, & the Environment* 13, no. 1 (2006): 2–6.

5. Van Jones, *The Green Collar Economy* (New York: Harper Collins, 2009).

6. Ben Block, "U.S. Election Brings Green Jobs in Focus," WorldWatch Institute, 2008, http://www.worldwatch.org/node/5923.

7. John McCain, "Press Release—Fact Sheet: John McCain's Jobs for America Economic Plan," July 7, 2008; Gerhard Peters and John T. Woolley, "American Presidency Project," http://www.presidency.ucsb.edu/ws/?pid=93853.

8. Ted Nordhaus and Michael Shellenberger, *Breakthrough: From the Death of Environmentalism to the Politics of Possibility* (New York: Houghton Mifflin Harcourt, 2007).

9. Van Jones, *Green Collar Economy.*

10. Peter Carvill, "Obama Seeks to Extend ITC Permanently," *PV Magazine,* February 5, 2015, http://www.pv-magazine.com/news/details/beitrag /obama-seeks-to-extend-itc-permanently_100018023.

11. U.S. Bureau of Labor Statistics, "Measuring Green Jobs," 2011.

12. Office of the White House, *A Retrospective Assessment of Clean Energy Investments in the Recovery Act,* February 2016, 31.

13. Ibid.

14. Pinderhughes, "Green Collar Jobs," 2.

15. Timothy Luke, "A Green New Deal: Why Green, How New, and What Is the Deal?" *Critical Policy Studies* 3, no. 1 (2009): 14–28.

16. In 2009, the U.S. Department of Treasury released a "Preliminary Analysis of the Jobs and Economic Impacts of Renewable Energy Projects Supported by the 1603 Treasury Grant Program." The report was authored by NREL staff and Marshall Goldberg of consulting firm MRG & Associates, a long-time expert witness at public utility proceedings on issues related to job creation. The 1603 program expired in 2011, though grants were still processed by the

Department of Energy through 2016. But the ITC too has come under attack from various quarters, including the American Legislative Exchange Council and the Edison Electric Institute, a think tank representing the viewpoints of investor-owned utilities. Ultimately the ITC and another tax credit used by the wind industry were extended by Congress. Democrats agreed to lift a decades-old oil export ban in exchange for Republican support for tax credits for the renewable energy industries.

17. Office of the White House, *Retrospective Assessment*, 31.

18. U.S. Bureau of Labor Statistics, "Measuring Green Jobs," http://www .bls.gov/green/, accessed June 2, 2012.

19. U.S. Bureau of Labor Statistics, "Careers in Solar Power," http://www .bls.gov/green/solar_power/, accessed June 2, 2012; more recent data are available from U.S. Department of Energy, "Solar Energy Jobs Outpace U.S. Economy," January 12, 2016, http://energy.gov/articles/solar-energy-jobs-outpace-us-economy.

20. Daniel Kammen, Kamal Kapadia, and Matthias Fripp, *Putting Renewables to Work: How Many Jobs Can the Clean Energy Industry Generate?* Renewable and Appropriate Energy Laboratory, University of California, Berkeley, 2004. The authors examined two deployment configurations and found utility-scale projects to produce 7.41 jobs/MW and distributed projects 10.56 jobs/MW.

21. International Renewable Energy Agency, "Renewable Energy and Jobs: Annual Review 2018," Abu Dhabi, 2018.

22. Kammen, Kapadia, and Fripp, *Putting Renewables to Work*.

23. U.S. Department of Energy, "U.S. Energy and Jobs Report," January 2017.

24. Eva Fleiß, Stefanie Hatzl, Sebastian Seebauer, and Alfred Posch, "Money, Not Morale: The Impact of Desires and Beliefs on Private Investment in Photovoltaic Citizen Participation Initiatives," *Journal of Cleaner Production* 141 (2017): 920–27.

25. Liridon Korcaj, Ulf J.J. Hahnel, and Hans Spada, "Intentions to Adopt Photovoltaic Systems Depends on Homeowners' Expected Personal Gains and Behavior of Peers," *Renewable Energy* 75 (2015): 407–15.

26. Chelsea Schelly, "Residential Solar Electricity Adoption: What Motivates, and What Matters? A Case Study of Early Adopters," *Energy Research & Social Science* 2 (2014): 183–91.

27. Ryan Wiser, D. Millstein, T. Mai, Jordan Macknick, and A. Carpenter, "The Environmental and Public Health Benefits of Achieving High Penetrations of Solar Energy in the United States," *Energy* 113 (2016): 472–86.

28. Brain Mayer, *Blue-Green Coalitions: Fighting for Safe Workplaces and Healthy Communities* (Ithaca, NY: Cornell University Press, 2009).

29. Kirsten Jenkins, Darren McCauley, Raphael Heffron, Hannes Stephan, and Robert Rehner, "Energy Justice: A Conceptual Review," *Energy Research & Social Science* 11 (2016): 174–82.

30. Nordhaus and Shellenberger, *Breakthrough*.

31. Matthew A. Cole, "Trade, the Pollution Haven Hypothesis and the Environmental Kuznets Curve: Examining the Linkages," *Ecological Economics* 48, no. 1 (2004): 71–81.

32. Paul D. Moskowitz, L. D. Hamilton, Samuel C. Morris, K. M. Novak, and Michael D. Rowe, *Photovoltaic Energy Technologies: Health and Environmental Effects*, BNL-51284, National Center for Analysis of Energy Systems, Brookhaven National Laboratory, Upton, NY, 1980.

33. Gary Gereffi, "Commodity Chains and Regional Divisions of Labor in East Asia," *Journal of Asian Business* 12 (1996): 75–112.

34. Ariana Eunjung Cha, "Solar Energy Firms Leave Waste Behind in China," *Washington Post*, March 9, 2008, 1.

35. Ibid.

36. Ibid.

37. Hong Yang, Xiannjin Huang, and Julian R. Thompson, "Correspondence: Tackle Pollution from Solar Panels," *Nature* 509 (2014): 563.

38. Ibid.

39. Shamsiah Ali Oettinger, "Survival of the Fittest and Cleanest," *PV Magazine*, March 28, 2011.

40. *Solar Industry Magazine*, "China's New Rules for Solar Polysilicon Factories Expected to Force Consolidation," January 26, 2011.

41. Ted Smith, David Sonnenfeld, and David Pellow, *Challenging the Chip: Labor Rights and Environmental Justice in the Global Electronics Industry* (Philadelphia, PA: Temple University Press, 2006).

42. Personal communication with Seb Nardicchia, Milpitas, California, June 1, 2010.

43. Paul Landsbergis, Janet Cahill, and Peter Schnall, "The Impact of Lean Production and Related New Systems of Work Organization on Worker Health," *Journal of Occupational Health Psychology* 4 (1999): 108–30.

44. Erica L. Plambeck and Terry A Taylor, "Sell the Plant? The Impact of Contract Manufacturing on Innovation, Capacity, and Profitability," *Management Science* 51 (2005): 133–50.

45. Jenny Chan and Ngai Pun, "Suicide as Protest for the New Generation of Chinese Migrant Workers: Foxconn, Global Capital, and the State," *Asia-Pacific Journal* 37 (2010): 2–10.

46. Naomi Klein, *No Logo* (Ottawa: Knopf Press, 1999).

47. Sharon LaFraniere, "Chinese Protesters Accuse Solar Panel Plant of Pollution," *New York Times*, September 19, 2011.

48. Ibid.

49. Oliver Morton, "A New Day Dawning? Silicon Valley Sunrise," *Nature* 443, no. 7107 (2006): 19–22.

50. Silicon Valley Toxics Coalition, "Green Jobs Platform for Solar," May 2009, http://svtc.org/our-work/solar/green-jobs-platform-for-solar/.

51. Paul Anastas and John Warner, *Green Chemistry: Theory and Practice* (Oxford University Press, 1998).

52. Adapted from "Worker Rights," Occupational Safety & Health Administration, U.S. Department of Labor, OSHA 3021-11R 2016, https://www.osha.gov/Publications/osha3021.pdf.

53. Silicon Valley Toxics Coalition, *Toward a Just and Sustainable Solar Energy Industry*, San Jose, California, 2009.

54. Julian Agyeman, Robert Doyle Bullard, and Bob Evans, *Just Sustainabilities: Development in an Unequal World* (Cambridge, MA: MIT Press, 2003).

55. Paul Robbins, "Cries along the Chain of Accumulation," *Geoforum* 54 (2011): 233–35.

56. Dustin Mulvaney, "Energy and Global Production Networks," in *The Palgrave Handbook of the International Political Economy of Energy*, ed. Thijs Van de Graaf, Benjamin K. Sovacool, Arunabha Ghosh, Florian Kern, and Michael T. Klare (New York: Springer, 2016), 621–40.

57. George Marcus, "Ethnography in/of the World System: The Emergence of Multi-Sited Ethnography," *Annual Review of Anthropology* 24, no. 1 (1995): 95–117.

58. James McCarthy, "First World Political Ecology: Lessons from the Wise Use Movement," *Environment and Planning A* 34, no. 7 (2002): 1281–1302.

59. There is a lot of commodity fetish focus in agrofood systems research. See e.g. Michael K. Goodman, "Reading Fair Trade: Political Ecological Imaginary and the Moral Economy of Fair Trade Foods," *Political Geography* 23, no. 7 (2004): 891–915.

60. Arjun Appadurai, *The Social Life of Things: Commodities in Cultural Perspective* (Cambridge: Cambridge University Press, 1988).

61. Gary Gereffi and M. Korzeniewicz, *Commodity Chains and Global Capitalism* (Westport, CT: Greenwood Press, 1994), 2.

62. William H. Friedland, "Reprise on Commodity Systems Methodology," *International Journal of Sociology Agriculture and Food* 9, no. 1 (2001): 82–103.

63. Gale Raj-Reichert, "Safeguarding Labour in Distant Factories: Health and Safety Governance in an Electronics Global Production Network," *Geoforum* 44 (2013): 23–31.

64. James G. Carrier, "Protecting the Environment the Natural Way: Ethical Consumption and Commodity Fetishism," *Antipode* 42, no. 3 (2010): 672–89.

65. Doreen Massey, "Power-Geometry and a Progressive Sense of Place," in *Mapping the Futures: Local Cultures, Global Change*, ed. Jon Bird, Barry Curtis, Tim Putnam, George Robertson, and Lisa Tickner (London: Routledge, 1993), 59–69.

66. Marcus, "Ethnography in/of the World System," 107.

67. Jenny Chan, Ngai Pun, and Mark Selden, "The Politics of Global Production: Apple, Foxconn and China's New Working Class," *New Technology, Work and Employment* 28 (2013): 100–15.

68. Marcus, "Ethnography in/of the World System," 102.

69. Ibid.

70. Susanne Freidberg, "On the Trail of the Global Green Bean: Methodological Considerations in Multi-Site Ethnography," *Global Networks* 1, no. 4 (2001), 353–368.

71. Marcus, "Ethnography in/of the World System," 96.

72. Jennifer Bair, *Frontiers of Commodity Chain Research* (Palo Alto, CA: Stanford University Press, 2008).

73. Immanuel Wallerstein, "The Rise and Future Demise of the World Capitalist System: Concepts for Comparative Analysis," *Comparative Studies in Society and History* 16, no. 4 (1974): 387–415.

74. David Harvey, *The Condition of Postmodernity: An Enquiry into the Origins of Cultural Change* (New York: Wiley, 1989).

75. William R. Caraher, Bret Weber, Kostis Kourelis, and Richard Rothaus, "The North Dakota Man Camp Project: The Archaeology of Home in the Bakken Oil Fields," *Historical Archaeology* 51, no. 2 (2017): 267–87.

76. Ramachandra Guha and Joan Martínez Alier, *Varieties of Environmentalism: Essays North and South* (London: Routledge, 2013).

77. Bair, *Frontiers of Commodity Chain Research.*

78. Sheila Jasanoff, "Technologies of Humility," *Nature* 450 (2007): 33.

79. Rachel Schurman and Dennis Takahashi-Kelso, *Engineering Trouble: Biotechnology and its Discontents* (Berkeley: University of California Press, 2003).

80. Kristen Schrader-Frechette, "Methodological Rules For Four Classes of Scientific Uncertainty," in *Scientific Uncertainty and Environmental Problem Solving,* edited by Jack Lemons (Cambridge: Blackwell Science, 1996), 12–39.

81. Ian Cook, "Follow the Thing: Papaya," *Antipode* 36, no. 4 (2004): 642–64.

82. Marcus, "Ethnography in/of the World System," 100.

83. Freidberg, "On the Trail," 354.

CHAPTER 2. GREEN NEW DEAL

1. Thomas Freidman, "The Power of Green," *New York Times,* April 15, 2007.

2. Office of the White House, "A Retrospective Assessment of Clean Energy Investments in the Recovery Act," February, 2016.

3. Numerous photographs published by Reuters, the Associated Press, and Getty Images situate the president in front of solar farms and or inside photovoltaic manufacturing facilities from 2009 to 2012.

4. Office of the White House, "Retrospective Assessment," 2.

5. Declan Butler, "Thin Films: Ready for Their Close-Up?" *Nature* 454 (2008): 558–59.

6. Fraunhofer Institute for Solar Energy, "Photovoltaics Report," Freiberg, Germany, August 27, 2018.

7. The author has tracked solar developments in an Excel database since 2008.

8. Shikhar Ghosh and Ramana Nanda, "Venture Capital Investment in the Clean Energy Sector," Entrepreneurial Management Working Paper 11–020, Harvard Business School, 2010.

9. United States Energy Policy Act of 2005, Public Law 109–58, August 8, 2005.

10. Loan Programs Office, "The Financing Force behind America's Clean Energy Economy," Department of Energy, 2011, http://lpo.energy.gov/.

11. Ibid.

12. Frank Rusco, Director of Natural Resources and Environment, "DOE Loan Programs," Testimony before the Subcommittee on Oversight and Investigations, Committee on Energy and Commerce, House of Representatives, GAO-14-645T, May 30, 2014.

13. U.S. Department of Energy, "Loan Guarantee Solicitation Announcement: Federal Loan Guarantees for Projects that Employ Innovative Energy Efficiency, Renewable Energy, and Advanced Transmission and Distribution Technologies," July 29, 2009.

14. Iris Kuo, "DOE Loan Chief: Where We'll Invest in 2011," *Venture Beat,* December 15, 2010, http://venturebeat.com/2010/12/15/doe-loan-chief-where-well-invest-in-2011-how-abound-solar-could-compete-with-china/.

15. Office of the White House, "Retrospective Assessment," 22.

16. National Renewable Energy Laboratory, "Innovation: Thin-Film Manufacturing Process Gives Edge to Photovoltaic Start-Up," NREL/FS-6A4-47570, Department of Energy, June 2010.

17. Council of State Governments, *Green Jobs Report,* 2009, http://knowledgecenter.csg.org/kc/system/files/Green_Jobs_Report_0.pdf.

18. Brian Eckhouse and E. Roston, "Trump Can't Kill Solyndra Loan Office That Outperforms Banks," *Bloomberg News,* November 29, 2016.

19. Governments, private companies, and nonprofits use the Jobs and Economic Development Impact model to approximate job creation and economic impacts. D. Steinberg, G. Porro, and M. Goldberg, *Preliminary Analysis of the Jobs and Economic Impacts of Renewable Energy Projects Supported by the §1603 Treasury Grant Program,* NREL/TP-6A20-52739, National Renewable Energy Laboratory, Golden, CO, 2012.

20. France Innovation Scientifique & Transfert, *Photovoltaic Thin Film Cells: Intellectual Property Overview,* FRINNOV, Paris, 2009.

21. Henry Waxman, House Energy and Commerce Committee Hearing, November 16, 2011. Congressman Waxman was co-sponsor of the Waxman-Markey cap-and-trade bill proposals, which were leading climate policies under consideration in the U.S.

22. Steven Chu, "The Solyndra Failure: Testimony of U.S. Department of Energy Secretary Steven Chu," House of Representatives, Committee on Energy and Commerce, 112th Congress, November 17, 2011.

23. Steven Chu, "Clean Energy Jobs and American Power Act: Statement of U.S. Department of Energy Secretary Steven Chu," Senate Committee on Environment and Public Works, 111th Congress. October 27, 2009.

24. House of Representatives, "Solyndra Failure."

25. Senate Committee, "Clean Energy Jobs."

26. In my time with SVTC I spoke off the record with three different Solyndra employees who noted challenges with the manufacturing process, during discussion about the Solar Scorecard in 2009 and 2010 at the Solyndra factory in Fremont, California, and SVTC's San José headquarters, in addition to several times on the phone. Solyndra employees declined to be interviewed on the record for this book in 2010, and given the national media attention on the Solyndra case, I did not seek follow-up interviews with former employees due to concern about human subjects. A visit to a San José e-waste processing

facility revealed several dumpsters of Solyndra's CIGS glass tubes filled with a mineral oil, some with cells added, others without. All the experts I conferred with during this research after this discovery confirmed my suspicion that the company was having trouble with quality control and manufacturing a consistent product. The media would later pick up the argument put forth by the Obama administration that China's trade policy was the reason for the failure of Solyndra. Most financial experts suggest that it was probably a combination of competition from China and manufacturing problems.

27. *Global Trade Atlas*, 2011.

28. Keith Bradsher, "Glut of Solar Panels Poses a New Threat to China," *New York Times*, October 4, 2012.

29. Kearney Alliance, "China's Solar Industry and the U.S. Anti-Dumping/Anti-Subsidy Trade Case," ChinaTrade.com, May 2012.

30. Jocelyn Durkey, "State Renewable Portfolio Standards and Goals," National Conference of State Legislatures, December 16, 2016.

31. Joseph Schumpeter, *Capitalism, Socialism, and Democracy* (New York: Harper and Brothers, 1942); see also Hermann Scheer, *The Solar Economy: Renewable Energy for a Sustainable Global Future* (New York: Routledge, 2013), 277.

32. Richard Foster and Sarah Kaplan, *Creative Destruction: Why Companies That Are Built to Last Underperform the Market—And How to Successfully Transform Them* (New York: Crown Business, 2011).

33. John E. Elliott, "Marx and Schumpeter on Capitalism's Creative Destruction: A Comparative Restatement," *Quarterly Journal of Economics* 95, no. 1 (1980): 45–68.

34. Gregory Unruh, "Understanding Carbon Lock-in," *Energy Policy* 28 (2000): 817–30.

35. North Carolina Clean Energy Technology Center, "DSIRE: Database of State Incentives for Renewables & Efficiency," http://www.dsireusa.org/.

36. Margaret Taylor, "Beyond Technology-Push and Demand-Pull: Lessons from California's Solar Policy," *Energy Economics* 30 (2008): 2829–54.

37. The ITC is slated to decline to lower percentages and eventually sunset for residential projects in 2022.

38. U.S. Government Accountability Office, "Electricity Generation Projects: Additional Data Could Improve Understanding of the Effectiveness of Tax Expenditures," GAO-15-302, April 2015, https://www.gao.gov/assets/670/669881.pdf.

39. Michael Weiss, "Everybody Loves Solar Energy, but . . ." *New York Times*, September 24, 1989.

40. California Energy Commission and California Public Utilities Commission, "History of Solar Energy in California," 2013, *Go Solar California*, http://www.gosolarcalifornia.ca.gov/about/gosolar/california.php.

41. Isaac Asimov, "Reason," in *I, Robot* (New York: Street and Smith, 1941), 319–36.

42. California Public Utilities Commission, *Renewable Energy Portfolio Standard: Annual Report*, November 2017.

43. Darrell Issa, of the 49th District of California in the U.S. House of Representatives, was the ranking member of the House Committee on Oversight and Government Reform, and often used the phrase "too big to fail" to describe solar projects, drawing on the metaphor used to describe the large investment banks bailed out through the Troubled Asset Relief Program a few years prior.

44. Interview (anonymous), Department of Energy Loan Guarantee Office.

45. Seema Mehta, "Romney Accuses Obama of 'Crony Capitalism' in Solyndra Trip," *Los Angeles Times,* May 31, 2012.

46. Frank Rusco, "DOE Loan Programs," Testimony by the Director of Natural Resources and Environment before the Subcommittee on Oversight and Investigations, Committee on Energy and Commerce, House of Representatives, GAO-14-645T, May 30, 2014; "DOE Loan Programs: Information on Implementation of GAO Recommendations and Program Costs," GAO-16-150T, Government Accountability Office, 2016.

47. Ibid.

48. Rusco, "DOE Loan Programs."

49. Ibid.

50. Department of Energy, "Loan Guarantee Solicitation," 52.

51. Timothy Luke, "A Green New Deal: Why Green, How New, and What Is the Deal?" *Critical Policy Studies* 3 no. 1 (2009): 14–28.

52. Bill Lee and Ed LiPuma, *Financial Derivatives and the Globalization of Risk* (Durham, NC: Duke University Press, 2004).

CHAPTER 3. INNOVATIONS IN PHOTOVOLTAICS

1. Quoted in John Perlin, *From Space to Earth: The Story of Solar Electricity* (Cambridge: Harvard University Press, 1999), 11.

2. Vasilis M. Fthenakis, Hyung Chul Kim, and Erik Alsema, "Emissions from Photovoltaic Life Cycles," *Environmental Science and Technology* 42 (2008): 2168–74.

3. Ibid., 2169.

4. Langdon Winner, "Upon Opening the Black Box and Finding It Empty: Social Constructivism and the Philosophy of Technology," *Science, Technology, & Human Values* 18, no. 3 (1993): 362–78.

5. Langdon Winner, *The Whale and the Reactor: A Search for Limits in an Age of High Technology* (Chicago, IL: University of Chicago Press, 1986).

6. Trevor Pinch and Wiebe Bijker, "The Social Construction of Facts and Artefacts: Or How the Sociology of Science and the Sociology of Technology Might Benefit Each Other," *Social Studies of Science* 14 (1984): 399–441.

7. Lewis Mumford, *Technics and Civilization* (Chicago, IL: University of Chicago Press, 2010).

8. Michel Callon, "Society in the Making: The Study of Technology as a Tool for Sociological Analysis," in *The Social Construction of Technological Systems: New Directions in the Sociology and History of Technology,* ed. Wiebe E. Bijker, Thomas Parke Hughes, and Trevor J. Pinch (Cambridge, MA: MIT Press, 1987), 83–103.

9. Ken Zweibel, "Thin Film Photovoltaics," Solar Energy Research Institute [now the National Renewable Energy Laboratory], Golden, CO, 1989.

10. Vasilis M. Fthenakis, "Sustainability of Photovoltaics: The Case for Thin-Film Solar Cells," *Renewable and Sustainable Energy Reviews* 13, no. 9 (2009): 2746–50.

11. Robert Margolis, R. Mitchell, and Ken Zweibel. "Lessons Learned from the Photovoltaic Manufacturing, Technology/Photovoltaic Manufacturing R&D and Thin-Film Photovoltaics Partnership Projects," Technical Report NREL/TP-520–39780, National Renewable Energy Laboratory, Golden, CO, 2006.

12. National Renewable Energy Laboratory, "Thin-Film Partnership," https://www.nrel.gov/pv/thin-film-partnership.html.

13. First Solar's 2017 annual report suggests over $10 billion from 2015 through 2017 alone: http://investor.firstsolar.com/financial-information.

14. Fthenakis, "Sustainability of Photovoltaics," 2748.

15. First Solar, "First Solar Announces Acquisition of Turner Renewable Energy," November 30, 2007, http://investor.firstsolar.com/news-releases/news-release-details/first-solar-announces-acquisition-turner-renewable-energy.

16. Hartmut Steinberger, "Health, Safety and Environmental Risks from the Operation of CdTe and CIS Thin-Film Modules," *Progress in Photovoltaics: Research and Applications* 6 (1998): 99–103.

17. Vasilis. M. Fthenakis, S. C. Morris, Paul D. Moskowitz, and D. L. Morgan. "Toxicity of Cadmium Telluride, Copper Indium Diselenide, and Copper Gallium Diselenide," *Progress in Photovoltaics: Research and Applications* 7, no. 6 (1999): 489–97. Daniel L. Morgan, Cassandra J. Shines, Shawn P. Jeter, Ralph E. Wilson, Michael P. Elwell, Herman C. Price, and Paul D. Moskowitz, "Acute Pulmonary Toxicity of Copper Gallium Diselenide, Copper Indium Diselenide, and Cadmium Telluride Intratracheally Instilled into Rats," *Environmental Research* 71, no. 1 (1995): 16–24.

18. Gunnar Nordberg, Robert Franciscus, Martin Herber, and Lorenzo Alessio, *Cadmium in the Human Environment: Toxicity and Carcinogenicity*, No. 118, International Agency for Research on Cancer, Lyon, 1992.

19. M. Kasuya, H. Teranishi, K. Aoshima, T. Katoh, H. Horiguchi, Y. Morikawa, M. Nishijo, and K. Iwata, "Water Pollution by Cadmium and the Onset of Itai-itai Disease," *Water Science and Technology* 25, no. 11 (1992): 149–56.

20. Mariska J. de Wild-Scholten, Karsten Wambach, Eric A. Alsema, and A. Jäger-Waldau, "Implications of European Environmental Legislation for Photovoltaic Systems," 20th European Photovoltaic Energy Conference, Barcelona, June 6–10, 2004.

21. Vasilis M. Fthenakis and Ken Zweibel, "CdTe PV: Real and Perceived EHS Risks," National Center for PV and Solar Program, Washington, DC, 2003.

22. Parikhit Sinha, Christopher J. Kriegner, William A. Schew, Swiatoslav W Kaczmar, Matthew Traister, and David J. Wilson, "Regulatory Policy Governing Cadmium-Telluride Photovoltaics: A Case Study Contrasting Life Cycle Management with the Precautionary Principle," *Energy Policy* 36 (2008): 381–87.

23. Mark Anthony, "Cadmium Telluride Casts Shadow on First Solar, *Seeking Alpha*, November 7, 2007, https://seekingalpha.com/article/55392-cadmium-telluride-casts-shadow-on-first-solar.

24. First Solar, U.S. Securities and Exchange Commission Filing, Form S-1, July 19, 2007.

25. Vasilis M. Fthenakis and Paul Moskowitz, "Thin-Film Photovoltaic Cells: Health and Environmental Issues in Their Manufacture, Use, and Disposal," *Progress in Photovoltaics* 3, no. 5 (1995): 295–306.

26. Vasilis M. Fthenakis, "Hazard Analysis for the Protection of Photovoltaic Manufacturing Facilities," in *Proceedings of 3rd World Conference on Photovoltaic Energy Conversion*, vol. 2 (IEEE, 2003), 2090–93.

27. Daniel L. Morgan, Cassandra J. Shines, Shawn P. Jeter, Mark E. Blazka, Michael R. Elwell, Ralph E. Wilson, Sandra M. Ward, Herman C. Price, and Paul D. Moskowitz, "Comparative Pulmonary Absorption, Distribution, and Toxicity of Copper Gallium Diselenide, Copper Indium Diselenide, and Cadmium Telluride in Sprague-Dawley Rats." *Toxicology and Applied Pharmacology* 147, no. 2 (1997): 399–410.

28. Vasilis M. Fthenakis, M. Fuhrmann, J. Heiser, A. Lanzirotti, J. Fitts, and W. Wang, "Emissions and Encapsulation of Cadmium in CdTe Photovoltaic Modules during Fires," *Progress in Photovoltaics: Research and Applications* 13, no. 8 (2005): 713–23.

29. Chao Zeng, Adriana Ramos-Ruiz, Jim A. Field, and Reyes Sierra-Alvarez. "Cadmium Telluride (CdTe) and Cadmium Selenide (CdSe) Leaching Behavior and Surface Chemistry in Response to pH and O_2," *Journal of Environmental Management* 154 (2015): 78–85.

30. William D. Cyrs, Heather J. Avens, Zachary A. Capshaw, Robert A. Kingsbury, Jennifer Sahmel, and Brooke E. Tvermoes, "Landfill Waste and Recycling: Use of a Screening-Level Risk Assessment Tool for End-of-Life Cadmium Telluride (CdTe) Thin-Film Photovoltaic (PV) Panels," *Energy Policy* 68 (2014): 524–33.

31. Surawut Chuangchote, Manaskorn Rachakornkij, Thantip Punmatharith, Chanathip Pharino, Chulalak Changul, Prapat Pongkiatkul, "Review of Environmental, Health and Safety of CdTe Photovoltaic Installations throughout Their Life-Cycle," Document No. SD-20, First Solar, Tempe, AZ, 2012.

32. T. Brouwers, "Bio-elution Test on Cadmium Telluride," ECTX Consultant, Liège, Belgium, 2010, cited in Swiatoslav Kaczmar, Evaluating the Read-Across Approach on Cdte Toxicity for CdTe photovoltaics," 32nd Annual Meeting, Society of Environmental Toxicology and Chemistry North America, November 2011, Boston, MA.

33. European Chemicals Agency, "Cadmium Telluride," 2017, https://echa.europa.eu/registration-dossier/-/registered-dossier/12227.

34. Chris Eberspacher, Charles F. Gay, and Paul D. Moskowitz, "Strategies for Enhancing the Commercial Viability of CdTe-Based Photovoltaics," *Solar Energy Materials and Solar Cells* 41/42 (1996): 637–53.

35. Jasmina Lovrić, Sung Ju Cho, Francoise Winnik, and Dusica Maysinger, "Unmodified Cadmium Telluride Quantum Dots Induce Reactive Oxygen Species Formation Leading to Multiple Organelle Damage and Cell Death," *Chemistry & Biology* 12, no. 11 (2005): 1227–34; Raphaël Schneider, Cécile Wolpert, Hélène Guilloteau, Lavinia Balan, Jacques Lambert, and Christophe Merlin, "The Exposure of Bacteria to CdTe-Core Quantum Dots: The Importance of

Surface Chemistry on Cytotoxicity," *Nanotechnology* 20, no. 22 (2009), 225101.

36. A. Ades and G. Kazantzis, "Lung Cancer in a Non-Ferrous Smelter: The Role of Cadmium," *British Journal of Industrial Medicine* 45 (1988): 435–42.

37. Vasilis M. Fthenakis, Hyung Chul Kim, and Wenming Wang, "Life Cycle Inventory Analysis of the Production of Metals used in Photovoltaics," BNL-77919, Brookhaven National Laboratory, Upton, NY, 2007.

38. Data from a presentation by Lisa Krueger at First Solar's Frankfurt Oder manufacturing facility, October 10, 2009.

39. Parikhit Sinha, Michael Fischman, Jim Campbell, Gaik Cheng Lee, and Lein Sim Lim, "Biomonitoring of CdTe PV Manufacturing and Recycling Workers," *43rd IEEE Photovoltaic Specialists Conference*, 2016, 3587–92.

40. U.S. Occupational Safety and Health Administration, "Cadmium: Toxic and Hazardous Substances Regulations, Part 1910.1027," 2016.

41. Non-Toxic Solar Alliance, "Position Paper on the Revision of the RoHS Directive," Brussels, Belgium, March 31, 2010.

42. Quoted in Veronica Webster, "Fighting for Clean Solar Energy in Europe," Bellona, May 4, 2010, http://bellona.org/news/renewable-energy/solar/2010-05-fighting-for-clean-solar-energy-in-europe.

43. *PV Magazine*, "Cadmium Won't Be Banned under RoHS, as Lobbyists Battle it Out," November 24, 2010.

44. Dominik Sollman and Christoph Podewils, "How Dangerous is Cadmium Telluride?" *Photon International,* September 2009.

45. Personal communication from a representative of SolarWorld Industries, Freiburg, Germany, September 12, 2009.

46. Dominik Sollman and Christoph Podewils, "How Dangerous is Cadmium Telluride?" *Photon Magazine,* September 2009.

47. *PV Magazine,* "Professor Green: I Would Like to See a Date Fixed for RoHS-Compliance," September 14, 2010.

48. Personal communication from a representative of SolarWorld Industries, Freiburg, Germany, September 12, 2009.

49. Eric Hafter, senior vice president of the Solar Group at Sharp, made this comment in a public talk at Solar Power International in Los Angeles in 2010. The same comment was made to *Greentech Media* in a 2010 interview.

50. Author interview with SolarWorld's Karsten Wambach, July 14, 2010.

51. Norwegian Geotechnical Institute, "Environmental Risks Regarding the Use and End-of-Life Disposal of CdTe PV Modules," Trondheim, Norway, April 16, 2010.

52. Ibid, 4.

53. Personal communication with Lisa Krueger, 2009, on a site visit to the manufacturing plant in Frankfurt Oder, Germany, September 1, 2009; personal communication with Andreas Wade of Calyxo at Q-Cells in Solar Valley, Germany, September 3, 2009.

54. M. Held, "Life Cycle Assessment of CdTe Module Recycling," 24th EU PVSEC Conference, Hamburg, Germany, 2009.

55. Personal communication with Andreas Wade, Calyxo headquarters, Germany, October 4, 2009.

56. Vasilis M. Fthenakis, Wenming Wang, and Hyung Kim, "Life Cycle Inventory Analysis of the Production of Metals Used in Photovoltaics," *Renewable and Sustainable Energy Reviews* 13 (2009): 493–517.

57. Yuangen Yang, Feili Li, Xiangyang Bi, Li Sun, Taoze Liu, Zhisheng Jin, and Congqiang Liu, "Lead, Zinc, and Cadmium in Vegetable Crops in a Zinc Smelting Region and Its Potential Human Toxicity," *Bulletin of Environmental Contamination and Toxicology* 87 (2011): 586–90.

58. This quote is from a Calyxo advertisement in 2008 that described the efforts to sustainably manage material flows.

59. Ibid.

60. Participant observation, research presentation, Solar Power International, October 10, 2009, Los Angeles, California, 2008.

61. Participant observation, Thin Film Solar Conference, San Francisco, California, April 28–29, 2009.

62. The site is no longer available.

63. Ken Zweibel, phone interview, June 14, 2009.

64. Marco Raugei and Vasilis M. Fthenakis, "Cadmium Flows and Emissions from CdTe PV: Future Expectations," *Energy Policy* 38 (2010): 5223–28.

65. Public talk by a representative from Solar Frontier, PARC, Palo Alto, California, August 30, 2010.

66. Interview with a representative from Avancis, München, Germany, September 8, 2009.

67. SEMATECH, "Silane Safety Improvement Project: S71 Final Report," Albany, NY, 1994.

68. *Forbes,* "Five Technologies Set to Change the Decade," January 7, 2009.

69. Keith Bradsher, "Solar Rises in Malaysia during Trade Wars over Panels," *New York Times,* December 14, 2017.

70. Mark Jaffe, "Colorado Orders Abound Solar to Clean Up Hazardous Waste at Four Sites," *Denver Post,* February 25, 2013.

71. Interview with Mark Chen, Abound Solar, Solar Power International, Los Angeles, September 10, 2009.

72. U.S. Geological Survey, "Tellurium: The Bright Future of Solar Energy," Fact Sheet 2014–3077, USGS Mineral Sources Program, April 2015. See also U.S. Geological Survey, "Tellurium," Mineral Commodity Summaries, January 2017, https://minerals.usgs.gov/minerals/pubs/commodity/selenium/mcs-2017-tellu.pdf.

73. Jason Dearen, "Solar Industry Grapples with Hazardous Wastes," Associated Press, February 10, 2013.

74. Steven Chu, "The Solyndra Failure: Testimony of U.S. Department of Energy Secretary Steven Chu," House of Representatives, Committee on Energy and Commerce, 112th Congress, November 17, 2011.

75. *CBS News,* "Solyndra Not Dealing with Toxic Waste at Milpitas Facility," April 28, 2012.

76. Dearen, "Solar Industry Grapples with Hazardous Wastes."

77. National Renewable Energy Center and Fundación Chile, "First Solar CdTe Photovoltaic Technology: Environmental, Health and Safety Assessment," October 2013.

78. Vasilis M. Fthenakis, "Could CdTe PV Modules Pollute the Environment?" National Photovoltaic Environmental Health and Safety Assistance Center, Brookhaven National Laboratory, Upton, NY, 1999.

79. Fthenakis, Kim, and Alsema, "Emissions from Photovoltaic Life Cycles," 2168.

80. Quoted in Vasilis M. Fthenakis and Ken Zweibel, "CdTe PV: Real and Perceived EHS Risks," National Center for PV and Solar Program, Washington, DC, 2003, 2.

81. Fthenakis, Kim, and Alsema, "Emissions from Photovoltaic Life Cycles."

82. International Energy Agency, "Methodology Guidelines on Life Cycle Assessment of Photovoltaic Electricity," Report IEA-PVPS, Paris, 2015.

83. Garvin A. Heath and Margaret K. Mann, "Background and Reflections on the Life Cycle Assessment Harmonization Project," *Journal of Industrial Ecology* 16, no. s1 (2012): S8–S11.

84. Theodore Porter, *Trust in Numbers: The Pursuit of Objectivity in Science and Public Life* (Princeton, NJ: Princeton University Press, 1995).

85. Wendy Espeland and Mitchell Stevens, "Commensuration as a Social Process," *Annual Review of Sociology* 24 (1998): 314.

86. Malcolm Slesser and Ian Hounam, "Solar Energy Breeders," *Nature* 262 (1976): 244–45.

87. Vasilis M. Fthenakis and Hyung Chul Kim, "Photovoltaics: Life-Cycle Analyses," *Solar Energy* 85, no. 8 (2011): 1609–28.

88. Michael Power, *The Audit Society: Rituals of Verification* (Oxford: Oxford University Press, 1997).

89. David N. Pellow, "Environmental Inequality Formation toward a Theory of Environmental Injustice," *American Behavioral Scientist* 43, no. 4 (2000), 590.

90. John Ehrenfeld, "Industrial Ecology: Paradigm Shift or Normal Science?" *American Behavioral Scientist* 44 (2000): 229–44.

91. Fthenakis, Kim, and Alsema, "Emissions from Photovoltaic Life Cycles," 2168.

92. Fthenakis, "Could CdTe PV Modules Pollute the Environment?"

93. Ken Zweibel and Vasilis Fthenakis, "Cadmium Facts and Handy Comparisons," National Renewable Energy Laboratory, Golden, CO, 2006.

94. Vasilis Fthenakis and Hyung Chul Kim, "Cu(InGa)Se$_2$ Thin-film Solar Cells: Comparative Life Cycle Analysis of Buffer Layers," 22nd European Photovoltaic Solar Energy Conference, Milan, September 3–7, 2007.

95. Dustin Mulvaney, "Are Green Jobs Just Jobs? Cadmium Narratives in the Life Cycle of Photovoltaics," *Geoforum* 54 (2014): 178–86.

96. U.S. Department of Labor and Occupational Safety and Health Administration, "Cadmium," OSHA Report 3136–06R, 2004.

97. Ibid.

98. Marianne Sullivan, *Tainted Earth: Smelters, Public Health, and the Environment* (New Brunswick, NJ: Rutgers University Press, 2014).

99. Porter, *Trust in Numbers*.

100. Alex Preda, "Socio-Technical Agency in Financial Markets: The Case of the Stock Ticker," *Social Studies of Science* 36 (2006): 753–82.

101. Alex Preda, "STS and Social Studies of Finance," in *The Handbook of Science and Technology Studies*, ed. E. Hackett, O. Amsterdamska, M. Lynch, and J. Wajcman (Cambridge, MA: MIT Press, 2007).

102. Donald Mackenzie, "Making Things the Same: Gases, Emission Rights and the Politics Of Carbon Markets," *Accounting, Organizations, Society* 34 (2009): 440–55.

103. Winner, *The Whale and the Reactor*.

104. John R. Bohland and Ken Smigielski, "First Solar's CdTe Module Manufacturing Experience: Environmental Health and Safety Results," *Conference Record of the Twenty-Eighth IEEE Photovoltaic Specialists Conference*, September 15–22, 2000.

105. Deborah A. Stone, *Policy Paradox: The Art of Political Decision Making* (New York: W. W. Norton, 1997).

106. Nikolas Rose, *Powers of Freedom: Reframing Political Thought* (Cambridge: Cambridge University Press, 1999), 199.

107. Goeffrey C. Bowker and Susan Leigh Star, *Sorting Things Out: Classification and Its Consequences* (Cambridge, MA: MIT Press, 2005).

108. Vasilis M. Fthenakis, "Prevention and Control of Accidental Releases of Hazardous Materials in PV Facilities," *Progress in Photovoltaics* 6 (1998): 91–98.

109. Colin Bailie and Michael McGehee, "High-efficiency Tandem Perovskite Solar Cells," *MRS Bulletin* 40 (2015): 681–85.

CHAPTER 4. RECYCLING AND PRODUCT STEWARDSHIP

1. Interview with Karsten Wambach at Sunicon/SolarWorld Industries, Freiburg, Germany, September 2009.

2. International Renewable Energy Association, "End-of-Life Management: Solar Photovoltaic Panels," Abu Dhabi, 2016.

3. Energy Information Agency, "Annual Energy Outlook 2014," Department of Energy, 2014.

4. Nancy M. Haegel, Robert Margolis, Tonio Buonassisi, David Feldman, Armin Froitzheim, Raffi Garabedian, and Martin Green, "Terawatt-Scale Photovoltaics: Trajectories and Challenges," *Science* 356, no. 6334 (2017): 141–43.

5. International Renewable Energy Association, *End-of-Life Management: Solar Photovoltaic Panels,* Abu Dhabi, 2016.

6. Björn A. Andersson and Staffan Jacobsson, "Monitoring and Assessing Technology Choice: The Case of Solar Cells," *Energy Policy* 28, no. 14 (2000): 1037–49.

7. Nicola Jones, "A Scarcity of Rare Metals is Hindering Green Technologies," *Yale Environment 360*, November 18, 2013.

8. U.S. Department of Energy, "SunShot Vision Study," Office of Energy Efficiency and Renewable Energy, February 2012, https://www1.eere.energy .gov/solar/pdfs/47927.pdf.

9. Valerie B. Grasso, "Rare Earth Elements in National Defense: Background, Oversight Issues, and Options For Congress," Congressional Research Service, Library of Congress, Washington, DC, 2013.

10. Ruth Michaelson, "The PV Race to Replace Silver," *PV Magazine,* August 15, 2011.

11. Jonathan Silver, "Department of Energy Talks Department of Energy: An Insider's Look Back at the Loan-Guarantee Program," *Solar Industry* 4, no. 6 (2011): 1, 12–15.

12. Henry E. Hilliard, "Silver Recycling in the United States in 2000," Circular 1196-N, U.S. Geological Survey, 2003.

13. Anja Müller, Karsten Wambach, and Eric Alsema, "Life Cycle Analysis of Solar Module Recycling Process," Symposium Proceedings, Materials Research Society, January 2006, 895.

14. Parikhit Sinha, Michael Fischman, Jim Campbell, Gaik Cheng Lee, and Lein Sim Lim, "Biomonitoring of CdTe PV Manufacturing and Recycling Workers," *43rd IEEE Photovoltaic Specialists Conference,* 2016, 3587–92.

15. Vasilis M. Fthenakis and W. Wang, "Extraction and Separation of Cd and Te from Cadmium Telluride Photovoltaic Manufacturing Scrap," *Progress in Photovoltaics: Research and Applications* 14, no. 4 (2006): 363–71.

16. Mike Redlinger, Alan Goodrich, Michael Woodhouse, Martin Lokanc, and Roderick Eggert, "The Present, Mid-Term, and Long-Term Supply Curves for Tellurium; and Updates in the Results from NREL's CdTe PV Module Manufacturing Cost Model," NREL Technical Report, 2013.

17. David Phillips, "How First Solar's Tellurium Deal Shows the Fragile Economics of Solar Panels," CBS News, November 29, 2010.

18. Redlinger, et al., "Present, Mid-Term, and Long-Term Supply Curves."

19. Oliver Ristau, "First Solar Invests in Mexican Tellurium Mine," *PV Magazine,* October 19, 2011.

20. R.R. Gay, "Status and Prospects for CIS-Based Photovoltaic," *Solar Energy Materials and Solar Cells* 47, nos. 1–4 (1997): 19–26.

21. Kristin J. Cummings, Walter E. Donat, David B. Ettensohn, Victor L. Roggli, Peter Ingram, and Kathleen Kreiss, "Pulmonary Alveolar Proteinosis in Workers at an Indium Processing Facility," *American Journal of Respiratory and Critical Care Medicine* 181, no. 5 (2010): 458–64.

22. Redlinger, et al., "Present, Mid-Term, and Long-Term Supply Curves," 15.

23. U.S. Geological Survey, "Platinum-Group Metals: Statistics and Information," 2013.

24. Cyrus A. Wadia, Paul Alivisatos, and Daniel M. Kammen, "Materials Availability Expands the Opportunity for Large-scale Photovoltaics Deployment," *Environmental Science & Technology* 43, no. 6 (2009): 2072–77.

25. SB 1020, Recycling: Hazardous Waste: Photovoltaic Panels: Collection and Recycling Programs, Introduced by Senator Monning, February 14, 2014, http://leginfo.legislature.ca.gov/faces/billNavClient.xhtml?bill_id=201320140SB1020.

26. Jasmina Lovrić, Sung Jo Cho, Francoise Winnik, and Dusica Maysinger, "Unmodified Cadmium Telluride Quantum Dots Induce Reactive Oxygen Species Formation Leading to Multiple Organelle Damage and Cell Death," *Chemical Biology* 12, no. 11 (2005): 1227–34.

27. Vasilis M. Fthenakis, S.C. Morris, Paul D. Moskowitz, and Daniel L. Morgan, "Toxicity of Cadmium Telluride, Copper Indium Diselenide, and Cop-

per Gallium Diselenide," *Progress in Photovoltaics: Research and Applications* 7, no. 6 (1999): 489–97.

28. Vasilis M. Fthenakis, "Life Cycle Impact Analysis of Cadmium in CdTe PV Production," *Renewable and Sustainable Energy Reviews* 8, no. 4 (2004): 303–34.

29. Adriana Ramos-Ruiz, Jean V. Wilkening, James A. Field, and Reyes Sierra-Alvarez, "Leaching of Cadmium and Tellurium from Cadmium Telluride (Cdte) Thin-Film Solar Panels under Simulated Landfill Conditions," *Journal of Hazardous Materials* 336 (2017): 57–64.

30. Hartmut Steinberger, "Health, Safety and Environmental Risks from the Operation of CdTe and CIS Thin-film Modules," *Progress in Photovoltaics* 6 (1998): 99–103.

31. Norwegian Geotechnical Institute, "Environmental Risks Regarding the Use and End-of-Life Disposal of CdTe PV Modules," Trondheim, Norway, April 16, 2010.

32. Vasilis M. Fthenakis, "End-of-Life Management and Recycling of PV Modules," *Energy Policy* 28, no. 14 (2000): 1051–58.

33. California Department of Toxic Substances Control, "Laws, Regulations, and Policies – Regulations," https://www.dtsc.ca.gov/LawsRegsPolicies /Regs/.

34. Andrew Michler, "Sonnenschiff: Solar City Produces 4x the Energy it Consumes," *Inhabit*, July 27, 2011, https://inhabitat.com/sonnenschiff-solar-city-produces-4x-the-energy-it-needs.

35. Joseph McCabe, "First Solar's Stewardship of Recycled CdTe Modules in Question," *GreenTech Media*, August 27, 2013, https://www.greentechmedia .com/articles/read/first-solars-stewardship-of-recycled-cdte-modules-in-question.

36. Jones, "Scarcity of Rare Metals."

37. European Commission, "Directive 2002/96/EC of the European Parliament and of the Council of 27 January 2003 on Waste Electrical & Electronic Equipment (WEEE)," January 27, 2003, http://eur-lex.europa.eu/legal-content /EN/TXT/?uri=CELEX:32017R0699.

38. This was according to a press release on the association PV Cycle's website (http;//www.pvcycle.org, accessed April 2, 2012), but is no longer posted.

39. European Commission, "Study on Photovoltaic Panels Supplementing the Impact Assessment for a Recast of the WEEE Directive," Final Report, April 14, 2011.

40. This was in the press released cited earlier.

41. Ibid.

42. Under European law, energy recovery from polymer incineration counts toward the recycled amount.

43. Based on data collected from the SVTC Solar Scorecard, http://www .solarscorecard.com.

44. Vasilis M. Fthenakis, "PV Recycling in the U.S.," European PV Solar Energy Conference and Exhibition and International Energy Agency Workshop on PV Life Management and Recycling, Amsterdam, the Netherlands, September 23, 2014.

45. SB 1020.

46. Edward Woodhouse and Steve Breyman, "Green Chemistry as Social Movement?" *Science, Technology, & Human Values* 30, no. 2 (2005): 199–222.

47. Jeannette M. Kadro and Anders Hagfeldt, "The End-of-Life of Perovskite PV," *Joule* 1, no. 1 (2017): 29–46.

CHAPTER 5. GREEN CIVIL WAR

1. Quoted in Daniel Stone, "Feinstein and the Mojave vs. Solar Power," *Newsweek*, January 12, 2010.

2. Michael Pasqualetti, "Wind Energy Landscapes: Society and Technology in the California Desert," *Society and Natural Resources* 14 (2001): 689–99.

3. Peter Alagona, *After the Grizzly: Endangered Species and the Politics of Place in California* (Berkeley: University of California Press, 2013).

4. U.S. Bureau of Land Management, "California Public Lands 2015, National System of Public Lands," Sacramento, 2015.

5. James Skillen, *The Nation's Largest Landlord: The Bureau of Land Management in the American West* (Lawrence: University Press of Kansas, 2009).

6. Cary Funk and Brian Kennedy, "The Politics of Climate," Pew Research Center, October 4, 2016, Washington, DC.

7. Juliet E. Carlisle, Stephanie L. Kane, David Solan, Madelaine Bowman, and Jeffrey C. Joe, "Public Attitudes Regarding Large-scale Solar Energy Development in the US," *Renewable and Sustainable Energy Reviews* no. 48 (2015): 835–47.

8. Christian Hunold and Steven Leitner, "'Hasta la Vista, Baby!' The Solar Grand Plan, Environmentalism, and Social Constructions of the Mojave Desert," *Environmental Politics* 20, no. 5 (2011): 687–704.

9. Juliet E. Carlisle, David Solan, Stephanie Kane, and Jeffrey Joe, "Utility-Scale Solar and Public Attitudes toward Siting: A Critical Examination of Proximity," *Land Use Policy* 58 (2016): 491–501.

10. Frank Wheat, *California Desert Miracle: The Fight for Desert Parks and Wilderness* (El Cajon, CA: Sunbelt, 1999).

11. "Green Civil War: Projects vs Preservation," *New York Times,* January 12, 2010.

12. Todd Woody, "It's Green against Green in Mojave Desert Solar Battle," *Yale Environment 360*, February 1, 2010.

13. U.S. Energy Policy Act of 2005, Public Law 109–58, August 8, 2005.

14. Bruce Pendery, "BLM's Retained Rights: How Requiring Environmental Protection Fulfills Oil and Gas Lease Obligations," *Environmental Law* 40 (2010): 599–685.

15. Jennifer F. Massouh, George D. Cannon, Suzanne M. Logan, and David L. Schwartz, "Real Promise or False Hope: DOE's Title XVII Loan Guarantee," *Electricity Journal* 22, no. 4 (2009): 53–67.

16. U.S. Department of the Interior, Order No. 3285, Renewable Energy Development by the Department of the Interior, March 11, 2009, https://www.blm.gov/or/energy/opportunity/files/order_3285.pdf.

17. Nate Seltenrich, "Oakland Invades the Desert," *East Bay Express,* December 8, 2010.

18. U.S. Federal Lands Management and Policy Act of 1976, §501(a)(4).

19. U.S. Department of the Interior, "Solar Energy Interim Rental Policy," Instruction Memorandum No. 2010–141, June 10, 2010, https://www.blm.gov /policy/im-2010-141.

20. *Associated Press,* "Green Light for Solar Power on Federal Lands: Interior Chief Sets Goal of 13 Commercial Projects by 2010 across West," June 30, 2009.

21. Janine Blaeloch, "Big Solar's Footprint on Public Lands," *Solar Done Right Newsletter,* December 12, 2011.

22. Ibid., 1.

23. Author's database, completed by tracking ROW status updates from the BLM.

24. Personal communication, Michael Picker, senior adviser to the governor for renewable energy facilities, Arrowhead Lake, California, October 15, 2011. Picker served under both Republican governor Arnold Schwarzenegger and Democratic governor Jerry Brown.

25. Until 2016, when a larger project opened in China, the largest in the world was the Topaz Solar Farm, a project that once sought, and was offered, but ultimately did not collect ARRA support.

26. U.S. National Environmental Policy Act of 1969, §102(C).

27. Council on Environmental Quality, Regulations for the implementation of the National Environmental Policy Act, 2005, §1502.1.

28. California Energy Commission, "American Reinvestment & Recovery Act (AARA) Solar Projects Progress at the Energy Commission," September 29, 2010.

29. Gibson Dunn, "Environmental Compliance Review Required for Projects Funded under the American Recovery and Reinvestment Act of 2009," Client Alert, May 7, 2009, https://www.gibsondunn.com/environmental-compliance-review-required-for-projects-funded-under-the-american-recovery-and-reinvestment-act-of-2009/.

30. Lacey Babcock, "Expedited Project Deadlines & NEPA Compliance: The Problematic Use of Categorical Exemptions," School of Design, Department of City and Regional Planning, University of Pennsylvania, May 7, 2010.

31. Anonymous, author conversation with BLM staff responsible for preparing a solar energy zone, March 2, 2010. Solar Energy Zones are discussed in more detail in the next chapter.

32. Karoun Demijian, "Facts that Disprove Conspiracy Theory about Harry Reid, Cliven Bundy," *Las Vegas Sun,* April 17, 2014.

33. Executive Order 13212, Actions to Expedite Energy-Related Projects, *Federal Register* 28357, May 18, 2001.

34. Ibid., 769.

35. U.S. Department of the Interior, "Salazar Approves Sixth and Largest Solar Project Ever on Public Lands" (press release), October 25, 2010.

36. Ibid.

37. Office of the White House, "A Retrospective Assessment of Clean Energy Investments in the Recovery Act," February, 2016, 31.

38. Ibid.

39. Todd Woody, "Solar Millennium Files for Bankruptcy as Solar Shakeout Continues," *Forbes,* December 21, 2011.

40. U.S. Bureau of Land Management, "Fast-Track Renewable Energy Projects", http://www.blm.gov/wo/st/en/prog/energy/renewable_energy/fast-track_renewable .html, accessed July 1, 2015.

41. Memorandum of Understanding between the Department of the Interior and the State of California on Renewable Energy, http://www.drecp.org /documents/docs/California_and_Dept_of_Interior_MOU_1-13-12.pdf.

42. David Myers and April Sall, interview, May 28, 2010.

43. Ibid.

44. Brett Prior, "Tessera / Stirling Sell Their Other Major Dish Project to PV Developer," *Greentech Media,* February 16, 2011.

45. James Navarro, "Right Idea, Wrong Place: Groups Sue Solar Project to Protect Imperiled Wildlife and Wild Lands," *Defenders of Wildlife Blog,* March 27, 2012.

46. Noaki Schwartz and Jason Dearen, "Native American Groups Sue to Stop Solar Projects," Associated Press, February 27, 2011.

47. Not all fast-tracked projects ultimately used ARRA funds. Some companies chose to use private finance; several other companies went out of business.

48. Gordon Walker and Noel Cass, "Carbon Reduction, 'the Public' and Renewable Energy: Engaging with Socio-Technical Configurations," *Area* 39, no. 4 (2007): 458–69.

49. Roopali Phadke, "Public Deliberation and the Geographies of Wind Justice," *Science as Culture* 22, no. 2 (2013): 247–55.

50. Carlisle et al., "Public Attitudes Regarding Large-Scale Solar Energy."

51. Dan van der Horst, "NIMBY or Not? Exploring the Relevance of Location and the Politics of Voiced Opinions in Renewable Energy Siting Controversies," *Energy Policy* 35, no. 5 (2007): 2705–14.

52. Phadke, "Public Deliberation."

53. Rene Germain, Donald W. Floyd, and Stephen V. Stehman, "Public Perceptions of the USDA Forest Service Public Participation Process," *Forest Policy and Economics* 3, no. 3 (2001): 113–24.

54. The expert panels' key recommendations included (1) incorporation of public participation in the planning and decision-making process instead of simply as a formal procedural requirement; (2) design of a public-participation process that is inclusive, collaborative, transparent, and communicated in good faith; and (3) that environmental assessments and decisions with substantial scientific content should be supported by collaborative, broadly based, integrated, and iterative analytic-deliberative processes. Tom Dietz and Paul Stern, *Public Participation In Environmental Assessment and Decision Making* (Washington, DC: National Academies Press, 2008).

55. Derek Bell, Tim Gray, and Claire Haggett, "The 'Social Gap' in Wind Farm Siting Decisions: Explanations and Policy Responses," *Environmental Politics* 14, no. 4 (2005): 460–77.

56. Pasqualetti, "Wind Energy Landscapes," 689–99.

57. Christian Brannstrom, Wendy Jepson, and Nicole Persons, "Social Perspectives on Wind-Power Development in West Texas," *Annals of the Association of American Geographers* 101, no. 4 (2011): 839–51.

58. Roopali Phadke, "Resisting and Reconciling Big Wind: Middle Landscape Politics in the New American West," *Antipode* 43 (2011): 754–76.

59. Maarten Wolsink, "Invalid Theory Impedes Our Understanding: A Critique on the Persistence of the Language of NIMBY," *Transactions of the Institute of British Geographers* 31, no. 1 (2006): 85–91.

60. Maarten Wolsink, "Entanglement of Interests and Motives: Assumptions behind the NIMBY-Theory on Facility Siting," *Urban Studies* 31, no. 6 (1994), 855.

61. Pasqualetti, "Wind Energy Landscapes," 690.

62. Ibid., 689.

63. Phadke, "Resisting and Reconciling," 754.

64. Jeremy Firestone and Willett Kempton, "Public Opinion about Large Offshore Wind Power: Underlying Factors," *Energy Policy* 35, no. 3 (2007): 1584–98.

65. Md. Munjur E. Moula, Johanna Maula, Mohamed Hamdy, Tingting Fang, Nusrat Jung, and Risto Lahdelma, "Researching Social Acceptability of Renewable Energy Technologies in Finland," *International Journal of Sustainable Built Environment* 2, no. 1 (2013): 89–98.

66. Kristy Michaud, Juliet E. Carlisle, and Eric R. A. N. Smith, "NIMBYism vs. Environmentalism in Attitudes toward Energy Development," *Environmental Politics* 17, no. 1 (2008): 20–39.

67. Jennifer M. Bernstein, "Climate Change, Industrial Solar, and the Globalized Local in Joshua Tree, California," *Yearbook of the Association of Pacific Coast Geographers* 78 (2016): 80–93.

68. Maarten Wolsink, "Planning of Renewables Schemes: Deliberative and Fair Decision-Making on Landscape Issues Instead of Reproachful Accusations of Non-Cooperation," *Energy Policy* 35, no. 5 (2007): 2692–2704.

69. van der Horst, "NIMBY or Not?"

70. Gordon Walker, Noel Cass, Kate Burningham, and Julie Barnett, "Renewable Energy and Sociotechnical Change: Imagined Subjectivities of 'the Public' and Their Implications," *Environment and Planning A* 42, no. 4 (2010): 931–47.

71. Phadke, "Public Deliberation," 255.

72. Kelvin Mason and Paul Milbourne, "Constructing a 'Landscape Justice' for Windfarm Development: The Case of Nant Y Moch, Wales." *Geoforum* 53 (2014): 104–15.

73. Myers and Sall used the term "green halo" in my interview with them in 2010.

74. Sethshtier, "Of Grizzlies and Tortoises," *High Country* News, October 29, 2010, https://www.hcn.org/blogs/range/of-grizzlies-and-tortoises.

75. Myers and Sall interview.

76. Ibid.

77. "Petition to List the Flat-Tail Horned Lizard (*Phrynosoma mcallii*) as Endangered under the California Endangered Species Act," Center for Biological Diversity before the California Fish and Game Commission, June 9, 2014.

78. Myers and Sall interview.

79. Ibid.

80. Steven R. Archer and Katharine I. Predick. "Climate Change and Ecosystems of the Southwestern United States," *Rangelands* 30, no. 3 (2008): 23–28.

81. J. M. Randall, S. S. Parker, J. Moore, B. Cohen, L. Crane, B. Christian, D. Cameron, J. MacKenzie, K. Klausmeyer, and S. Morrison, "Mojave Desert Ecoregional Assessment," Nature Conservancy, San Francisco, CA, 2010, https://www.scienceforconservation.org/assets/downloads/Mojave_Desert_Ecoregional_Assessment_2010.pdf.

82. John Berger, *Charging Ahead: The Business of Renewable Energy and What It Means for America* (Berkeley: University of California Press, 1998).

83. *New York Times,* "RFK Jr., Enviros Clash over Mojave Solar Proposal," September 9, 2009.

84. Ibid.

85. Ibid.

86. Robert F. Kennedy Jr., "An Ill Wind off Cape Cod," *New York Times,* December 16, 2005.

87. *New York Times,* "RFK Jr., Enviros Clash," 1.

88. Todd Woody, "Desert Vistas vs. Solar Power," *New York Times,* December 21, 2009.

89. Basin and Range Watch, "Calico Solar Project (SES Solar 1) Updates," http://www.basinandrangewatch.org/CalicoUpdates.html

90. Myers and Sall interview.

91. U.S. Bureau of Land Management, "Solar Energy Program, Western Solar Plan," http://blmsolar.anl.gov/.

92. Associated Press, "Native American Groups Sue to Stop Solar Projects," *San Francisco Chronicle,* February 28, 2011.

93. Tiffany Hsu, "Imperial Valley Solar Project Sued by Indian Tribe," *Los Angeles Times* blogs, November 4, 2010, http://latimesblogs.latimes.com/greenspace/2010/11/imperial-valley-solar-project-sued-by-indian-tribe.html.

94. Louis Sahagun, "Discovery of Indian Artifacts Complicates Genesis Solar Project," *Los Angeles Times,* April 24, 2012, http://articles.latimes.com/2012/apr/24/local/la-me-solar-bones-20120424.

95. Chris Clarke, "Documentary Explores Conflict between Tribes, Energy Developers," KCET, March 8, 2013.

96. Alfredo Figueroa, interview with the author, Blythe, California, March 15, 2011.

97. Ibid.

98. Californians for Renewable Energy, "Complaint and Protest against the California-Based Roncentrating Solar Power (CSP) Developer Genesis Solar LLC, the California Energy Commission (CEC), the United States Department of Interior Bureau of Land Management (BLM) and the United States Department of Energy (US DOE)," 58693, https://efiling.energy.ca.gov/Lists/DocketLog.aspx?docketnumber=09-AFC-08.

99. Ibid.

100. Sahagun, "Discovery of Indian Artifacts," 1.

101. Karen Eugenie Rignall, "Solar Power, State Power, and the Politics of Energy Transition in Pre-Saharan Morocco," *Environment and Planning A* 48, no. 3 (2016): 540–57.

102. Julie Cart, "Environmentalists Feeling Burned by Rush to Build Solar Projects," *Los Angeles Times,* April 6, 2012.

103. Solar Energy Industries Association, "Concentrating Solar Power," https://www.seia.org/initiatives/concentrating-solar-power.

104. Randall et al., "Mojave Desert Ecoregional Assessment."

105. Frederick Turner, Philip Medica and Craig Lyons, "Reproduction and Survival of the Desert Tortoise in Ivanpah Valley, California," *COPEIA* 4 (1984): 811–20.

106. Craig Stanford, *The Last Tortoise: A Tale of Extinction in Our Lifetime* (Cambridge, MA: Harvard University Press, 2010).

107. Mike Stark, "Big Money for the Lowly Desert Tortoise," Associated Press, February 22, 2009.

108. Cameron Barrows, "Sensitivity to Climate Change for Two Reptiles at the Mojave-Sonoran Desert Interface," *Journal of Arid Environments* 75 no. 7 (2011): 629–35.

109. On December 14, 2010, the USFWS determined that the Sonoran population of desert tortoises warranted listing under the Endangered Species Act, but that higher conservation priorities and budget limitations precluded listing (Alagona, *After the Grizzly*).

110. Interview with Laura Cunningham, Amargosa Valley, Nevada, 2009.

111. Alagona, *After the Grizzly,* 149–74.

112. U.S President Barack Obama, "Solar Power to Power a Clean Economy," weekly radio address, October 10, 2010.

113. Bill O'Brien, "Follow the Sun: Cone Drive's Heliostats will Help California Solar Company," *Traverse City Record-Eagle*, February 13, 2011.

114. The reply the interior secretary to this letter by can be found here: https://www.doi.gov/sites/doi.gov/files/migrated/foia/os/upload/2012-00113ci.pdf.

115. U.S. Fish and Wildlife Service, "Desert Tortoise Recovery Plan: Mojave Population," June 1994.

116. Barrows, "Sensitivity to Climate Change," 629.

117. Center for Biological Diversity, "Public Comment on Final EIS for Ivanpah Solar Electric Generating System," Tucson, AZ, 2010.

118. Judith Lewis Mernit, "Here Comes the Sun," *Audubon,* September-October, 2011.

119. Jim Andre, interview with author, March 2010.

120. Basin and Range Watch, "Ivanpah Valley ACEC Nomination Petition," Las Vegas Bureau of Land Management Field Office, October 23, 2011.

121. U.S. Bureau of Land Management, "Biological Assessment for the Ivanpah Solar Electric Generating System," July 2010.

122. Center for Biological Diversity, "Questions and Answers Related to the Settlement Agreement between the Center for Biological Diversity and Bright-Source Energy, Inc. Regarding the Ivanpah Solar Electric Generating System

Project in the Mojave Desert," October 2010, https://www.biologicaldiversity.org/species/reptiles/desert_tortoise/pdfs/CBD_Ivanpah_Factsheet_8-8-2011.pdf.

123. Western Watersheds Project, "Ivanpah Solar Thermal Threatens Thousands of Desert Tortoise," undated, https://www.westernwatersheds.org/newsmedia/online-messenger/ivanpah-solar-thermal-threatens-thousands-desert-tortoise.htm. The Sierra Club lawsuit against the Calico site was thrown out by the courts in April 2011.

124. Jessica Corso, "9th Circ. Buries Enviro Suit over $2.2B Mojave Solar Project," *Law 360*, May 5, 2015.

125. U.S. Bureau of Land Management, "Revised Biological Assessment for the Ivanpah Solar Electric Generating System," April 19, 2011.

126. U.S. Bureau of Land Management, "Biological Assessment for the Ivanpah Solar Electric Generating System," December 7, 2009.

127. Center for Biological Diversity, "Disastrous Desert Tortoise Translocation Suspended" (press release), Tucson, AZ, October 10, 2008.

128. U.S. Bureau of Land Management, "Revised Biological Assessment."

129. Julie Cart, "Saving Desert Tortoises Is a Costly Hurdle for Solar Projects," *Los Angeles Times*, March 4, 2012.

130. Nancy Pfund, "Opinion: Donuts, Renewable Energy and What They Say about America," *San Jose Mercury News*, April 25, 2012.

131. David Denelski, "It's Not Easy Being Green: Ivanpah Solar Plant Near Nevada Burns a Lot of Natural Gas, Making It a Greenhouse Gas Emitter under State Law," *Orange County Register*, October 15, 2015.

132. Joe Ryan, "NRG's Massive California Solar Plant Finally Making Enough Power," *Bloomberg News*, February 1, 2017.

133. Interview with Defenders of Wildlife attorney Joshua Basofin, March 20, 2010.

134. *Associated Press*, "Mojave Desert Solar Plant Cooking 'Alarming' Number of Birds in Mid-air," August 18, 2014.

135. Judith Lewis Mernit, "Is Big Desert Solar Killing Birds in Southern California?" *High Country News*, August 18, 2013.

136. H. T. Harvey & Associates, "Ivanpah Solar Electric Generating System Avian & Bat Monitoring Plan: October 2013–2014 Annual Report," 2015.

137. Valerie Kuletz, *Tainted Desert: Environmental and Social Ruin in the American West* (New York: Routledge, 1998).

138. Janine Blaeloch, "Solar Power, and Tortoises Too," *Los Angeles Times*, October 16, 2012.

139. Grey Brechin, *Imperial San Francisco: Urban Power, Earthly Ruin* (Berkeley: University of California Press, 2006).

140. Steve Lerner, *Sacrifice Zones: The Frontlines of Toxic Chemical Exposure in the United States* (Cambridge, MA: MIT Press, 2010).

141. Julia Fox, "Mountaintop Removal in West Virginia: An Environmental Sacrifice Zone," *Organization & Environment* 12, no. 2 (1999): 163–83.

142. The adjective "dismal" might be added to cost–benefit analysis. Harvard president and Obama-administration treasury secretary Lawrence Summers notoriously said it would be wise to send hazardous waste to developing countries because their people will die of other causes before the onset of the

health effects. John Bellamy Foster, "'Let Them Eat Pollution': Capitalism and the World Environment," *Monthly Review* 44, no. 8 (1993): 10–21.

143. Fox, "Mountaintop Removal," 167.

144. BrightSource, "Ivanpah Project Receives Federal Permits; Approved to Commence Construction" (press release), October 7, 2010.

145. Ken Salazar, remarks at Solar Power International Conference, Los Angles, California, October 13, 2010.

146. John McPhee, *Basin and Range* (New York: Macmillan, 1981).

147. Edward Abbey, *Desert Solitaire: A Season in the Wilderness* (New York: Random House, 1968).

148. Mark David Spence, *Dispossessing the Wilderness: Indian Removal and the Making of the National Parks* (Oxford University Press, 1999).

149. Donald Worster, *Under Western Skies: Nature and History in the American West* (Oxford University Press, 1992).

150. Skillen, *Nation's Largest Landlord*.

151. George Cameron Coggins and Parthenia Blessing Evans, "Multiple Use, Sustained Yield Planning on the Public Lands," *University of Colorado Law Review* 53 (1981): 411–31.

152. Mike Davis, "The Dead West: Ecocide in Marlboro Country," *New Left Review* (1993): 49–59.

153. Rebecca Solnit, *Savage Dreams: A Journey into the Landscape Wars of the American West* (Berkeley: University of California Press, 1994), 62.

154. Davis, "Dead West," 58.

155. Ibid., 61.

156. Solnit, *Savage Dreams*.

157. Beth A. Newingham, Cheryl H. Vanier, Therese N. Charlet, Kiona Ogle, Stanley D. Smith, and Robert S. Nowak, "No Cumulative Effect of 10 Years Of Elevated [CO_2] on Perennial Plant Biomass Components in the Mojave Desert," *Global Change Biology* 19, no. 7 (2013): 2168–81.

158. Jayne Belnap, "Recovery Rates of Cryptobiotic Crusts: Inoculant Use and Assessment Methods," *Great Basin Naturalist* 53 (1993): 89–95.

159. Ray Rasker, "An Exploration into the Economic Impact of Industrial Development versus Conservation on Western Public Lands," *Society and Natural Resources* 19, no. 3 (2006): 191–207.

160. Andrew Hansen, Ray Rasker, Bruce Maxwell, Jay Rotella, Jerry Johnson, Andrea Wright Parmenter, Ute Langner, Warren Cohen, Rick Lawrence, and Matthew Kraska, "Ecological Causes and Consequences of Demographic Change in the New West: As Natural Amenities Attract People and Commerce to the Rural West, the Resulting Land-Use Changes Threaten Biodiversity, Even in Protected Areas, and Challenge Efforts to Sustain Local Communities and Ecosystems," *BioScience* 52, no. 2 (2002): 151–62.

161. Scott Streater, "Fast-Tracked Solar Project Could Speed Mojave Desert's Demise," *New York Times*, November 12, 2009.

162. Ronnie Lipschutz and Dustin Mulvaney, "The Road Not Taken, Round II: Centralized vs. Distributed Energy Strategies and Human Security," in *International Handbook of Energy Security*, ed. H. Dyer and Julia Trombett (Northampton, MA: Edward Elgar, 2013), 483–506.

163. Davis, "Dead West," 58; Solnit, *Savage Dreams.*

164. Bill Powers and Sheila Bowers, "Distributed Solar PV: Why It Should Be the Centerpiece of U.S. Solar Energy Policy," Solar Done Right, September 10, 2010.

165. Wilderness Society, "Solar Facility Will Improve Toxic Lands while Providing Much Needed Jobs and Energy to Local Economy," (press release), May 13, 2010.

166. Michael Brune, "Zombie Attacks on Rooftop Solar," *Sierra Club Blog,* March 17, 2015.

167. Interview with Kevin Emmerich, Amargosa Valley, Nevada, March 2009.

168. John Kennedy, "First Solar Project Doesn't Jeopardize Protected Tortoise," *Law 360,* March 31, 2015.

169. Basin and Range Watch, "First Solar Hits Another Desert Tortoise Jackpot! 152 Relocated," December 16, 2014; Silver State South Solar, "Table 1. Summary of SSS Desert Tortoise Translocations," unpublished data acquired from the BLM by Basin and Range Watch, http://www.basinandrangewatch .org/Silver-State-S-Tortoises-Moved-2014%20Sheet1.pdf.

170. Rebecca Solnit, "Are We Missing the Big Picture on Climate Change?" *New York Times Magazine,* December 2, 2014, 13.

171. Solnit, "Missing the Big Picture," 13.

172. Ibid., 13.

173. Clark Miller and Jennifer Richter, "Social Planning for Energy Transitions," *Current Sustainable/Renewable Energy Reports* 1, no. 3 (2014): 77–84.

174. Hunold and Leitner, "'Hasta la Vista, Baby!'," 701.

175. Ibid., 702.

176. Anonymous interview with concerned community member, April 14, 2012.

177. Jeanne Nienaber Clarke and Daniel McCool, *Staking Out the Terrain: Power and Performance among Natural Resource Agencies* (Albany, NY: SUNY Press, 1996).

178. U.S. Department of the Interior, "Interior Department Finalizes Rule Providing a Foundation for the Future of BLM's Renewable Energy Program" (press release), November 11, 2016.

CHAPTER 6. THE WESTERN SOLAR PLAN

1. Jeff Lovich and Joshua R. Ennen, "Wildlife Conservation and Solar Energy Development in the Desert Southwest, United States," *BioScience* 61, no. 12 (2011): 982–92.

2. Damon Turney and Vasilis M. Fthenakis, "Environmental Impacts from the Installation and Operation of Large-Scale Solar Power Plants," *Renewable and Sustainable Energy Reviews* 15, no. 6 (2011): 3261–70.

3. Vaclav Smil, "On Energy and Land: Switching from Fossil Fuels to Renewable Energy Will Change Our Patterns of Land Use," *American Scientist* 72, no. 1 (1984), 18.

4. Martin J. Pasqualetti and Byron A. Miller, "Land Requirements for the Solar and Coal Options," *Geographical Journal* 150, no. 2 (1984): 192–212.

5. Robert I. McDonald, Joseph Fargione, Joe Kiesecker, William M. Miller, and Jimmie Powell, "Energy Sprawl or Energy Efficiency: Climate Policy Impacts on Natural Habitat for the United States of America," *PLoS One* 4, no. 8 (2009): e6802–13.

6. Sean Ong, Clinton Campbell, Paul Denholm, Robert Margolis, and Garvin Heath, "Land-Use Requirements for Solar Power Plants in the United States," NREL/TP-6A20–56290, National Renewable Energy Laboratory, Golden, CO, 2013.

7. Rebecca R. Hernandez, Madison K. Hoffacker, and Christopher B. Field, "Land-Use Efficiency of Big Solar," *Environmental Science & Technology* 48, no. 2 (2014): 1315–23.

8. Turney and Fthenakis, "Environmental Impacts."

9. Ibid., 3263.

10. Rebecca R. Hernandez, Madison K. Hoffacker, Michelle L. Murphy-Mariscal, Grace C. Wu, and Michael F. Allen, "Solar Energy Development Impacts on Land Cover Change and Protected Areas," *Proceedings of the National Academy of Sciences* 112, no. 44 (2015): 13579–84.

11. David M. Stoms, Stephanie L. Dashiell, and Frank W. Davis, "Siting Solar Energy Development to Minimize Biological Impacts," *Renewable Energy* 57 (2013): 289–98.

12. Matthew L. Farnsworth, Brett G. Dickson, Luke J. Zachmann, Ericka E. Hegeman, Amanda R. Cangelosi, Thomas G. Jackson Jr, and Amanda F. Scheib, "Short-Term Space-Use Patterns of Translocated Mojave Desert Tortoise in Southern California," *PloS one* 10, no. 9 (2015): e0134250.

13. Jennifer M. Germano, Kimberleigh J. Field, Richard A. Griffiths, Simon Clulow, Jim Foster, Gemma Harding, and Ronald R. Swaisgood, "Mitigation-Driven Translocations: Are We Moving Wildlife in the Right Direction?" *Frontiers in Ecology and the Environment* 13, no. 2 (2015): 100–05.

14. U.S. Department of Energy, "First Solar Topaz Farm Draft Environmental Impact Statement," 2011, http://energy.gov/lpo/about-us/environmental-compliance/environmental-impact-statements/eis-0458-first-solar-topaz.

15. Louis Sahagun, "Problems Cast Shadows of Doubt on Solar Project," *Los Angeles Times*, February 11, 2012.

16. Louis Sahagun, "Canine Distemper in Kit Foxes Spreads in Mojave Desert," *Los Angeles Times*, April 18, 2012.

17. Public comment made at EIS meeting for the Panoche Solar Farm, Hollister, California, March 12, 2011.

18. Rebecca A. Kagan, Tabitha C. Viner, Pepper W. Trail, and Edgard O. Espinoza, "Avian Mortality at Solar Energy Facilities in Southern California: A Preliminary Analysis," National Fish and Wildlife Forensics Laboratory, Ashland, OR.

19. Michael D. McCrary, Robert L. McKernan, Ralph W. Schreiber, William D. Wagner, and Terry C. Sciarrotta, "Avian Mortality at a Solar Energy Power Plant," *Journal of Field Ornithology* (1986): 135–41.

20. H.T. Harvey & Associates, "Ivanpah Solar Electric Generating System Avian & Bat Monitoring Plan: October 2013–2014 Annual Report," April 2015.

21. This public comment from U.S. Fish and Wildlife Service division chief Sorenson to the CEC was posted on the Basin and Range Watch website.

22. Ibid.

23. American Bird Conservancy, "Position Paper: Solar Energy," The Plains, VA, 2015.

24. Jennifer A. Smith and James F. Dwyer, "Avian Interactions with Renewable Energy Infrastructure: An Update," The Condor 118, no. 2 (2016): 411–23.

25. Chris Clarke, "Water Birds Turning up Dead at Solar Projects in the Desert," KCET, July 17, 2013. https://www.kcet.org/redefine/water-birds-turning-up-dead-at-solar-projects-in-the-desert.

26. Gábor Horváth, Miklós Blahó, Ádám Egri, György Kriska, István Seres, and Bruce Robertson, "Reducing the Maladaptive Attractiveness of Solar Panels to Polarotactic Insects," Conservation Biology 24, no. 6 (2010): 1644–53.

27. Leroy J. Walston Jr., Katherine E. Rollins, Kirk E. LaGory, Karen P. Smith, and Stephanie A. Meyers, "A Preliminary Assessment of Avian Mortality at Utility-Scale Solar Energy Facilities in the United States," Renewable Energy no. 92 (2016): 405–14.

28. Benjamin Means, "Prohibiting Conduct, Not Consequences: The Limited Reach of the Migratory Bird Treaty Act," Michigan Law Review 97, no. 3 (1998): 823–42.

29. Interview with ecologist Jeff Aardahl, March 18, 2010.

30. E.O. Wilson and Thomas Lovejoy, "A Mojave Solar Project in the Bighorns' Way," New York Times, September 11, 2015.

31. John Wehausen and Clinton W. Epps, "Protecting Desert Bighorn Sheep Migration Corridors in Mojave Desert: Guest Commentary," Daily Bulletin, January 12, 2015.

32. Eugene P. Odum, Fundamentals of Ecology (Toronto: W.B. Saunders, 1993).

33. Jeff Lovich and David Bainbridge, "Anthropogenic Degradation of the Southern California Desert Ecosystem and Prospects for Natural Recovery and Restoration," Environmental Management 24, no. 3 (1999): 309–26.

34. Interview with Jim Andre at the Sweeney Granite Mountains Desert Center, near Kelso, California.

35. California Senate Bill 34, Solar Thermal and Photovoltaic Power Plants: Siting: California Endangered Species Act: Mitigation Measures, 2009–2010, http://leginfo.legislature.ca.gov/faces/billNavClient.xhtml?bill_id=200920100SB34.

36. Dustin Mulvaney, "Identifying the Roots of Green Civil War over Utility-Scale Solar Energy Projects on Public Lands across the American Southwest," Journal of Land Use Science 12, no. 6 (2017): 493–515.

37. Bureau of Land Management (BLM), "BLM Announces Scoping Meeting and Notice of Intent to Prepare Environmental Assessment for Solar Project Water Request" (press release), 2014.

38. George B. Frisvold and Tatiana Marquez, "Water Requirements for Large-Scale Solar Energy Projects in the West," *Journal of Contemporary Water Research & Education* 151, no. 1 (2013): 106–16.

39. Rebecca Nelson and Debra Perrone, "The Role of Permitting Regimes in Western United States Groundwater Management," *Groundwater* 54, no. 6 (2016): 761–64.

40. Francesco Zanatta, "Automatic Washing Device for Continuous Surfaces, in Particular Solar Thermal Collectors, Photovoltaic Panels, Continuous Glazed Building Walls and Similar Surfaces," U.S. Patent 9,192,966, issued November 24, 2015.

41. Herman K. Trabish, "Construction Halted at First Solar's 230 MW Antelope Valley Site," *Greentech Media*, April 22, 2013.

42. Ibid.

43. Ibid.

44. Ibid.

45. This was mentioned during a short presentation at an "open house" meeting with the local public for the proposed 550 MW Topaz Solar Farm in 2009.

46. Trabish, "Construction Halted."

47. William H. Schlesinger, "Carbon Storage in the Caliche of Arid Soils: A Case Study from Arizona," *Soil Science* 133, no. 4 (1982): 247–56.

48. Penelope Serrano-Ortiz, Marilyn Roland, Sergio Sanchez-Moral, Ivan A. Janssens, Francisco Domingo, Yves Goddéris, and Andrew S. Kowalski, "Hidden, Abiotic CO_2 Flows and Gaseous Reservoirs in the Terrestrial Carbon Cycle: Review and Perspectives," *Agricultural and Forest Meteorology* 150, no. 3 (2010): 321–29.

49. Michael F. Allen and Alan McHughen, "Solar Power in the Desert: Are the Current Large-Scale Solar Developments Really Improving California's Environment?" Center for Conservation Biology, University of California, Riverside, 2011.

50. Joseph Fargione, Jason Hill, David Tilman, Stephen Polasky, and Peter Hawthorne, "Land Clearing and the Biofuel Carbon Debt," *Science* 319, no. 5867 (2008): 1235–38.

51. Greg Barron-Gafford, Rebecca L. Minor, Nathan A. Allen, Alex D. Cronin, Adria E. Brooks, and Mitchell A. Pavao-Zuckerman, "The Photovoltaic Heat Island Effect: Larger Solar Power Plants Increase Local Temperatures," *Scientific Reports* 6 (2016): 35070.

52. One letter from University of California, Santa Cruz, ecologist Barry Sinervo detailed several hypotheses based on his prior research on reptile thermoregulation.

53. Valéry Masson, Marion Bonhomme, Jean-Luc Salagnac, Xavier Briottet, and Aude Lemonsu, "Solar Panels Reduce both Global Warming and Urban Heat Island," *Frontiers in Environmental Science* 2, no. 14 (2014): 1–10.

54. Clifford K. Ho, Cianan A. Sims, and Joshua M. Christian, "Evaluation of Glare at the Ivanpah Solar Electric Generating System," *Energy Procedia* 69 (2015): 1296–1305.

55. Steven A. Sumner and Peter M. Layde. "Expansion of Renewable Energy Industries and Implications for Occupational Health," *Journal of the American Medical Association* 302, no. 7 (2009): 787–89.

56. Andrew Gray, "Valley Fever Hits Carrizo Plain Solar Farms Hard," KYET, May 2, 2013.

57. Sarah Zhang, "A Huge Solar Plant Caught on Fire, and That's the Least of Its Problems," *Wired,* May 23, 2016.

58. Kera Abraham, "Will Central California's Proposed Place in the Sun Ruin Life for Sustainable Farmers?" *Monterey County Weekly,* February 25, 2010.

59. Sarah McBride, "With Solar Power, It's Green on Green," Reuters, January 10, 2011.

60. California Department of Conservation, "The Williamson Act and Solar Energy," presentation to Fresno County, August 2012.

61. Jason A. Wilken, Gail Sondermeyer, Dennis Shusterman, Jennifer McNary, Duc J. Vugia, Ann McDowell, Penny Borenstein, et al., "Coccidioidomycosis among Workers Constructing Solar Power Farms, California, USA, 2011–2014," *Emerging Infectious Diseases* 21, no. 11 (2015): 1997.

62. Ibid.

63. Confidential interview with a property owner in California Valley.

64. Chris Clarke, "Solar Energy Development in the Carrizo Plain Draws Lawsuits," KCET, April 23, 2011, http://www.kcet.org/news/redefine/revisit /commentary/carrizo-plain-solar-draws-lawsuit.html.

65. U.S. Department of the Interior, "BLM Initiates Environmental Analysis of Solar Energy Development" (press release), May 29, 2008.

66. U.S. Department of the Interior, "Secretary Salazar, Gov. Schwarzenegger Sign Initiative to Expedite Renewable Energy Development," October 12, 2009.

67. Personal communication, anonymous BLM district staff, Las Cruces, New Mexico, March 23, 2010.

68. Carl Walters, *Adaptive Management of Renewable Resources* (New York: Macmillan, 1986).

69. Southern California Edison, Presentation to the DRECP Stakeholder Meeting, August 17, 2011.

70. U.S. Bureau of Land Management, "BLM Rejects Solar Development in Silurian Valley," News Release CA-SO-15–06, November 20, 2014.

71. Wehausen and Epps, "Protecting Desert Bighorn Sheep Migration."

72. U.S Bureau of Land Management, "Desert Renewable Energy Conservation Plan: Record of Decision," September 2016.

73. Leroy J. Walston, Shruti K. Mishra, Heidi M. Hartmann, Ihor Hlohowskyj, James McCall, and Jordan Macknick, "Examining the Potential for Agricultural Benefits from Pollinator Habitat at Solar Facilities in the United States," *Environmental Science and Technology* 52, no. 13 (2018): 7566–76.

74. Madison K. Hoffacker, Michael Allen, and Rebecca Hernandez, "Land-Sparing Opportunities for Solar Energy Development in Agricultural Landscapes: A Case Study of the Great Central Valley, CA, United States," *Environmental Science and Technology* 51, no. 24 (2017): 14472–82.

75. Madison K. Hoffacker, Michael F. Allen, and Rebecca R. Hernandez, "Land-Sparing Opportunities for Solar Energy Development in Agricultural Landscapes: A Case Study of the Great Central Valley, CA, United States," *Environmental Science and Technology* 51, no. 24 (2017): 14472–82.

76. Public comment to the DRECP from Basin and Range Watch, January 30, 2015.

77. Kara A. Moore-O'Leary, Rebecca R. Hernandez, Dave S. Johnston, Scott R. Abella, Karen E. Tanner, Amanda C. Swanson, Jason Kreitler, and Jeffrey E. Lovich, "Sustainability of Utility-Scale Solar Energy: Critical Ecological Concepts," *Frontiers in Ecology and the Environment* 15, no. 7 (2017): 385–94.

CHAPTER 7. BREAKTHROUGH TECHNOLOGIES AND SOLAR TRADE WARS

1. Steven Chu, "The Solyndra Failure: Testimony of U.S. Department of Energy Secretary Steven Chu," Committee on Energy and Commerce, House of Representatives, 112th Congress, November 17, 2011.

2. Carl Pope and Bjorn Lomborg, "Debate: The State of Nature," *Foreign Affairs,* October 22, 2009.

3. Stephen Pacala and Robert Socolow, "Stabilization Wedges: Solving the Climate Problem for the Next 50 Years with Current Technologies," *Science* 305, no. 5686 (2004): 968–72.

4. Varun Sivaram, *Taming the Sun: Innovations to Harness Solar Energy and Power the Planet* (Cambridge: MIT Press, 2018).

5. Declan Butler, "Thin Films: Ready for Their Close-up?" *Nature* 454 (2008): 558–59.

6. Jonathan Silver, "Department of Energy Talks Department of Energy: An Insider's Look Back at the Loan-Guarantee Program," *Solar Industry* 4, no. 6 (2011): 1, 12–15.

7. Bloomberg New Energy Finance, "League Table Results Book," 2011, https://www.bnef.com/assets/pdfs/league-table-results-book-2011.pdf.

8. Reed Landberg and Brian Eckhouse, "Trump Solar Duties Strike at $161 billion China-Led Industry," *Bloomberg,* January 23, 2018.

9. Nassim Nicholas Taleb, *The Black Swan: The Impact of the Highly Improbable* (New York: Random House, 2007).

10. Nilima Choudhury, "Cleantech Venture Investment in 2011 Reaches Highest Level since 2008," *PV-Tech,* January 6, 2013.

11. Steven Johnson, *Where Good Ideas Come From: The Seven Patterns of Innovation* (New York: Penguin, 2011).

12. Michael J. Roberts, Joseph B. Lassiter III, and Ramana Nanda, "U.S. Department of Energy & Recovery Act Funding: Bridging the 'Valley of Death'," Case 810–144, Harvard Business School, June 2010.

13. World Intellectual Property Organization, "Photovoltaic Thin Film Cells," FRINNOV, Paris, 2009.

14. William M. Adams, *Green Development: Environment and Sustainability in the Third World* (New York: Routledge, 2003).

15. Kathleen McAfee, "Selling Nature to Save It? Biodiversity and Green Developmentalism," *Environment And Planning D: Society and Space* 17, no. 2 (1999): 133–54.

16. Nikolas Rose, *Power of Freedom: Reframing Political Thought* (Cambridge, UK: Cambridge University Press, 1999).

17. David Harvey, *A Brief History of Neoliberalism* (Oxford: Oxford University Press, 2007).

18. David Hess, *Good Green Jobs in a Global Economy: Making and Keeping New Industries in the United States* (Cambridge, MA: MIT Press, 2012).

19. Paul Krugman, "Building a Green Economy," *New York Times Magazine*, April 7, 2010.

20. Fraunhofer Institute for Solar Energy, "Photovoltaics Report," Freiberg, Germany, November 17, 2016.

21. *Venture Beat*, "DOE Loan Chief: Where We'll Invest in 2011," December 15, 2010, http://venturebeat.com/2010/12/15/doe-loan-chief-where-well-invest-in-2011-how-abound-solar-could-compete-with-china/.

22. Marc Lifsher, "Unlike Solyndra, Other California Solar Projects Appear on Track," *Los Angeles Times*, October 15, 2011, http://articles.latimes.com /2011/oct/15/business/la-fi-1015-solar-loans-20111015.

23. Kris Bevill, "Getting Better: A Conversation with the U.S. DOE's Jonathan Silver, Head of Loan Programs," *Ethanol Producer Magazine*, August 15, 2011.

24. Ibid., 22.

25. Dana Hull, "The Man behind Solyndra's Rise and Fall: Chris Gronet," *Mercury News*, November 26, 2011.

26. Solyndra, "Method of Depositing Materials on a Non-planar Surface," U.S. patent no. 7563725 B2, 2007.

27. Hearing before the Committee on Oversight and Government Reform, 112th Congress, March 20, 2012, https://www.gpo.gov/fdsys/pkg/CHRG-112hhrg74042/html/CHRG-112hhrg74042.htm.

28. Anita Hamilton, "Best Inventions of 2008," *Time*, October 28, 2008.

29. Kevin Bullis, "Better Solar for Big Buildings: A Startup Is Selling Cylindrical Solar Cells That Can Generate More Power than Conventional Panels," *MIT Technology Review*, October 7, 2008.

30. Jim Snyder, "FBI Raid on Solyndra May Herald Escalation of Watchdog Probe," *Bloomberg News*, September 10, 2011.

31. U.S. President Barack Obama, "Remarks by the President on the Economy at Solyndra, Inc., Fremont, California," White House press release, May 26, 2010.

32. Steven Chu, "The Solyndra Failure: Testimony of U.S. Department of Energy Secretary Steven Chu," Committee on Energy and Commerce, House of Representatives, 112th Congress, November 17, 2011.

33. Ibid.

34. U.S. Securities and Exchange Commission, "Solyndra, Inc.," Form S-1, December 18, 2009.

35. Robert Dydo, "Manufacturing Costs at First Solar No Longer Offer Advantage over the Chinese Solar Companies," *Seeking Alpha,* March 2, 2012, https://seekingalpha.com/article/407391-manufacturing-costs-at-first-solar-no-longer-offer-advantage-over-the-chinese-solar-companies.

36. Natural Resources Committee, H.R. 2915, "Chairmen Seek to Halt Risky WAPA Loans by Energy Department, Give Notice on Further Oversight and Document Requests," 112th Congress, November 9, 2011, http://naturalresources.house.gov/News/DocumentSingle.aspx?DocumentID=268112.

37. Ibid.

38. Ibid.

39. Alison Vekshin and Mark Chediak, "Solyndra's $733 Million Plant Had Whistling Robots, Spa Showers," Bloomberg News, September 28, 2011.

40. Committee on Energy and Commerce, "Solyndra Failure."

41. U.S. Geological Survey, "Selenium and Tellurium Statistics and Information," Mineral Commodity Studies, 2017.

42. *PV Tech,* "Venture Capital Solar Investments Fall by Almost 50%," January 10, 2013, http://www.pv-tech.org/news/venture_capital_solar_investments_fall_by_almost_50.

43. Committee on Energy and Commerce, "Solyndra Failure."

44. Quoted in Keith Bradsher, "For Solar Panel Industry, a Volley of Trade Cases," *New York Times,* October 11, 2012.

45. An example of a crony capitalism argument from the other side is the role of Dick Cheney in directing oil, gas, and defense policy to the benefit of his former firm, Halliburton.

46. U.S. Department of Energy, "Energy Department Finalizes Loan Guarantee to Support California Solar Generation Project" (press release), September 30, 2011.

47. John McArdle, "Issa, DOE Official Clash over Obama's Early Role in Loan Guarantee," *Energy and Environment News,* September 22, 2011, https://www.eenews.net/stories/1059954060.

48. Statement of Jens Meyerhoff, First Solar, before the Committee on Energy and Natural Resources, U.S. Senate, September 23, 2010, https://www.energy.senate.gov/public/index.cfm/files/serve?File_id=3EC7189C-C026-A3EF-7D7C-FE79B83040A4.

49. Email from Dong Kim, U.S. Department of Energy, June 23, 2011, quoted in Mathew Mosk, "GOP Says Energy Dept. Cut Corners to Lend Az. Solar Firm $1.6 Billion," *ABC News,* March 19, 2012.

50. Letter from Jens Meyerhoff, First Solar, to Jonathan Silver, director of the Loan Programs Office, U.S. DOE, May 18, 2011.

51. U.S. House of Representatives Committee on Oversight and Government Reform, "The Department of Energy's Disastrous Management of Loan Guarantee Programs," Staff Report, U.S. House of Representatives, 112th Congress, March 20, 2012.

52. Eric Wesoff, "First Solar Acquires GE's PrimeStar Solar IP," *Greentech Media,* August 6, 2013.

53. Jonathan Gifford, "A Disruptive Partnership," *PV Magazine,* May 5, 2014.

54. Ron Wyden, "China's Subsidy of Solar Panels," *New York Times,* October 28, 2012.

55. Keith Bradsher, "U.S. Solar Panel Makers Say China Violated Trade Rules," *New York Times,* October 19, 2011.

56. Quoted in ibid., 1.

57. Kan Sichao, "Chinese Photovoltaic Market and Industry Outlook," *Institute of Energy Economics Journal,* April 2010.

58. Kevin Bullis, "Solar's Great Leap Forward," *MIT Technology Review,* June 22, 2010.

59. Ibid.

60. Yong-hau Wang, Guo-liang Luo, and Yi-wei Guo, "Why Is There Overcapacity in China's PV Industry in Its Early Growth Stage?" *Renewable Energy* 72 (2014): 188–94.

61. Keith Bradsher, "Trade War in Solar Takes Shape," *New York Times,* November 9, 2011.

62. Ron Wyden, "China's Grab for Green Jobs," October 14, 2011, https://www.wyden.senate.gov/download/staff-report-chinas-grab-for-green-jobs_-examination-of-the-surge-of-solar-goods-exports-from-china-.

63. Michael Hobday, "East versus Southeast Asian Innovation Systems: Comparing OEM- and TNC-led Growth in Electronics," *Technology, Learning, And Innovation: Experiences of Newly Industrializing Economies* (2000): 129–69.

64. Jenny Chan and Pun Ngai, "Suicide as Protest for the New Generation of Chinese Migrant Workers: Foxconn, Global Capital, and the State," *Asia-Pacific Journal* 37 no. 2 (2010): 2–10.

65. *PV Magazine,* "Foxconn Finalizes Sharp Acquisition," August 15, 2016, https://www.pv-magazine.com/2016/08/15/foxconn-finalizes-sharp-acquisition_100025762/.

66. U.S. Department of Commerce, *China: Competition from State-Owned Enterprises,* International Trade Administration, June, 17, 2016.

67. The author's database of manufacturing showed LDK in the top ten of cell and module manufacturers in 2011.

68. Phil Milford and Michael Bathon, "LDK Files Bankruptcy in U.S. Court on China Solar Glut," *Bloomberg News,* October 21, 2014.

69. U.S. Census Bureau, "U.S. Trade in Goods with China, 2012," https://www.census.gov/foreign-trade/balance/c5700.html.

70. Robert Scott, *The China Toll,* Economic Policy Institute, Washington, DC, August 23, 2012.

71. Usha Haley and George Haley, *Subsidies to Chinese Industry: State Capitalism, Business Strategy, and Trade Policy* (Oxford: Oxford University Press, 2013).

72. First Solar, "First Solar and Ordos Take Key Step Forward in 2 GW Project" (press release), November 17, 2009.

73. William Pentland, "First Solar's Dangerous Game," *Forbes,* May 10, 2011.

74. U.S.-China Economic and Security Review Commission, "China's Intellectual Property Rights and Indigenous Innovation Policy," hearing before the U.S.-China Economic and Security Review Commission, May 4, 2011.

75. Diane Caldwell, "Solar Company Seeks Stiff U.S. Tariffs to Deter Chinese Spying: SolarWorld Americas Says Hackers in China Stole Documents," *New York Times*, September 1, 2014.

76. Morris Goldstein, "Currency Manipulation and Enforcing the Rules of the International Monetary System," in *Reforming the IMF for the 21st Century*, edited by Edwin M. Truman (Peterson Institute for Ecological Economics, Washington, DC, 2006): 150–51.

77. Keith Bradsher, "On Clean Energy, China Skirts Rules," *New York Times*, September 8, 2010.

78. Personal communication with Karsten Wambach, SolarWorld Industries, Freiburg, Germany, October 5, 2010.

79. *Business Wire,* "Solar Bubble to Burst in 2009 as Supply Exceeds Demand," March 20, 2008.

80. Lux Research, "Solar State of the Market Q1 2008: The End of the Beginning," February 29, 2008.

81. Leondra Walet, "Ban on Scrap Polysilicon to Boost China Solar Sector," *Reuters,* August 27, 2009.

82. Ibid., 1.

83. Vivian L., "China Solar: Who Survives China's Polysilicon Shakeout?" *Greentech Media,* April 12, 2012.

84. Ibid.

85. Ayush Verma, "Solar Stars: China's Top 5 Manufacturers," *IAmRenew,* April 9, 2018, https://www.iamrenew.com/green-energy/solar-stars-chinas-top-5-solar-manufacturers/.

86. World Trade Organization, "Anti-dumping, Subsidies, Safeguards," Article VI of the General Agreement on Tariffs and Trade, Geneva, 1994.

87. *Renewable Energy World,* "President Obama: 'Questionable Competitive Practices Coming out of China'," November 2, 2011.

88. The Obama administration filed suit with the WTO, and China was forced to lift the export restrictions in 2015. World Trade Organization, "China: Measures Related to the Exportation of Rare Earths, Tungsten and Molybdenum," Dispute Settlement DS431, 2015.

89. U.S. Department of Energy, "2010 Critical Materials Strategy Summary," 2010.

90. SolarWorld Industries, "Petition for the Imposition of Antidumping and Countervailing Duties against Crystalline Silicon Photovoltaic Cells from the People's Republic of China," October 19, 2011.

91. World Trade Organization, "Anti-dumping."

92. Dana Hull, "U.S. Solar Industry Divided over Trade Action against China," *San Jose Mercury News,* February 15, 2012.

93. Speech at the Conference on the Renaissance of American Manufacturing, March 27, 2012.

94. Bradsher, "China Skirts Rules."

95. Ibid.

96. Kearney Alliance, "China's Solar Industry and the U.S. Anti-Dumping/Anti-Subsidy Trade Case," May 19, 2012, http://www.chinasolartrade.com.

97. In 2017 CSUN announced that it would build a 1 GW crystalline silicon manufacturing facility in Sacramento, California, to begin production in 2019.

98. Eric Wesoff, "Jigar Shah's Letter to Gordon Brinser of SolarWorld," *Greentech Media,* December 20, 2011.

99. Coalition for Affordable Solar Energy, Press Release, November 7, 2011.

100. Danny Kennedy, "Saudi Arabia Makes Big Bet on Solar," *Forbes,* February 10, 2012.

101. Quoted in Jeff Himmelman, "The Secret to Solar Power," *New York Times Magazine,* August 9, 2012.

102. Quoted in *Solar Industry Magazine,* "Solar Energy Industries Association Weighs in on Investigation into China's RE Practices," September 16, 2010.

103. Ibid.

104. Quoted in Stephen Lacey, "Are We in a Solar Trade War with China?" *Grist,* October 19, 2011.

105. U.S. International Trade Commission, "Crystalline Silicon Photovoltaic Cells and Modules From China: Investigation Nos. 701-TA-481 and 731-TA-1190," Washington, DC, 2012.

106. U.S. Department of Commerce, "Antidumping and Countervailing Duty Operations, Referral of Potential Evasion Concerns to the Department of Homeland Security," memo, April 11, 2013.

107. Keith Bradsher, "China Bends to U.S. Complaint on Solar Panels but Plans Retaliation," *New York Times,* November 21, 2011.

108. Leslie Hook, "China Imposes Tariffs on Polysilicon Exports from U.S. and South Korea," *Financial Times,* July 18, 2013.

109. Christian Roselund, "Hemlock to Close Tennessee Polysilicon Site," *PV Magazine,* December 19, 2014.

110. Ibid.

111. Ibid.

112. Ibid.

113. World Trade Organization, "Certain Measures Affecting the Renewable Energy Generation Sector," Dispute DS412, 2014.

114. *Economic Times,* "India to Achieve Target of 100 GW of Solar Power by 2022," December 21, 2017.

115. Office of the U.S. Trade Representative, "United States Challenges India's Restrictions on U.S. Solar Exports," February 2013.

116. Tom Kenning, "Both Cases Aired in India's Solar Anti-Dumping Investigation," *PV Tech,* December 13, 2017, https://www.pv-tech.org/news/both-sides-air-their-case-for-indias-solar-anti-dumping-investigation.

117. Yuki Naguchi, "After Solyndra, Other Energy Loans Draw Scrutiny," *National Public Radio,* October 3, 2011.

118. Hess, *Good Green Jobs,* 2.

119. Adrian Smith and Rob Raven, "What is Protective Space? Reconsidering Niches in Transitions to Sustainability," *Research Policy* 41, no. 6 (2012): 1025–36.

120. Karl Marx, *Grundrisse* (New York: Penguin, 1993).

121. Joseph A. Schumpeter, *Capitalism, Socialism, and Democracy* (New York: Harper and Brothers, 1942).

CHAPTER 8. SOLAR POWER AND A JUST TRANSITION

1. Quoted in John Perlin, *From Space to Earth: The Story of Solar Electricity* (Cambridge, MA: Harvard University Press, 1999).

2. U. Gangopadhyay, S.K. Dhungel, K. Kim, U. Manna, P.K. Basu, H.J. Kim, B. Karunagaran, K.S. Lee, J.S. Yoo, and J. Yi, "Novel Low Cost Chemical Texturing for Very Large Area Industrial Multi-Crystalline Silicon Solar Cells," *Semiconductor Science and Technology* 20, no. 9 (2005): 938.

3. D.S. Strebkov, A. Pinov, V.V. Zadde, E.N. Lebedev, E.P. Belov, N.K. Efimov, and S.I. Kleshevnikova, "Chlorine Free Technology for Solar-Grade Silicon Manufacturing," Workshop on Crystalline Silicon Solar Cells and Modules, Winter Park, CO, August 8–11, 2004.

4. Jan A. Staessen, A. Amery, R.R. Lauwerys, Harry A. Roels, G. Ide, and G. Vyncke. "Renal Function and Historical Environmental Cadmium Pollution from Zinc Smelters," *The Lancet* 343, no. 8912 (1994): 1523–27.

5. International Finance Corporation, *Utility-Scale Solar Photovoltaic Power Plants: A Project Developer's Guide*, Washington, DC, 2015.

6. Jim Lazar, "Value of Solar and Grid Benefits Studies," Regulatory Assistance Project, July 21, 2016, https://www.raponline.org/wp-content/uploads/2016/08/rap-lazar-euci-value-of-solar-studies-2016-july-21-2016.pdf.

7. Julian Agyeman, Robert Doyle Bullard, and Bob Evans, *Just Sustainabilities: Development in an Unequal World* (Cambridge, MA: MIT Press, 2003).

8. S.A. Sumner and P.M. Layde, "Expansion of Renewable Energy Industries and Implications for Occupational Health," *Journal of the American Medical Association* 302 (2009): 787–89.

9. Kersty Hobson, "Closing the Loop or Squaring the Circle? Locating Generative Spaces for the Circular Economy," *Progress in Human Geography* 40, no. 1 (2016): 88–104.

10. Vasilis M.Fthenakis, Hyung Chul Kim, and Eric Alsema, "Emissions from Photovoltaic Life Cycles," *Environmental Science and Technology* 42, no. 2 (2008): 2168.

11. Ibid.

12. R. Mis-Fernández, I. Rimmaudo, V. Rejón, E. Hernandez-Rodriguez, I. Riech, A. Romeo, and J.L. Peña, "Deep Study of $MgCl_2$ as Activator in CdS/CdTe Solar Cells," *Solar Energy* 155 (2017): 620–26.

13. Y.S. Tsuo, J.M. Gee, P. Menna, D.S. Strebkov, A. Pinov, and V. Zadde, *Environmentally Benign Silicon Solar Cell Manufacturing*, NREL/CP-590-23902, National Renewable Energy Laboratory, Golden, CO, 1998.

14. Desert Renewable Energy Conservation Plan (DRECP) Independent Science Advisors, "Recommendations of Independent Science Advisors for the California Desert Renewable Energy Conservation Plan," DRECP-1000-2010-008-F, October 2010.

15. Karen Eugenie Rignall, "Solar Power, State Power, and the Politics of Energy Transition in Pre-Saharan Morocco," *Environment and Planning A* 48, no. 3 (2016): 540–57.

16. Andreas Wade, Parikhit Sinha, Karen Drozniak, Dustin Mulvaney, and Jessica Slomka, "Ecodesign, Ecolabeling and Green Procurement Policies: Enabling More Sustainable Photovoltaics?" *Proceedings of the IEEE Photovoltaic Specialist Conference and World Conference on Photovoltaic Electricity Conversion*, June 16, 2018.

Bibliography

Abbey, Edward, *Desert Solitaire: A Season in the Wilderness* (New York: Random House, 1968).

Abraham, Kera, "Will Central California's Proposed Place in the Sun Ruin Life for Sustainable Farmers?" *Monterey County Weekly*, February 25, 2010.

Adams, William M., *Green Development: Environment and Sustainability in the Third World* (New York: Routledge, 2003).

Ades, A., and G. Kazantzis, "Lung Cancer in a Non-Ferrous Smelter: The Role of Cadmium," *British Journal of Industrial Medicine* 45 (1988): 435–42.

Agyeman, Julian, Robert Doyle Bullard, and Bob Evans, *Just Sustainabilities: Development in an Unequal World* (Cambridge, MA: MIT Press, 2003).

Alagona, Peter, *After the Grizzly: Endangered Species and the Politics of Place in California* (Berkeley: University of California Press, 2013).

Allen, Michael F., and Alan McHughen, "Solar Power in the Desert: Are the Current Large-Scale Solar Developments Really Improving California's Environment?" Center for Conservation Biology, University of California, Riverside, 2011.

Anastas, Paul, and John Warner, *Green Chemistry: Theory and Practice* (Oxford University Press, 1998).

Andersson, Björn A., and Staffan Jacobsson, "Monitoring and Assessing Technology Choice: The Case of Solar Cells," *Energy Policy* 28, no. 14 (2000): 1037–49.

Anthony, Mark, "Cadmium Telluride Casts Shadow on First Solar," *Seeking Alpha*, November 7, 2007, https://seekingalpha.com/article/55392-cadmium-telluride-casts-shadow-on-first-solar.

Appadurai, Arjun, *The Social Life of Things: Commodities in Cultural Perspective* (Cambridge: Cambridge University Press, 1988).

Archer, Steven R., and Katharine I. Predick. "Climate Change and Ecosystems of the Southwestern United States," *Rangelands* 30, no. 3 (2008): 23–28.

Asimov, Isaac, "Reason," in *I, Robot* (New York: Street and Smith, 1941), 319–36.

Associated Press, "Green Light for Solar Power on Federal Lands: Interior Chief Sets Goal of 13 Commercial Projects by 2010 across West," June 30, 2009.

Associated Press, "Mojave Desert Solar Plant Cooking 'Alarming' Number of Birds in Mid-air," August 18, 2014.

Babcock, Lacey, "Expedited Project Deadlines & NEPA Compliance: The Problematic Use of Categorical Exemptions," School of Design, Department of City and Regional Planning, University of Pennsylvania, May 7, 2010.

Bailie, Colin, and Michael McGehee, "High-Efficiency Tandem Perovskite Solar Cells," *MRS Bulletin* 40 (2015): 681–85.

Bair, Jennifer, *Frontiers of Commodity Chain Research* (Palo Alto, CA: Stanford University Press, 2008).

Barron-Gafford, Greg, Rebecca L. Minor, Nathan A. Allen, Alex D. Cronin, Adria E. Brooks, and Mitchell A. Pavao-Zuckerman, "The Photovoltaic Heat Island Effect: Larger Solar Power Plants Increase Local Temperatures," *Scientific Reports* 6 (2016): 35070.

Barrows, Cameron, "Sensitivity to Climate Change for Two Reptiles at the Mojave-Sonoran Desert Interface," *Journal of Arid Environments* 75, no. 7 (2011): 629–35.

Basin and Range Watch, "Ivanpah Valley ACEC Nomination Petition," Las Vegas Bureau of Land Management Field Office, October 23, 2011.

Bell, Derek, Tim Gray, and Claire Haggett, "The 'Social Gap' in Wind Farm Siting Decisions: Explanations and Policy Responses," *Environmental Politics* 14, no. 4 (2005): 460–77.

Belnap, Jayne, "Recovery Rates of Cryptobiotic Crusts: Inoculant Use and Assessment Methods," *Great Basin Naturalist* 53 (1993): 89–95.

Berger, John, *Charging Ahead: The Business of Renewable Energy and What It Means for America* (Berkeley: University of California Press, 1998).

Bernstein, Jennifer M., "Climate Change, Industrial Solar, and the Globalized Local in Joshua Tree, California," *Yearbook of the Association of Pacific Coast Geographers* 78 (2016): 80–93.

Bevill, Kris, "Getting Better: A Conversation with the U.S. DOE's Jonathan Silver, Head of Loan Programs," *Ethanol Producer Magazine*, August 15, 2011.

Bickerstaff, Karen, Gordon Walker, and Harriet Bulkeley, *Energy Justice in a Changing Climate: Social Equity and Low-Carbon Energy* (London: Zed, 2013).

Blaeloch, Janine, "Big Solar's Footprint on Public Lands," *Solar Done Right Newsletter*, December 12, 2011.

Blaeloch, Janine, "Solar Power, and Tortoises Too," *Los Angeles Times*, October 16, 2012.

Block, Ben, "U.S. Election Brings Green Jobs in Focus," WorldWatch Institute, 2008, http://www.worldwatch.org/node/5923.

Bloomberg New Energy Finance, "League Table Results Book," 2011, https://www.bnef.com/assets/pdfs/league-table-results-book-2011.pdf.

Bohland, John R., and Ken Smigielski, "First Solar's CdTe Module Manufacturing Experience: Environmental Health and Safety Results," *Conference Record of the Twenty-Eighth IEEE Photovoltaic Specialists Conference,* September 15–22, 2000.

Bowker, Goeffrey C., and Susan Leigh Star, *Sorting Things Out: Classification and Its Consequences* (Cambridge, MA: MIT Press, 2005).

Bradsher, Keith, "China Bends to U.S. Complaint on Solar Panels but Plans Retaliation," *New York Times,* November 21, 2011.

Bradsher, Keith, "For Solar Panel Industry, a Volley of Trade Cases," *New York Times,* October 11, 2012.

Bradsher, Keith, "Glut of Solar Panels Poses a New Threat to China," *New York Times,* October 4, 2012.

Bradsher, Keith, "On Clean Energy, China Skirts Rules," *New York Times,* September 8, 2010.

Bradsher, Keith, "Solar Rises in Malaysia during Trade Wars over Panels," *New York Times,* December 14, 2017.

Bradsher, Keith, "Trade War in Solar Takes Shape," *New York Times,* November 9, 2011.

Bradsher, Keith, "U.S. Solar Panel Makers Say China Violated Trade Rules," *New York Times,* October 19, 2011.

Brannstrom, Christian, Wendy Jepson, and Nicole Persons, "Social Perspectives on Wind-Power Development in West Texas," *Annals of the Association of American Geographers* 101, no. 4 (2011): 839–51.

Brechin, Grey, *Imperial San Francisco: Urban Power, Earthly Ruin* (Berkeley: University of California Press, 2006).

BrightSource, "Ivanpah Project Receives Federal Permits; Approved to Commence Construction" (press release), October 7, 2010.

Brouwers, T., "Bio-elution Test on Cadmium Telluride," ECTX Consultant, Liège, Belgium, 2010.

Brune, Michael, "Zombie Attacks on Rooftop Solar," *Sierra Club Blog,* March 17, 2015.

Bullis, Kevin, "Better Solar for Big Buildings: A Startup Is Selling Cylindrical Solar Cells That Can Generate More Power than Conventional Panels," *MIT Technology Review,* October 7, 2008.

Bullis, Kevin, "Solar's Great Leap Forward," *MIT Technology Review,* June 22, 2010.

Business Wire, "Solar Bubble to Burst in 2009 as Supply Exceeds Demand," March 20, 2008.

Butler, Declan, "Thin Films: Ready for Their Close-Up?" *Nature* 454 (2008): 558–59.

Caldwell, Diane, "Solar Company Seeks Stiff U.S. Tariffs to Deter Chinese Spying: SolarWorld Americas Says Hackers in China Stole Documents," *New York Times,* September 1, 2014.

California Department of Conservation, "The Williamson Act and Solar Energy," presentation to Fresno County, August 2012.

California Department of Toxic Substances Control, "Laws, Regulations, and Policies: Regulations," https://www.dtsc.ca.gov/LawsRegsPolicies/Regs/.

California Energy Commission, "American Reinvestment & Recovery Act (AARA) Solar Projects Progress at the Energy Commission," September 29, 2010.

California Energy Commission and California Public Utilities Commission, "History of Solar Energy in California," 2013, Go Solar California, http://www.gosolarcalifornia.ca.gov/about/gosolar/california.php.

California Independent System Operator, "California ISO Solar Production Soars to New Record" (press release), July 14, 2016.

California Public Utilities Commission, Renewable Energy Portfolio Standard: Annual Report, November 2017.

California Senate Bill 34, Solar Thermal and Photovoltaic Power Plants: Siting: California Endangered Species Act: Mitigation Measures, 2009–2010, http://leginfo.legislature.ca.gov/faces/billNavClient.xhtml?bill_id=200920108SB34.

California Senate Bill 1020, Recycling: Hazardous Waste: Photovoltaic Panels: Collection and Recycling Programs, Introduced by Senator Monning, February 14, 2014, http://leginfo.legislature.ca.gov/faces/billNavClient.xhtml?bill_id=201320140SB1020.

Callon, Michel, "Society in the Making: The Study of Technology as a Tool for Sociological Analysis," in The Social Construction of Technological Systems: New Directions in the Sociology and History of Technology, ed. Wiebe E. Bijker, Thomas Parke Hughes, and Trevor J. Pinch (Cambridge, MA: MIT Press, 1987), 83–103.

Caraher, William R., Bret Weber, Kostis Kourelis, and Richard Rothaus, "The North Dakota Man Camp Project: The Archaeology of Home in the Bakken Oil Fields," Historical Archaeology 51, no. 2 (2017): 267–87.

Carlisle, Juliet E., Stephanie L. Kane, David Solan, Madelaine Bowman, and Jeffrey C. Joe, "Public Attitudes Regarding Large-Scale Solar Energy Development in the US," Renewable and Sustainable Energy Reviews 48 (2015): 835–47.

Carlisle, Juliet E., David Solan, Stephanie Kane, and Jeffrey Joe, "Utility-Scale Solar and Public Attitudes toward Siting: A Critical Examination of Proximity," Land Use Policy 58 (2016): 491–501.

Carrier, James G., "Protecting the Environment the Natural Way: Ethical Consumption and Commodity Fetishism," Antipode 42, no. 3 (2010): 672–89.

Cart, Julie, "Environmentalists Feeling Burned by Rush to Build Solar Projects," Los Angeles Times, April 6, 2012.

Cart, Julie, "Saving Desert Tortoises Is a Costly Hurdle for Solar Projects," Los Angeles Times, March 4, 2012.

Carvill, Peter, "Obama Seeks to Extend ITC Permanently," PV Magazine, February 5, 2015, http://www.pv-magazine.com/news/details/beitrag/obama-seeks-to-extend-itc-permanently_100018023.

Castranova, Vincent, and Val Vallyathan, "Silicosis and Coal Workers' Pneumoconiosis," Environmental Health Perspectives 108, no. 4 (2000): 675–84.

CBS News, "Solyndra Not Dealing with Toxic Waste at Milpitas Facility," April 28, 2012.

Center for Biological Diversity, "Public Comment on Final EIS for Ivanpah Solar Electric Generating System," Tucson, AZ, 2010.

Center for Biological Diversity, "Questions and Answers Related to the Settlement Agreement between the Center for Biological Diversity and BrightSource Energy, Inc. Regarding the Ivanpah Solar Electric Generating System Project in the Mojave Desert," October 2010, https://www.biologicaldiversity.org/species/reptiles/desert_tortoise/pdfs/CBD_Ivanpah_Factsheet_8-8-2011.pdf.

Cha, Ariana Eunjung, "Solar Energy Firms Leave Waste behind in China," Washington Post, March 9, 2008, 1.

Chan, Jenny, and Ngai Pun, "Suicide as Protest for the New Generation of Chinese Migrant Workers: Foxconn, Global Capital, and the State," Asia-Pacific Journal 8, no. 37 (2010): 2–10.

Chan, Jenny, Ngai Pun, and Mark Selden, "The Politics of Global Production: Apple, Foxconn and China's New Working Class," New Technology, Work and Employment 28 (2013): 100–15.

Chen, Jenq-Renn, Hsiao-Yun Tsai, Shang-Kay Chen, Huan-Ren Pan, Shuai-Ching Hu, Chun-Cheng Shen, Chia-Ming Kuan, Yu-Chen Lee, and Chih-Chin Wu, "Analysis of a Silane Explosion in a Photovoltaic Fabrication Plant," Process Safety Progress 25, no. 3 (2006): 237–44.

Choudhury, Nilima, "Cleantech Venture Investment in 2011 Reaches Highest Level since 2008," PV-Tech, January 6, 2013.

Chu, Steven, "Clean Energy Jobs and American Power Act: Statement of U.S. Department of Energy Secretary Steven Chu," Senate Committee on Environment and Public Works, 111th Congress, October 27, 2009.

Chu, Steven, "The Solyndra Failure: Testimony of U.S. Department of Energy Secretary Steven Chu," House of Representatives, Committee on Energy and Commerce, 112th Congress, November 17, 2011.

Chuangchote, Surawut, Manaskorn Rachakornkij, Thantip Punmatharith, Chanathip Pharino, Chulalak Changul, Prapat Pongkiatkul, "Review of Environmental, Health and Safety of CdTe Photovoltaic Installations throughout Their Life-Cycle," Document SD-20, First Solar, Tempe, AZ, 2012.

Clarke, Chris, "Documentary Explores Conflict between Tribes, Energy Developers," KCET, March 8, 2013, https://www.kcet.org/socal-focus/documentary-explores-conflict-between-tribes-energy-developers.

Clarke, Chris, "Solar Energy Development in the Carrizo Plain Draws Lawsuits," KCET, April 23, 2011, http://www.kcet.org/news/redefine/revisit/commentary/carrizo-plain-solar-draws-lawsuit.html.

Clarke, Chris, "Water Birds Turning Up Dead at Solar Projects in the Desert," KCET, July 17, 2013. https://www.kcet.org/redefine/water-birds-turning-up-dead-at-solar-projects-in-the-desert.

Clarke, Jeanne Nienaber, and Daniel McCool, Staking Out the Terrain: Power and Performance among Natural Resource Agencies (Albany, NY: SUNY Press, 1996).

Coalition for Affordable Solar Energy, Press Release, November 7, 2011.

Coggins, George Cameron, and Parthenia Blessing Evans, "Multiple Use, Sustained Yield Planning on the Public Lands," University of Colorado Law Review 53 (1981): 411–31.

Cole, Matthew A., "Trade, the Pollution Haven Hypothesis and the Environmental Kuznets Curve: Examining the Linkages," *Ecological Economics* 48, no. 1 (2004): 71–81.

Cook, Ian, "Follow the Thing: Papaya," *Antipode* 36, no. 4 (2004): 642–64.

Corso, Jessica, "9th Circ. Buries Enviro Suit over $2.2B Mojave Solar Project," *Law 360*, May 5, 2015.

Council of State Governments, *Green Jobs Report*, 2009, http://knowledge center.csg.org/kc/system/files/Green_Jobs_Report_0.pdf.

Crosby, Alfred W., *Children of the Sun* (New York: Norton, 2006).

Cummings, Kristin J., Walter E. Donat, David B. Ettensohn, Victor L. Roggli, Peter Ingram, and Kathleen Kreiss, "Pulmonary Alveolar Proteinosis in Workers at an Indium Processing Facility," *American Journal of Respiratory and Critical Care Medicine* 181, no. 5 (2010): 458–64.

Cyrs, William D., Heather J. Avens, Zachary A. Capshaw, Robert A. Kingsbury, Jennifer Sahmel, and Brooke E. Tvermoes, "Landfill Waste and Recycling: Use of a Screening-Level Risk Assessment Tool for End-of-Life Cadmium Telluride (CdTe) Thin-Film Photovoltaic (PV) Panels," *Energy Policy* 68 (2014): 524–33.

Dannenmaier, Eric, "Executive Exclusion and the Cloistering of the Cheney Energy Task Force," *New York University Environmental Law Journal* 16 (2008): 329–80.

Davis, Mike, "The Dead West: Ecocide in Marlboro Country," *New Left Review* (1993): 49–59.

Dearen, Jason, "Solar Industry Grapples with Hazardous Wastes," Associated Press, February 10, 2013.

Demijian, Karoun, "Facts that Disprove Conspiracy Theory about Harry Reid, Cliven Bundy," *Las Vegas Sun*, April 17, 2014.

Denelski, David, "It's Not Easy Being Green: Ivanpah Solar Plant Near Nevada Burns a Lot of Natural Gas, Making It a Greenhouse Gas Emitter under State Law," *Orange County Register*, October 15, 2015.

Desert Renewable Energy Conservation Plan Independent Science Advisors, *Recommendations of Independent Science Advisors for the California Desert Renewable Energy Conservation Plan*, DRECP-1000-2010-008-F, October 2010.

de Wild-Scholten, Mariska J., Karsten Wambach, Eric A. Alsema, and A. Jäger-Waldau, "Implications of European Environmental Legislation for Photovoltaic Systems," 20th European Photovoltaic Energy Conference, Barcelona, June 6–10, 2004.

Dietz, Tom, and Paul Stern, *Public Participation In Environmental Assessment and Decision Making* (Washington, DC: National Academies Press, 2008).

Durkey, Jocelyn, "State Renewable Portfolio Standards and Goals," National Conference of State Legislatures, December 16, 2016.

Dydo, Robert, "Manufacturing Costs at First Solar No Longer Offer Advantage over the Chinese Solar Companies," *Seeking Alpha*, March 2, 2012, https://seekingalpha.com/article/407391-manufacturing-costs-at-first-solar-no-longer-offer-advantage-over-the-chinese-solar-companies.

Eberspacher, Chris, Charles F. Gay, and Paul D. Moskowitz, "Strategies for Enhancing the Commercial Viability of CdTe-Based Photovoltaics," *Solar Energy Materials and Solar Cells* 41/42 (1996): 637–53.

Eckhouse, Brian, and E. Roston, "Trump Can't Kill Solyndra Loan Office That Outperforms Banks," *Bloomberg News,* November 29, 2016.

Economic Times, "India to Achieve Target of 100 GW of Solar Power by 2022," December 21, 2017.

Ehrenfeld, John, "Industrial Ecology: Paradigm Shift or Normal Science?" *American Behavioral Scientist* 44 (2000): 229–44.

Elliott, John E., "Marx and Schumpeter on Capitalism's Creative Destruction: A Comparative Restatement," *Quarterly Journal of Economics* 95, no. 1 (1980): 45–68.

Energy Information Agency, "Annual Energy Outlook 2014," Department of Energy, 2014.

Energy Information Agency, "Annual Energy Outlook 2017," U.S. Department of Energy, 2017.

Espeland, Wendy, and Mitchell Stevens, "Commensuration as a Social Process," *Annual Review of Sociology* 24 (1998): 314.

European Chemicals Agency, "Cadmium Telluride," 2017, https://echa.europa.eu/registration-dossier/-/registered-dossier/12227.

Fargione, Joseph, Jason Hill, David Tilman, Stephen Polasky, and Peter Hawthorne, "Land Clearing and the Biofuel Carbon Debt," *Science* 319, no. 5867 (2008): 1235–38.

Farnsworth, Matthew L., Brett G. Dickson, Luke J. Zachmann, Ericka E. Hegeman, Amanda R. Cangelosi, Thomas G. Jackson Jr, and Amanda F. Scheib, "Short-Term Space-Use Patterns of Translocated Mojave Desert Tortoise in Southern California," *PloS ONE* 10, no. 9 (2015): e0134250.

Firestone, Jeremy, and Willett Kempton, "Public Opinion about Large Offshore Wind Power: Underlying Factors," *Energy Policy* 35, no. 3 (2007): 1584–98.

First Solar, "First Solar and Ordos Take Key Step Forward in 2 GW Project" (press release), November 17, 2009.

First Solar, "First Solar Announces Acquisition of Turner Renewable Energy," November 30, 2007, http://investor.firstsolar.com/news-releases/news-release-details/first-solar-announces-acquisition-turner-renewable-energy.

First Solar, U.S. Securities and Exchange Commission Filing, Form S-1, July 19, 2007.

Fleiß, Eva, Stefanie Hatzl, Sebastian Seebauer, and Alfred Posch, "Money, Not Morale: The Impact of Desires and Beliefs on Private Investment in Photovoltaic Citizen Participation Initiatives," *Journal of Cleaner Production* 141 (2017): 920–27.

Forbes, "Five Technologies Set to Change the Decade," January 7, 2009.

Foster, Richard, and Sarah Kaplan, *Creative Destruction: Why Companies That Are Built to Last Underperform the Market—And How to Successfully Transform Them* (New York: Crown Business, 2011).

Fox, Julia, "Mountaintop Removal in West Virginia: An Environmental Sacrifice Zone," *Organization & Environment* 12, no. 2 (1999): 163–83.

France Innovation Scientifique & Transfert, *Photovoltaic Thin Film Cells: Intellectual Property Overview,* FRINNOV, Paris, 2009.

Fraunhofer Institute for Solar Energy, *Photovoltaics Report,* Freiberg, Germany, November 17, 2016.

Fraunhofer Institute for Solar Energy, *Photovoltaics Report,* Freiberg, Germany, August 27, 2018.

Freidberg, Susanne, "It's Complicated: Corporate Sustainability and the Uneasiness of Life Cycle Assessment." *Science as Culture* 24, no. 2 (2015): 157–82.

Freidberg, Susanne, "On the Trail of the Global Green Bean: Methodological Considerations in Multi-Site Ethnography," *Global Networks* 1, no. 4 (2001), 353–368.

Friedland, William H., "Reprise on Commodity Systems Methodology," *International Journal of Sociology Agriculture and Food* 9, no. 1 (2001): 82–103.

Freidman, Thomas, "The Power of Green," *New York Times,* April 15, 2007.

Frisvold, George B., and Tatiana Marquez, "Water Requirements for Large-Scale Solar Energy Projects in the West," *Journal of Contemporary Water Research & Education* 151, no. 1 (2013): 106–16.

Fthenakis, Vasilis M., "Could CdTe PV Modules Pollute the Environment?" National Photovoltaic Environmental Health and Safety Assistance Center, Brookhaven National Laboratory, Upton, NY, 1999.

Fthenakis, Vasilis M. "End-of-Life Management and Recycling of PV Modules," *Energy Policy* 28, no. 14 (2000): 1051–58.

Fthenakis, Vasilis M., "Hazard Analysis for the Protection of Photovoltaic Manufacturing Facilities," in *Proceedings of 3rd World Conference on Photovoltaic Energy Conversion,* vol. 2 (IEEE, 2003), 2090–93.

Fthenakis, Vasilis M. "Life Cycle Impact Analysis of Cadmium in CdTe PV Production," *Renewable and Sustainable Energy Reviews* 8, no. 4 (2004): 303–34.

Fthenakis, Vasilis M., "Prevention and Control of Accidental Releases of Hazardous Materials in PV Facilities," *Progress in Photovoltaics* 6 (1998): 91–98.

Fthenakis, Vasilis M, "PV Recycling in the U.S.," European PV Solar Energy Conference and Exhibition and International Energy Agency Workshop on PV Life Management and Recycling, Amsterdam, The Netherlands, September 23, 2014.

Fthenakis, Vasilis M., "Sustainability of Photovoltaics: The Case for Thin-Film Solar Cells," *Renewable and Sustainable Energy Reviews* 13, no. 9 (2009): 2746–50.

Fthenakis, Vasilis M., M. Fuhrmann, J. Heiser, A. Lanzirotti, J. Fitts, and Wen-ming Wang, "Emissions and Encapsulation of Cadmium in CdTe Photovoltaic Modules during Fires," *Progress in Photovoltaics: Research and Applications* 13, no. 8 (2005): 713–23.

Fthenakis, Vasilis M., and Hyung Chul Kim, "Photovoltaics: Life-Cycle Analyses." *Solar Energy* 85, no. 8 (2011): 1609–28.

Fthenakis, Vasilis M., and Hyung Chul Kim, "Cu(InGa)Se$_2$ Thin-film Solar Cells: Comparative Life Cycle Analysis of Buffer Layers," 22nd European Photovoltaic Solar Energy Conference, Milan, September 3–7, 2007.

Fthenakis, Vasilis M., Hyung Chul Kim, and Erik Alsema, "Emissions from Photovoltaic Life Cycles," *Environmental Science and Technology* 42, no. 6 (2008): 2168–74.

Fthenakis, Vasilis M., Hyung Chul Kim, and Wenming Wang, "Life Cycle Inventory Analysis of the Production of Metals used in Photovoltaics," BNL-77919, Brookhaven National Laboratory, Upton, NY, 2007.

Fthenakis, Vasilis M., S. C. Morris, Paul D. Moskowitz, and Daniel L. Morgan, "Toxicity of Cadmium Telluride, Copper Indium Diselenide, and Copper Gallium Diselenide," *Progress in Photovoltaics: Research and Applications* 7, no. 6 (1999): 489–97.

Fthenakis, Vasilis M., and Paul Moskowitz, "Thin-Film Photovoltaic Cells: Health and Environmental Issues in Their Manufacture, Use, and Disposal," *Progress in Photovoltaics* 3, no. 5 (1995): 295–306.

Fthenakis, Vasilis M., and Wenming Wang, "Extraction and Separation of Cd and Te from Cadmium Telluride Photovoltaic Manufacturing Scrap," *Progress in Photovoltaics: Research and Applications* 14, no. 4 (2006): 363–71.

Fthenakis, Vasilis M., Wenming Wang, and Hyung Kim, "Life Cycle Inventory Analysis of the Production of Metals Used in Photovoltaics," *Renewable and Sustainable Energy Reviews* 13 (2009): 493–517.

Fthenakis, Vasilis M., and Ken Zweibel, "CdTe PV: Real and Perceived EHS Risks," National Center for PV and Solar Program, Washington, DC, 2003.

Funk, Cary, and Brian Kennedy, "The Politics of Climate," Pew Research Center, October 4, 2016, Washington, DC.

Gangopadhyay, U., S. K. Dhungel, K. Kim, U. Manna, P. K. Basu, H. J. Kim, B. Karunagaran, K. S. Lee, J. S. Yoo, and J. Yi, "Novel Low Cost Chemical Texturing for Very Large Area Industrial Multi-Crystalline Silicon Solar Cells," *Semiconductor Science and Technology* 20, no. 9 (2005): 938.

Gereffi, Gary, "Commodity Chains and Regional Divisions of Labor in East Asia," *Journal of Asian Business* 12 (1996): 75–112.

Gereffi, Gary, and M. Korzeniewicz, *Commodity Chains and Global Capitalism* (Westport, CT: Greenwood Press, 1994).

Germain, Rene, Donald W. Floyd, and Stephen V. Stehman, "Public Perceptions of the USDA Forest Service Public Participation Process," *Forest Policy and Economics* 3, no. 3 (2001): 113–24.

Germano, Jennifer M., Kimberleigh J. Field, Richard A. Griffiths, Simon Clulow, Jim Foster, Gemma Harding, and Ronald R. Swaisgood, "Mitigation-Driven Translocations: Are We Moving Wildlife in the Right Direction?" *Frontiers in Ecology and the Environment* 13, no. 2 (2015): 100–05.

Ghosh, Shikhar, and Ramana Nanda, "Venture Capital Investment in the Clean Energy Sector," Entrepreneurial Management Working Paper 11–020, Harvard Business School, 2010.

Gifford, Jonathan, "A Disruptive Partnership," *PV Magazine*, May 5, 2014.

Goldstein, Morris, "Currency Manipulation and Enforcing the Rules of the International Monetary System," in *Reforming the IMF for the 21st Century*, ed. Edwin M. Truman (Peterson Institute for Ecological Economics, Washington, DC, 2006): 150–51.

Goodman, Michael K., "Reading Fair Trade: Political Ecological Imaginary and the Moral Economy of Fair Trade Foods," *Political Geography* 23, no. 7 (2004): 891–915.

Grasso, Valerie B., "Rare Earth Elements in National Defense: Background, Oversight Issues, and Options For Congress," Congressional Research Service, Library of Congress, Washington, DC, 2013.

Gray, Andrew, "Valley Fever Hits Carrizo Plain Solar Farms Hard," KYET, May 2, 2013.

Groom, Nichola, "Solar Company Evergreen Files for Bankruptcy," *Reuters,* August 15, 2011.

Guha, Ramachandra, and Joan Martínez Alier, *Varieties of Environmentalism: Essays North and South* (London: Routledge, 2013).

Haegel, Nancy M., Robert Margolis, Tonio Buonassisi, David Feldman, Armin Froitzheim, Raffi Garabedian, and Martin Green, "Terawatt-Scale Photovoltaics: Trajectories and Challenges," *Science* 356, no. 6334 (2017): 141–43.

Haley, Usha, and George Haley, *Subsidies to Chinese Industry: State Capitalism, Business Strategy, and Trade Policy* (Oxford: Oxford University Press, 2013).

Hamilton, Anita, "Best Inventions of 2008," *Time*, October 28, 2008.

Hansen, Andrew, Ray Rasker, Bruce Maxwell, Jay Rotella, Jerry Johnson, Andrea Wright Parmenter, Ute Langner, Warren Cohen, Rick Lawrence, and Matthew Kraska, "Ecological Causes and Consequences of Demographic Change in the New West," *BioScience* 52, no. 2 (2002): 151–62.

Harvey, David, *A Brief History of Neoliberalism* (Oxford: Oxford University Press, 2007).

Harvey, David, *The Condition of Postmodernity: An Enquiry into the Origins of Cultural Change* (New York: Wiley, 1989).

Hearing before the Committee on Oversight and Government Reform, 112th Congress, March 20, 2012, https://www.gpo.gov/fdsys/pkg/CHRG-112hhrg74042/html/CHRG-112hhrg74042.htm.

Heath, Garvin A., and Margaret K. Mann, "Background and Reflections on the Life Cycle Assessment Harmonization Project," *Journal of Industrial Ecology* 16, no. S1 (2012): S8–S11.

Held, M., "Life Cycle Assessment of CdTe Module Recycling," 24th EU PVSEC Conference, Hamburg, Germany, 2009.

Hernandez, Rebecca R., Madison K. Hoffacker, and Christopher B. Field, "Land-Use Efficiency of Big Solar," *Environmental Science & Technology* 48, no. 2 (2014): 1315–23.

Hernandez, Rebecca R., Madison K. Hoffacker, Michelle L. Murphy-Mariscal, Grace C. Wu, and Michael F. Allen, "Solar Energy Development Impacts on Land Cover Change and Protected Areas," *Proceedings of the National Academy of Sciences* 112, no. 44 (2015): 13579–84.

Hess, David, *Good Green Jobs in a Global Economy: Making and Keeping New Industries in the United States* (Cambridge, MA: MIT Press, 2012).

Hilliard, Henry E., "Silver Recycling in the United States in 2000," Circular 1196-N, U.S. Geological Survey, 2003.

Himmelman, Jeff, "The Secret to Solar Power," *New York Times Magazine,* August 9, 2012.

Ho, Clifford K., Cianan A. Sims, and Joshua M. Christian, "Evaluation of Glare at the Ivanpah Solar Electric Generating System," *Energy Procedia* 69 (2015): 1296–1305.

Hobday, Michael, "East versus Southeast Asian Innovation Systems: Comparing OEM- and TNC-led Growth in Electronics," *Technology, Learning, And Innovation: Experiences of Newly Industrializing Economies* (2000): 129–69.

Hobson, Kersty, "Closing the Loop or Squaring the Circle? Locating Generative Spaces for the Circular Economy," *Progress in Human Geography* 40, no. 1 (2016): 88–104.

Hoffacker, Madison K., Michael F. Allen, and Rebecca R. Hernandez, "Land-Sparing Opportunities for Solar Energy Development in Agricultural Landscapes: A Case Study of the Great Central Valley, CA, United States," *Environmental Science and Technology* 51, no. 24 (2017): 14472–82.

Hook, Leslie, "China Imposes Tariffs on Polysilicon Exports from U.S. and South Korea," *Financial Times,* July 18, 2013.

Horváth, Gábor, Miklós Blahó, Ádám Egri, György Kriska, István Seres, and Bruce Robertson, "Reducing the Maladaptive Attractiveness of Solar Panels to Polarotactic Insects," *Conservation Biology* 24, no. 6 (2010): 1644–53.

Hsu, Tiffany, "Imperial Valley Solar Project Sued by Indian Tribe," *Los Angeles Times* blogs, November 4, 2010, http://latimesblogs.latimes.com/greenspace /2010/11/imperial-valley-solar-project-sued-by-indian-tribe.html.

H. T. Harvey & Associates, "Ivanpah Solar Electric Generating System Avian & Bat Monitoring Plan: October 2013–2014 Annual Report," April 2015.

Hull, Dana, "The Man behind Solyndra's Rise and Fall: Chris Gronet," *Mercury News,* November 26, 2011.

Hull, Dana, "U.S. Solar Industry Divided over Trade Action against China," *San Jose Mercury News,* February 15, 2012.

Hunold, Christian, and Steven Leitner, "'Hasta la Vista, Baby!' The Solar Grand Plan, Environmentalism, and Social Constructions of the Mojave Desert," *Environmental Politics* 20, no. 5 (2011): 687–704.

Intergovernmental Panel on Climate Change, *Climate Change 2013: The Physical Science Basis* (5th Assessment Report), Geneva, 2014.

International Energy Agency, "Methodology Guidelines on Life Cycle Assessment of Photovoltaic Electricity," Report IEA-PVPS, Paris, 2015.

International Finance Corporation, "Utility-Scale Solar Photovoltaic Power Plants: A Project Developers Guide," Washington, DC, 2015.

International Renewable Energy Agency, *End-of-Life Management: Solar Photovoltaic Panels,* Abu Dhabi, 2016.

International Renewable Energy Agency, "Renewable Energy and Jobs Annual Review," Abu Dhabi, 2014.

Jaffe, Mark, "Colorado Orders Abound Solar to Clean Up Hazardous Waste at Four Sites," *Denver Post,* February 25, 2013.

Jasanoff, Sheila, "Technologies of Humility," *Nature* 450 (2007): 33.

Jenkins, Kirsten, Darren McCauley, Raphael Heffron, Hannes Stephan, and Robert Rehner, "Energy Justice: A Conceptual Review," *Energy Research & Social Science* 11 (2016): 174–82.

Johnson, Steven, *Where Good Ideas Come From: The Seven Patterns of Innovation* (New York: Penguin, 2011).

Jones, Nicola, "A Scarcity of Rare Metals is Hindering Green Technologies," *Yale Environment 360,* November 18, 2013.

Jones, Van, *The Green Collar Economy* (New York: Harper Collins, 2009).

Kaczmar, Swiatoslav, "Evaluating the Read-Across Approach on CdTe toxicity for CdTe Photovoltaics," Society of Environmental Toxicology and Chemistry North America 32nd Annual Meeting, November 13–17, 2011, Boston, MA.

Kadro, Jeannette M., and Anders Hagfeldt, "The End-of-Life of Perovskite PV," *Joule* 1, no. 1 (2017): 29–46.

Kagan, Rebecca A., Tabitha C. Viner, Pepper W. Trail, and Edgard O. Espinoza, "Avian Mortality at Solar Energy Facilities in Southern California: A Preliminary Analysis," National Fish and Wildlife Forensics Laboratory, Ashland, OR.

Kahn, Debra, "U.S. Clean Tech Outpaced by China—Chu," *Energy & Environment News,* March 9, 2010.

Kammen, Daniel, Kamal Kapadia, and Matthias Fripp, *Putting Renewables to Work: How Many Jobs Can the Clean Energy Industry Generate?* Renewable and Appropriate Energy Laboratory, University of California, Berkeley, 2004.

Kasuya, M., H. Teranishi, K. Aoshima, T. Katoh, H. Horiguchi, Y. Morikawa, M. Nishijo, and K. Iwata, "Water Pollution by Cadmium and the Onset of Itai-itai Disease," *Water Science and Technology* 25, no. 11 (1992): 149–56.

Kearney Alliance, "China's Solar Industry and the U.S. Anti-Dumping/Anti-Subsidy Trade Case," May 19, 2012, http://www.chinasolartrade.com.

Kennedy, John, "First Solar Project Doesn't Jeopardize Protected Tortoise," *Law 360,* March 31, 2015.

Kennedy, Robert F., Jr, "An Ill Wind off Cape Cod," *New York Times,* December 16, 2005.

Kenning, Tom, "Both Cases Aired in India's Solar Anti-Dumping Investigation," *PV Tech,* December 13, 2017, https://www.pv-tech.org/news/both-sides-air-their-case-for-indias-solar-anti-dumping-investigation.

Kenning, Tom, "Solar Accounts for 1% of Global Electricity, How Long Will the Next 1% Take?" *PV Tech,* June 29, 2015, https://www.pv-tech.org/editors-blog/solar_accounts_for_1_of_global_electricity_how_long_will_the_next_1_take.

Klein, Naomi, *No Logo* (Ottawa: Knopf Press, 1999).

Korcaj, Liridon, Ulf J. J. Hahnel, and Hans Spada, "Intentions to Adopt Photovoltaic Systems Depends on Homeowners' Expected Personal Gains and Behavior of Peers," *Renewable Energy* 75 (2015): 407–15.

Krugman, Paul, "Building a Green Economy," *New York Times Magazine,* April 7, 2010.

Kuletz, Valerie, *Tainted Desert: Environmental and Social Ruin in the American West* (New York: Routledge, 1998).

Kuo, Iris, "DOE Loan Chief: Where We'll Invest in 2011," *Venture Beat,* December 15, 2010, http://venturebeat.com/2010/12/15/doe-loan-chief-where-well-invest-in-2011-how-abound-solar-could-compete-with-china/.

L., Vivian, "China Solar: Who Survives China's Polysilicon Shakeout?" *Greentech Media*, April 12, 2012.

Lacey, Stephen, "Are We in a Solar Trade War with China?" *Grist*, October 19, 2011.

LaFraniere, Sharon, "Chinese Protesters Accuse Solar Panel Plant of Pollution," *New York Times*, September 19, 2011.

Landberg, Reed, and Brian Eckhouse, "Trump Solar Duties Strike at $161 Billion China-Led Industry," *Bloomberg*, January 23, 2018.

Landsbergis, Paul, Janet Cahill, and Peter Schnall, "The Impact of Lean Production and Related New Systems of Work Organization on Worker Health," *Journal of Occupational Health Psychology* 4 (1999): 108–30.

Lazar, Jim, "Value of Solar and Grid Benefits Studies," Regulatory Assistance Project, July 21, 2016, https://www.raponline.org/wp-content/uploads/2016/08/rap-lazar-euci-value-of-solar-studies-2016-july-21-2016.pdf.

Lecuyer, Christophe, and David Brock, *Makers of the Microchip: A Documentary History of Fairchild Semiconductor* (Cambridge, MA: MIT Press, 2010).

Lee, Bill, and Ed LiPuma, *Financial Derivatives and the Globalization of Risk* (Durham, NC: Duke University Press, 2004).

Lerner, Steve, *Sacrifice Zones: The Frontlines of Toxic Chemical Exposure in the United States* (Cambridge, MA: MIT Press, 2010).

Lifsher, Marc, "Unlike Solyndra, Other California Solar Projects Appear on Track," *Los Angeles Times*, October 15, 2011, http://articles.latimes.com/2011/oct/15/business/la-fi-1015-solar-loans-20111015.

Lipschutz, Ronnie, and Dustin Mulvaney, "The Road Not Taken, Round II: Centralized vs. Distributed Energy Strategies and Human Security," in *International Handbook of Energy* Security, ed. H. Dyer and Julia Trombett (Northampton, MA, Edward Elgar, 2013), 483–506.

Loan Programs Office, Department of Energy, "The Financing Force behind America's Clean Energy Economy," 2011.

Lovich, Jeffrey E., and David Bainbridge, "Anthropogenic Degradation of the Southern California Desert Ecosystem and Prospects for Natural Recovery and Restoration," *Environmental Management* 24, no. 3 (1999): 309–26.

Lovich, Jeffrey E., and Joshua R. Ennen, "Wildlife Conservation and Solar Energy Development in the Desert Southwest, United States," *BioScience* 61, no. 12 (2011): 982–92.

Lovrić, Jasmina, Sung Jo Cho, Francoise Winnik, and Dusica Maysinger, "Unmodified Cadmium Telluride Quantum Dots Induce Reactive Oxygen Species Formation Leading to Multiple Organelle Damage and Cell Death," *Chemical Biology* 12, no. 11 (2005): 1227–34.

Luke, Timothy, "A Green New Deal: Why Green, How New, and What Is the Deal?" *Critical Policy Studies* 3, no. 1 (2009): 14–28.

Lux Research, "Solar State of the Market Q1 2008: The End of the Beginning," February 29, 2008.

MacKenzie, Donald, "Making Things the Same: Gases, Emission Rights and the Politics Of Carbon Markets," *Accounting, Organizations, Society* 34 (2009): 440–55.

Marcus, George, "Ethnography in/of the World System: The Emergence of Multi-Sited Ethnography," *Annual Review of Anthropology* 24, no. 1 (1995): 95–117.

Margolis, Robert, R. Mitchell, and Ken Zweibel, "Lessons Learned from the Photovoltaic Manufacturing, Technology/Photovoltaic Manufacturing R&D and Thin-Film Photovoltaics Partnership Projects," Technical Report NREL/TP-520-39780, National Renewable Energy Laboratory, Golden, CO, 2006.

Martin, Chris, "China Flooded U.S. with Solar Panels before Trump's Tariffs," *Bloomberg*, February 14, 2018, https://www.bloomberg.com/news/articles /2018-02-16/china-flooded-u-s-with-solar-panels-before-trump-s-tariffs.

Marx, Karl, *Grundrisse* (New York: Penguin, 1993).

Mason, Kelvin, and Paul Milbourne, "Constructing a 'Landscape Justice' for Windfarm Development: The Case of Nant Y Moch, Wales." *Geoforum* 53 (2014): 104–15.

Massey, Doreen, "Power-Geometry and a Progressive Sense of Place," in *Mapping the Futures: Local Cultures, Global Change*, ed. Jon Bird, Barry Curtis, Tim Putnam, George Robertson, and Lisa Tickner (London: Routledge, 1993), 59–69.

Masson, Valéry, Marion Bonhomme, Jean-Luc Salagnac, Xavier Briottet, and Aude Lemonsu, "Solar Panels Reduce both Global Warming and Urban Heat Island," *Frontiers in Environmental Science* 2, no. 14 (2014): 1–10.

Massouh, Jennifer F., George D. Cannon, Suzanne M. Logan, and David L. Schwartz, "Real Promise or False Hope: DOE's Title XVII Loan Guarantee," *Electricity Journal* 22, no. 4 (2009): 53–67.

Mayer, Brian, *Blue-Green Coalitions: Fighting for Safe Workplaces and Healthy Communities* (Ithaca, NY: Cornell University Press, 2009).

McAfee, Kathleen, "Selling Nature to Save It? Biodiversity and Green Developmentalism," *Environment And Planning D: Society and Space* 17, no. 2 (1999): 133–54.

McArdle, John, "Issa, DOE Official Clash over Obama's Early Role in Loan Guarantee," *Energy and Environment News*, September 22, 2011, https:// www.eenews.net/stories/1059954060.

McBride, Sarah, "With Solar Power, It's Green on Green," Reuters, January 10, 2011.

McCabe, Joseph, "First Solar's Stewardship of Recycled CdTe Modules in Question," *GreenTech Media*, August 27, 2013, https://www.greentechmedia .com/articles/read/first-solars-stewardship-of-recycled-cdte-modules-in-question.

McCain, John, "Press Release—Fact Sheet: John McCain's Jobs for America Economic Plan," July 7, 2008.

McCarthy, James, "First World Political Ecology: Lessons from the Wise Use Movement," *Environment and Planning A* 34, no. 7 (2002): 1281–1302.

McDonald, Robert I., Joseph Fargione, Joe Kiesecker, William M. Miller, and Jimmie Powell, "Energy Sprawl or Energy Efficiency: Climate Policy Impacts on Natural Habitat for the United States of America," *PLoS ONE* 4, no. 8 (2009): e6802–13.

McPhee, John, *Basin and Range* (New York: Macmillan, 1981).

Means, Benjamin, "Prohibiting Conduct, Not Consequences: The Limited Reach of the Migratory Bird Treaty Act," *Michigan Law Review* 97, no. 3 (1998): 823–42.

Mehta, Seema, "Romney Accuses Obama of 'Crony Capitalism' in Solyndra Trip," *Los Angeles Times,* May 31, 2012.

Memorandum of Understanding between the Department of the Interior and the State of California on Renewable Energy, http://www.drecp.org/documents/docs/California_and_Dept_of_Interior_MOU_1-13-12.pdf.

Mernit, Judith Lewis, "Here Comes the Sun," *Audubon,* September-October, 2011.

Mernit, Judith Lewis, "Is Big Desert Solar Killing Birds in Southern California?" *High Country News,* August 18, 2013.

Michaelson, Ruth, "The PV Race to Replace Silver," *PV Magazine,* August 15, 2011.

Michaud, Kristy, Juliet E. Carlisle, and Eric R.A.N. Smith, "NIMBYism vs. Environmentalism in Attitudes toward Energy Development," *Environmental Politics* 17, no. 1 (2008): 20–39.

Michler, Andrew, "Sonnenschiff: Solar City Produces 4x the Energy it Consumes," *Inhabit,* July 27, 2011, https://inhabitat.com/sonnenschiff-solar-city-produces-4x-the-energy-it-needs.

Milford, Phil, and Michael Bathon, "LDK Files Bankruptcy in U.S. Court on China Solar Glut," *Bloomberg News,* October 21, 2014.

Miller, Clark, and Jennifer Richter, "Social Planning for Energy Transitions," *Current Sustainable/Renewable Energy Reports* 1, no. 3 (2014): 77–84.

Mis-Fernández, R., I. Rimmaudo, V. Rejón, E. Hernandez-Rodriguez, I. Riech, A. Romeo, and J.L. Peña, "Deep Study of $MgCl_2$ as Activator in CdS/CdTe Solar Cells," *Solar Energy* 155 (2017): 620–26.

Mooney, Chris, "Why Big Solar and Environmentalists are Clashing over the California Desert," *Washington Post,* August 15, 2016.

Moore-O'Leary, Kara A., Rebecca R. Hernandez, Dave S. Johnston, Scott R. Abella, Karen E. Tanner, Amanda C. Swanson, Jason Kreitler, and Jeffrey E. Lovich, "Sustainability of Utility-Scale Solar Energy: Critical Ecological Concepts," *Frontiers in Ecology and the Environment* 15, no. 7 (2017): 385–94.

Morgan, Daniel L., Cassandra J. Shines, Shawn P. Jeter, Mark E. Blazka, Michael R. Elwell, Ralph E. Wilson, Sandra M. Ward, Herman C. Price, and Paul D. Moskowitz, "Comparative Pulmonary Absorption, Distribution, and Toxicity of Copper Gallium Diselenide, Copper Indium Diselenide, and Cadmium Telluride in Sprague-Dawley Rats," *Toxicology and Applied Pharmacology* 147, no. 2 (1997): 399–410.

Morgan, Daniel L., Cassandra J. Shines, Shawn P. Jeter, Ralph E. Wilson, Michael P. Elwell, Herman C. Price, and Paul D. Moskowitz, "Acute Pulmonary Toxicity of Copper Gallium Diselenide, Copper Indium Diselenide, and Cadmium Telluride Intratracheally Instilled into Rats," *Environmental Research* 71, no. 1 (1995): 16–24.

Morton, Oliver, "A New Day Dawning? Silicon Valley Sunrise," *Nature* 443, no. 7107 (2006): 19–22.

Moskowitz, Paul D., Vasilis M. Fthenakis, L. D. Hamilton, and J. C. Lee, "Public Health Issues in Photovoltaic Energy Systems: An Overview of Concerns," *Solar Cells* 19, no. 3 (1987): 287–99.

Moskowitz, Paul D., L. D. Hamilton, Samuel C. Morris, K. M. Novak, and Michael D. Rowe, *Photovoltaic Energy Technologies: Health and Environmental Effects*, BNL-51284, National Center for Analysis of Energy Systems, Brookhaven National Laboratory, Upton, NY, 1980.

Moula, Md. Munjur E., Johanna Maula, Mohamed Hamdy, Tingting Fang, Nusrat Jung, and Risto Lahdelma, "Researching Social Acceptability of Renewable Energy Technologies in Finland," *International Journal of Sustainable Built Environment* 2, no. 1 (2013): 89–98.

Mufson, Steven, "The Green Machine," *Washington Post*, December 9, 2008. http://www.washingtonpost.com/wp-dyn/content/article/2008/12/08/AR2008120803569.html?sid=ST2009090601541.

Müller, Anja, Karsten Wambach, and Eric Alsema, "Life Cycle Analysis of Solar Module Recycling Process," *Materials Research Society Symposium Proceedings*, 895, January 2006.

Mulvaney, Dustin, "Are Green Jobs Just Jobs? Cadmium Narratives in the Life Cycle of Photovoltaics," *Geoforum* 54 (2014): 178–86.

Mulvaney, Dustin, "Energy and Global Production Networks," in *The Palgrave Handbook of the International Political Economy of Energy*, ed. Thijs Van de Graaf, Benjamin K. Sovacool, Arunabha Ghosh, Florian Kern, and Michael T. Klare (New York: Springer, 2016), 621–40.

Mulvaney, Dustin, "Identifying the Roots of Green Civil War over Utility-Scale Solar Energy Projects on Public Lands across the American Southwest," *Journal of Land Use Science* 12, no. 6 (2017): 493–515.

Mumford, Lewis, *Technics and Civilization* (Chicago, IL: University of Chicago Press, 2010).

Naguchi, Yuki, "After Solyndra, Other Energy Loans Draw Scrutiny," *National Public Radio*, October 3, 2011.

National Renewable Energy Center and Fundación Chile, "First Solar CdTe Photovoltaic Technology: Environmental, Health and Safety Assessment," October 2013.

National Renewable Energy Laboratory, "Innovation: Thin-Film Manufacturing Process Gives Edge to Photovoltaic Start-Up," NREL/FS-6A4–47570, Department of Energy, June 2010.

National Renewable Energy Laboratory, "Thin-Film Partnership," https://www.nrel.gov/pv/thin-film-partnership.html.

Natural Resources Committee, H.R. 2915, "Chairmen Seek to Halt Risky WAPA Loans by Energy Department, Give Notice on Further Oversight and Document Requests," 112th Congress, November 9, 2011, http://naturalresources.house.gov/News/DocumentSingle.aspx?DocumentID=268112.

Navarro, James, "Right Idea, Wrong Place: Groups Sue Solar Project to Protect Imperiled Wildlife and Wild Lands," *Defenders of Wildlife Blog*, March 27, 2012.

Nelson, Rebecca, and Debra Perrone, "The Role of Permitting Regimes in Western United States Groundwater Management," *Groundwater* 54, no. 6 (2016): 761–64.

Newell, Peter, and Dustin Mulvaney, "The Political Economy of the 'Just Transition'," *Geographical Journal* 179, no. 2 (2013): 132–40.

Newingham, Beth A., Cheryl H. Vanier, Therese N. Charlet, Kiona Ogle, Stanley D. Smith, and Robert S. Nowak, "No Cumulative Effect of 10 Years Of Elevated [CO_2] on Perennial Plant Biomass Components in the Mojave Desert," *Global Change Biology* 19, no. 7 (2013): 2168–81.

New York Times, "American Inventor Uses Egypt's Sun for Power; Appliance Concentrates the Heat Rays and Produces Steam, Which Can Be Used to Drive Irrigation Pumps in Hot Climates," July 2, 1916.

New York Times, "RFK Jr., Enviros Clash over Mojave Solar Proposal," September 9, 2009.

Nordberg, Gunnar, Robert Franciscus, Martin Herber, and Lorenzo Alessio, *Cadmium in the Human Environment: Toxicity and Carcinogenicity,* No. 118, International Agency for Research on Cancer, Lyon, 1992.

Nordhaus, Ted, and Michael Shellenberger, *Breakthrough: From the Death of Environmentalism to the Politics of Possibility* (New York: Houghton Mifflin Harcourt, 2007).

Nordhaus, Ted, and Michael Shellenberger, "The Death of Environmentalism: Global Warming Politics in a Post-Environmental World," Breakthrough Institute, 2004, http://www.thebreakthrough.org/images/Death_of_Environmentalism.pdf.

Nordhaus, Ted, and Michael Shellenberger, "Green Jobs for Janitors," *New Republic,* October 6, 2010.

North Carolina Clean Energy Technology Center, "DSIRE: Database of State Incentives for Renewables & Efficiency," http://www.dsireusa.org/.

Norwegian Geotechnical Institute, "Environmental Risks Regarding the Use and End-of-Life Disposal of CdTe PV Modules," Trondheim, Norway, April 16, 2010.

O'Brien, Bill, "Follow the Sun: Cone Drive's Heliostats will Help California Solar Company," *Traverse City Record-Eagle,* February 13, 2011.

Odum, Eugene P., *Fundamentals of Ecology* (Toronto: W.B. Saunders, 1993).

Oettinger, Shamsiah Ali, "Survival of the Fittest and Cleanest," *PV Magazine* March 28, 2011.

Office of the U.S. Trade Representative, "United States Challenges India's Restrictions on U.S. Solar Exports," February 2013.

Office of the White House, "A Retrospective Assessment of Clean Energy Investments in the Recovery Act," February 2016.

Ong, Sean, Clinton Campbell, Paul Denholm, Robert Margolis, and Garvin Heath, "Land-Use Requirements for Solar Power Plants in the United States," NREL/TP-6A20–56290, National Renewable Energy Laboratory, Golden, CO, 2013.

Pacala, Stephen, and Robert Socolow, "Stabilization Wedges: Solving the Climate Problem for the Next 50 Years with Current Technologies," *Science* 305, no. 5686 (2004): 968–72.

Pasqualetti, Michael, "Wind Energy Landscapes: Society and Technology in the California Desert," *Society and Natural Resources* 14 (2001): 689–99.

Pasqualetti, Martin J., and Byron A. Miller, "Land Requirements for the Solar and Coal Options," *Geographical Journal* 150, no. 2 (1984): 192–212.

Pellow, David N., "Environmental Inequality Formation toward a Theory of Environmental Injustice," *American Behavioral Scientist* 43, no. 4 (2000), 581–601.

Pendery, Bruce, "BLM's Retained Rights: How Requiring Environmental Protection Fulfills Oil and Gas Lease Obligations," *Environmental Law* 40 (2010): 599–685.

Pentland, William, "First Solar's Dangerous Game," *Forbes,* May 10, 2011.

Perkins, John, *Changing Energy: The Transition to a Sustainable Future* (Oakland: University of California Press, 2018).

Perlin, John, *From Space to Earth: The Story of Solar Electricity* (Cambridge, MA: Harvard University Press, 1999).

Perlin, John, "Silicon Solar Cell Turns 50," NREL/BR-520–33947, National Renewable Energy Laboratory, Golden, CO, 2004.

Peters, Gerhard, and John T. Woolley, "American Presidency Project," http://www.presidency.ucsb.edu/ws/?pid=93853.

Pfund, Nancy, "Opinion: Donuts, Renewable Energy and What They Say about America," *San Jose Mercury News,* April 25, 2012.

Phadke, Roopali, "Public Deliberation and the Geographies of Wind Justice," *Science as Culture* 22, no. 2 (2013): 247–55.

Phadke, Roopali, "Resisting and Reconciling Big Wind: Middle Landscape Politics in the New American West," *Antipode* 43 (2011): 754–76.

Phillips, David, "How First Solar's Tellurium Deal Shows the Fragile Economics of Solar Panels," CBS News, November 29, 2010.

Pinch, Trevor, and Wiebe Bijker, "The Social Construction of Facts and Artefacts: Or How the Sociology of Science and the Sociology of Technology Might Benefit Each Other," *Social Studies of Science* 14 (1984): 399–441.

Pinderhughes, Rachel, "Green Collar Jobs: Workforce Opportunities in the Growing Green Economy," *Race, Poverty, & the Environment* 13, no. 1 (2006): 2–6.

Plambeck, Erica L., and Terry A Taylor, "Sell the Plant? The Impact of Contract Manufacturing on Innovation, Capacity, and Profitability," *Management Science* 51 (2005): 133–50.

Pope, Carl, and Bjorn Lomborg, "Debate: The State of Nature," *Foreign Affairs,* October 22, 2009.

Porter, Theodore, *Trust in Numbers: The Pursuit of Objectivity in Science and Public Life* (Princeton, NJ: Princeton University Press, 1995).

Power, Michael, *The Audit Society: Rituals of Verification* (Oxford: Oxford University Press, 1997).

Powers, Bill, and Sheila Bowers, "Distributed Solar PV: Why It Should Be the Centerpiece of U.S. Solar Energy Policy," Solar Done Right, September 10, 2010.

Preda, Alex, "Socio-Technical Agency in Financial Markets: The Case of the Stock Ticker," *Social Studies of Science* 36 (2006): 753–82.

Preda, Alex, "STS and Social Studies of Finance," in *The Handbook of Science and Technology Studies*, ed. E. Hackett, O. Amsterdamska, M. Lynch, and J. Wajcman (Cambridge, MA: MIT Press, 2007).

Prior, Brett, "Tessera / Stirling Sell Their Other Major Dish Project to PV Developer," *Greentech Media*, February 16, 2011.

PV Magazine, "Cadmium Won't Be Banned under RoHS, as Lobbyists Battle it Out," November 24, 2010.

PV Magazine, "Foxconn Finalizes Sharp Acquisition," August 15, 2016, https://www.pv-magazine.com/2016/08/15/foxconn-finalizes-sharp-acquisition_100025762/.

PV Magazine, "Professor Green: I Would Like to See a Date Fixed for RoHS-Compliance," September 14, 2010.

PV Tech, "Venture Capital Solar Investments Fall by Almost 50%," January 10, 2013, http://www.pv-tech.org/news/venture_capital_solar_investments_fall_by_almost_50.

Raj-Reichert, Gale, "Safeguarding Labour in Distant Factories: Health and Safety Governance in an Electronics Global Production Network," *Geoforum* 44 (2013): 23–31.

Ramos-Ruiz, Adriana, Jean V. Wilkening, James A. Field, and Reyes Sierra-Alvarez, "Leaching of Cadmium and Tellurium from Cadmium Telluride (Cdte) Thin-Film Solar Panels under Simulated Landfill Conditions," *Journal of Hazardous Materials* 336 (2017): 57–64.

Randall, J.M., S.S. Parker, J. Moore, B. Cohen, L. Crane, B. Christian, D. Cameron, J. MacKenzie, K. Klausmeyer and S. Morrison, "Mojave Desert Ecoregional Assessment," Nature Conservancy, San Francisco, California, 2010.

Rasker, Ray, "An Exploration into the Economic Impact of Industrial Development versus Conservation on Western Public Lands," *Society and Natural Resources* 19, no. 3 (2006): 191–207.

Raugei, Marco, and Vasilis M. Fthenakis, "Cadmium Flows and Emissions from CdTe PV: Future Expectations," *Energy Policy* 38 (2010): 5223–28.

Redlinger, Mike, Alan Goodrich, Michael Woodhouse, Martin Lokanc, and Roderick Eggert, "The Present, Mid-Term, and Long-Term Supply Curves for Tellurium; and Updates in the Results from NREL's CdTe PV Module Manufacturing Cost Model," Technical Report, National Renewable Energy Laboratory, 2013.

Renewable Energy World, "President Obama: 'Questionable Competitive Practices Coming out of China'," November 2, 2011.

Rignall, Karen Eugenie, "Solar Power, State Power, and the Politics of Energy Transition in Pre-Saharan Morocco," *Environment and Planning A* 48, no. 3 (2016): 540–57.

Ristau, Oliver, "First Solar Invests in Mexican Tellurium Mine," *PV Magazine*, October 19, 2011.

Robbins, Paul, "Cries along the Chain of Accumulation," *Geoforum* 54 (2011): 233–35.

Robbins, Paul, *Political Ecology: A Critical Introduction* (Oxford: Blackwell, 2011).

Roberts, Michael J., Joseph B. Lassiter III, and Ramana Nanda, "U.S. Department of Energy & Recovery Act Funding: Bridging the 'Valley of Death'," Case 810–144, Harvard Business School, June 2010.

Rose, Nikolas, *Powers of Freedom: Reframing Political Thought* (Cambridge, UK: Cambridge University Press, 1999).

Roselund, Christian, "Hemlock to Close Tennessee Polysilicon Site," *PV Magazine,* December 19, 2014.

Roth, Sammy, "World's Largest Solar Plant Opens in California Desert," *USA Today,* February 10, 2015.

Rusco, Frank, "DOE Loan Programs," Testimony by the Director of Natural Resources and Environment before the Subcommittee on Oversight and Investigations, Committee on Energy and Commerce, House of Representatives, GAO-14-645T, May 30, 2014.

Ryan, Joe, "NRG's Massive California Solar Plant Finally Making Enough Power," *Bloomberg News,* February 1, 2017.

Sahagun, Louis, "Canine Distemper in Kit Foxes Spreads in Mojave Desert," *Los Angeles Times,* April 18, 2012.

Sahagun, Louis, "Discovery of Indian Artifacts Complicates Genesis Solar Project," *Los Angeles Times,* April 24, 2012, http://articles.latimes.com /2012/apr/24/local/la-me-solar-bones-20120424.

Sahagun, Louis, "Problems Cast Shadows of Doubt on Solar Project," *Los Angeles Times,* February 11, 2012.

Scheer, Hermann, *The Solar Economy: Renewable Energy for a Sustainable Global Future* (New York: Routledge, 2013).

Schelly, Chelsea, "Residential Solar Electricity Adoption: What Motivates, and What Matters? A Case Study of Early Adopters," *Energy Research & Social Science* 2 (2014): 183–91.

Schlesinger, William H., "Carbon Storage in the Caliche of Arid Soils: A Case Study from Arizona," *Soil Science* 133, no. 4 (1982): 247–56.

Schneider, Raphaël, Cécile Wolpert, Hélène Guilloteau, Lavinia Balan, Jacques Lambert, and Christophe Merlin, "The Exposure of Bacteria to CdTe-Core Quantum Dots: The Importance of Surface Chemistry on Cytotoxicity," *Nanotechnology* 20, no. 22 (2009), 225101.

Schrader-Frechette, Kristen, "Methodological Rules For Four Classes of Scientific Uncertainty," in *Scientific Uncertainty and Environmental Problem Solving,* ed. Jack Lemons (Cambridge: Blackwell Science, 1996), 12–39.

Schumpeter, Joseph, *Capitalism, Socialism, and Democracy* (New York: Harper and Brothers, 1942).

Schurman, Rachel, and Dennis Takahashi-Kelso, *Engineering Trouble: Biotechnology and its Discontents* (Berkeley: University of California Press, 2003).

Schwartz, Noaki, and Jason Dearen, "Native American Groups Sue to Stop Solar Projects," Associated Press, February 27, 2011.

Scott, Robert, *The China Toll,* Economic Policy Institute, Washington, DC, August 23, 2012.

Seltenrich, Nate, "Oakland Invades the Desert," *East Bay Express,* December 8, 2010.

SEMATECH, "Silane Safety Improvement Project: S71 Final Report," Albany, NY, 1994.

Serrano-Ortiz, Penelope, Marilyn Roland, Sergio Sanchez-Moral, Ivan A. Janssens, Francisco Domingo, Yves Goddéris, and Andrew S. Kowalski, "Hidden, Abiotic CO_2 Flows and Gaseous Reservoirs in the Terrestrial Carbon Cycle: Review and Perspectives," *Agricultural and Forest Meteorology* 150, no. 3 (2010): 321–29.

Sichao, Kan, "Chinese Photovoltaic Market and Industry Outlook," *Institute of Energy Economics Journal*, April 2010.

Silicon Valley Toxics Coalition, *Toward a Just and Sustainable Solar Energy Industry*, San Jose, CA, 2009.

Silver, Jonathan. "Department of Energy Talks Department of Energy: An Insider's Look Back at the Loan-Guarantee Program," *Solar Industry* 4, no. 6 (2011): 1, 12–15.

Sinha, Parikhit, Michael Fischman, Jim Campbell, Gaik Cheng Lee, and Lein Sim Lim, "Biomonitoring of CdTe PV Manufacturing and Recycling Workers," *43rd IEEE Photovoltaic Specialists Conference*, 2016, 3587–92.

Sinha, Parikhit, Christopher J. Kriegner, William A. Schew, Swiatoslav W. Kaczmar, Matthew Traister, and David J. Wilson, "Regulatory Policy Governing Cadmium-Telluride Photovoltaics: A Case Study Contrasting Life Cycle Management with the Precautionary Principle," *Energy Policy* 36 (2008): 381–87.

Sivaram, Varun, *Taming the Sun: Innovations to Harness Solar Energy and Power the Planet* (Cambridge: MIT Press, 2018).

Skillen, James, *The Nation's Largest Landlord: The Bureau of Land Management in the American West* (Lawrence: University Press of Kansas, 2009).

Slesser, Malcolm, and Ian Hounam, "Solar Energy Breeders," *Nature* 262 (1976): 244–45.

Smil, Vaclav, "On Energy and Land: Switching from Fossil Fuels to Renewable Energy Will Change Our Patterns of Land Use," *American Scientist* 72, no. 1 (1984): 15–21.

Smith, Adrian, and Rob Raven, "What is Protective Space? Reconsidering Niches in Transitions to Sustainability," *Research Policy* 41, no. 6 (2012): 1025–36.

Smith, Ted, David Allan Sonnenfeld, and David N. Pellow, *Challenging the Chip: Labor Rights and Environmental Justice in the Global Electronics Industry* (Philadelphia, PA: Temple University Press, 2006).

Smith, Willoughby, "Effect of Light on Selenium during the Passage of an Electric Current," *Nature* 7, no. 303 (1873): 303.

Snyder, Jim, "FBI Raid on Solyndra May Herald Escalation of Watchdog Probe," *Bloomberg News,* September 10, 2011.

Solar Industry Magazine, "China's New Rules for Solar Polysilicon Factories Expected to Force Consolidation," January 26, 2011.

Solar Industry Magazine, "Solar Energy Industries Association Weighs in on Investigation into China's RE Practices," September 16, 2010.

SolarWorld Industries, "Petition for the Imposition of Antidumping and Countervailing Duties against Crystalline Silicon Photovoltaic Cells from the People's Republic of China," October 19, 2011.

Sollman, Dominik, and Christoph Podewils, "How Dangerous is Cadmium Telluride?" *Photon International,* September 2009.

Solnit, Rebecca, "Are We Missing the Big Picture on Climate Change?" *New York Times Magazine,* December 2, 2014, 13.

Solnit, Rebecca, *Savage Dreams: A Journey into the Landscape Wars of the American West* (Berkeley: University of California Press, 1994).

Solyndra, "Method of Depositing Materials on a Non-planar Surface," U.S. patent no. 7563725 B2, 2007.

Sovacool, Benjamin K., Matthew Burke, Lucy Baker, Chaitanya Kumar Kotikalapudi, and Holle Wlokas, "New Frontiers and Conceptual Frameworks for Energy Justice," *Energy Policy* 105, no. 6 (2017): 677–91.

Spence, Mark David, *Dispossessing the Wilderness: Indian Removal and the Making of the National Parks* (Oxford University Press, 1999).

Staessen, Jan A., A. Amery, R. R. Lauwerys, Harry A. Roels, G. Ide, and G. Vyncke, "Renal Function and Historical Environmental Cadmium Pollution from Zinc Smelters," *The Lancet* 343, no. 8912 (1994): 1523–27.

Stanford, Craig, *The Last Tortoise: A Tale of Extinction in Our Lifetime* (Cambridge, MA: Harvard University Press, 2010).

Stark, Mike, "Big Money for the Lowly Desert Tortoise," *Associated Press,* February 22, 2009.

Statement of Jens Meyerhoff, First Solar, Before the Committee on Energy and Natural Resources, U.S. Senate, September 23, 2010, https://www.energy.senate.gov/public/index.cfm/files/serve?File_id=3EC7189C-C026-A3EF-7D7C-FE79B83040A4.

Steinberg, D., G. Porro, and M. Goldberg, *Preliminary Analysis of the Jobs and Economic Impacts of Renewable Energy Projects Supported by the §1603 Treasury Grant Program,* NREL/TP-6A20–52739, National Renewable Energy Laboratory, Golden, CO, 2012.

Steinberger, Hartmut, "Health, Safety and Environmental Risks from the Operation of CdTe and CIS Thin-film Modules," *Progress in Photovoltaics* 6 (1998): 99–103.

Stoms, David M., Stephanie L. Dashiell, and Frank W. Davis, "Siting Solar Energy Development to Minimize Biological Impacts," *Renewable Energy* 57 (2013): 289–98.

Stone, Deborah A., *Policy Paradox: The Art of Political Decision Making* (New York: W. W. Norton, 1997).

Streater, Scott, "Fast-Tracked Solar Project Could Speed Mojave Desert's Demise," *New York Times,* November 12, 2009.

Strebkov, D. S., A. Pinov, V. V. Zadde, E. N. Lebedev, E. P. Belov, N. K. Efimov, and S. I. Kleshevnikova, "Chlorine Free Technology for Solar-Grade Silicon Manufacturing," Workshop on Crystalline Silicon Solar Cells and Modules, Winter Park, CO, August 8–11, 2004.

Sullivan, Marianne, *Tainted Earth: Smelters, Public Health, and the Environment* (New Brunswick, NJ: Rutgers University Press, 2014).

Sumner, Steven A., and Peter M. Layde, "Expansion of Renewable Energy Industries and Implications for Occupational Health," *Journal of the American Medical Association* 302, no. 7 (2009): 787–89.

Taleb, Nassim Nicholas, *The Black Swan: The Impact of the Highly Improbable* (New York: Random House, 2007).

Taylor, Margaret, "Beyond Technology-Push and Demand-Pull: Lessons from California's Solar Policy," *Energy Economics* 30 (2008): 2829–54.

Thompson, Neil, and Jennifer Ballen, "First Solar," Sloan School of Management, Massachusetts Institute of Technology, 2017.

Trabish, Herman K., "Construction Halted at First Solar's 230 MW Antelope Valley Site," *Greentech Media,* April 22, 2013.

Tsuo, Y. S., J. M. Gee, P. Menna, D. S. Strebkov, A. Pinov, and V. Zadde, *Environmentally Benign Silicon Solar Cell Manufacturing,* NREL/CP-590–23902, National Renewable Energy Laboratory, Golden, CO, 1998.

Turner, Frederick, Philip Medica, and Craig Lyons, "Reproduction and Survival of the Desert Tortoise in Ivanpah Valley, California," *Copeia* 4 (1984): 811–20.

Turney, Damon, and Vasilis M. Fthenakis, "Environmental Impacts from the Installation and Operation of Large-Scale Solar Power Plants," *Renewable and Sustainable Energy Reviews* 15, no. 6 (2011): 3261–70.

Unruh, Gregory, "Understanding Carbon Lock-in," *Energy Policy* 28 (2000): 817–30.

U.S. Bureau of Labor Statistics, "Careers in Solar Power," http://www.bls.gov/green/solar_power/, accessed June 2, 2012.

U.S. Bureau of Labor Statistics, "Measuring Green Jobs," 2011.

U.S. Bureau of Land Management, "Biological Assessment for the Ivanpah Solar Electric Generating System," December 7, 2009.

U.S. Bureau of Land Management, "BLM Rejects Solar Development in Silurian Valley," News Release CA-SO-15-06, November 20, 2014.

U.S. Bureau of Land Management, "California Public Lands 2015, National System of Public Lands," Sacramento, 2015.

U.S Bureau of Land Management, "Desert Renewable Energy Conservation Plan: Record of Decision," September 2016.

U.S. Bureau of Land Management, "Fast-Track Renewable Energy Projects," http://www.blm.gov/wo/st/en/prog/energy/renewable_energy/fast-track_renewable.html, accessed July 1, 2015.

U.S. Bureau of Land Management, "Revised Biological Assessment for the Ivanpah Solar Electric Generating System," April 19, 2011.

U.S Bureau of Land Management, "Solar Energy Program, Western Solar Plan," http://blmsolar.anl.gov/.

U.S. Census Bureau, "U.S. Trade in Goods with China, 2012," https://www.census.gov/foreign-trade/balance/c5700.html.

U.S.-China Economic and Security Review Commission, "China's Intellectual Property Rights and Indigenous Innovation Policy," hearing before the U.S.-China Economic and Security Review Commission, May 4, 2011.

U.S. Department of Commerce, "Antidumping and Countervailing Duty Operations, Referral of Potential Evasion Concerns to the Department of Homeland Security," memo, April 11, 2013.

U.S. Department of Commerce, *China: Competition from State-Owned Enterprises,* International Trade Administration, June, 17, 2016.

U.S. Department of Energy, "2010 Critical Materials Strategy Summary," 2010.

U.S. Department of Energy, "Energy Department Finalizes Loan Guarantee to Support California Solar Generation Project" (press release), September 30, 2011.

U.S. Department of Energy, "First Solar Topaz Farm Draft Environmental Impact Statement," 2011, http://energy.gov/lpo/about-us/environmental-compliance/environmental-impact-statements/eis-0458-first-solar-topaz.

U.S. Department of Energy, "Loan Guarantee Solicitation Announcement: Federal Loan Guarantees for Projects that Employ Innovative Energy Efficiency, Renewable Energy, and Advanced Transmission and Distribution Technologies," July 29, 2009.

U.S. Department of Energy, "Solar Energy Jobs Outpace U.S. Economy," January 12, 2016, http://energy.gov/articles/solar-energy-jobs-outpace-us-economy.

U.S. Department of Energy, "SunShot Vision Study," February 2012, Office of Energy Efficiency and Renewable Energy, https://www1.eere.energy.gov/solar/pdfs/47927.pdf.

U.S. Department of Labor and Occupational Safety and Health Administration, "Cadmium," OSHA Report 3136–06R, 2004.

U.S. Department of the Interior, "BLM Initiates Environmental Analysis of Solar Energy Development" (press release), May 29, 2008.

U.S. Department of the Interior, "Draft Solar Programmatic Environmental Impact Statement," May 29, 2008.

U.S. Department of the Interior, "Interior Department Finalizes Rule Providing a Foundation for the Future of BLM's Renewable Energy Program" (press release), November 11, 2016.

U.S. Department of the Interior, "Salazar Approves Sixth and Largest Solar Project Ever on Public Lands" (press release), October 25, 2010.

U.S. Department of the Interior, "Secretary Salazar, Gov. Schwarzenegger Sign Initiative to Expedite Renewable Energy Development," October 12, 2009.

U.S. Energy Policy Act of 2005, Public Law 109–58, August 8, 2005.

U.S. Federal Lands Management and Policy Act of 1976, §501(a)(4).

U.S. Fish and Wildlife Service, "Desert Tortoise Recovery Plan: Mojave Population," June 1994.

U.S. Geological Survey, "Platinum-Group Metals: Statistics and Information," 2013.

U.S. Geological Survey, "Selenium and Tellurium Statistics and Information," Mineral Commodity Summaries, 2017.

U.S. Geological Survey, "Tellurium," Mineral Commodity Summaries, January 2017, https://minerals.usgs.gov/minerals/pubs/commodity/selenium/mcs-2017-tellu.pdf.

U.S. Geological Survey, "Tellurium: The Bright Future of Solar Energy," Fact Sheet 2014-3077, USGS Mineral Sources Program, April 2015.

U.S. Government Accountability Office, "Electricity Generation Projects: Additional Data Could Improve Understanding of the Effectiveness of Tax Expenditures," GAO-15–302, April 2015, https://www.gao.gov/assets/670/669881.pdf.

U.S. House of Representatives Committee on Oversight and Government Reform, "The Department of Energy's Disastrous Management of Loan Guarantee Programs," Staff Report, U.S. House of Representatives, 112th Congress, March 20, 2012.

U.S. International Trade Commission, "Crystalline Silicon Photovoltaic Cells and Modules From China: Investigation Nos. 701-TA-481 and 731-TA-1190," Washington, DC, 2012.

U.S. National Environmental Policy Act of 1969, §102(C).

U.S. Occupational Safety and Health Administration, "Cadmium: Toxic and Hazardous Substances Regulations, Part 1910.1027," 2016.

U.S. President Barack Obama, "Remarks by the President on the Economy at Solyndra, Inc., Fremont, California," White House press release, May 26, 2010.

U.S President Barack Obama, "Solar Power to Power a Clean Economy," weekly radio address, October 10, 2010.

U.S. President Barack Obama, "State of the Union," 2011.

U.S. Securities and Exchange Commission, "Solyndra, Inc.," Form S-1, December 18, 2009.

Valiullin, Adel, and Valentin Tarabarin, "Archimedes' Burning Mirrors: Myth or Reality?" *History of Mechanism and Machine Science* 11 (2010): 387–96.

van der Horst, Dan, "NIMBY or Not? Exploring the Relevance of Location and the Politics of Voiced Opinions in Renewable Energy Siting Controversies," *Energy Policy* 35, no. 5 (2007): 2705–14.

Vekshin, Alison, and Mark Chediak, "Solyndra's $733 Million Plant Had Whistling Robots, Spa Showers," *Bloomberg News*, September 28, 2011.

Verma, Ayush, "Solar Stars: China's Top 5 Manufacturers," *IAmRenew,* April 9, 2018, https://www.iamrenew.com/green-energy/solar-stars-chinas-top-5-solar-manufacturers/.

Wade, Andreas, Parikhit Sinha, Karen Drozniak, Dustin Mulvaney, and Jessica Slomka, "Ecodesign, Ecolabeling and Green Procurement Policies – enabling more Sustainable Photovoltaics?" *Proceedings of the IEEE Photovoltaic Specialist Conference and World Conference on Photovoltaic Electricity Conversion,* June 16, 2018.

Wadia, Cyrus A., Paul Alivisatos, and Daniel M. Kammen, "Materials Availability Expands the Opportunity for Large-scale Photovoltaics Deployment," *Environmental Science & Technology* 43, no. 6 (2009): 2072–77.

Walet, Leondra, "Ban on Scrap Polysilicon to Boost China Solar Sector," *Reuters,* August 27, 2009.

Walker, Gordon, and Noel Cass, "Carbon Reduction, 'the Public' and Renewable Energy: Engaging with Socio-Technical Configurations," *Area* 39, no. 4 (2007): 458–69.

Walker, Gordon, Noel Cass, Kate Burningham, and Julie Barnett, "Renewable Energy and Sociotechnical Change: Imagined Subjectivities of 'the Public' and Their Implications," *Environment and Planning A* 42, no. 4 (2010): 931–47.

Wallerstein, Immanuel, "The Rise and Future Demise of the World Capitalist System: Concepts for Comparative Analysis," *Comparative Studies in Society and History* 16, no. 4 (1974): 387–415.

Walston, Leroy J., Shruti K. Mishra, Heidi M. Hartmann, Ihor Hlohowskyj, James McCall, and Jordan Macknick, "Examining the Potential for Agricultural Benefits from Pollinator Habitat at Solar Facilities in the United States," *Environmental Science and Technology* 52, no. 13 (2018): 7566–76.

Walston, Leroy J., Jr, Katherine E. Rollins, Kirk E. LaGory, Karen P. Smith, and Stephanie A. Meyers, "A Preliminary Assessment of Avian Mortality at Utility-Scale Solar Energy Facilities in the United States," *Renewable Energy* no. 92 (2016): 405–14.

Walters, Carl, *Adaptive Management of Renewable Resources* (New York: Macmillan, 1986).

Wang, Yong-hau, Guo-liang Luo, and Yi-wei Guo, "Why Is There Overcapacity in China's PV Industry in Its Early Growth Stage?" *Renewable Energy* 72 (2014): 188–94.

Webster, Veronica, "Fighting for Clean Solar Energy in Europe," Bellona, May 4, 2010, http://bellona.org/news/renewable-energy/solar/2010-05-fighting-for-clean-solar-energy-in-europe.

Wehausen, John, and Clinton W. Epps, "Protecting Desert Bighorn Sheep Migration Corridors in Mojave Desert: Guest Commentary," *Daily Bulletin*, January 12, 2015.

Weiss, Michael, "Everybody Loves Solar Energy, but . . ." *New York Times*, September 24, 1989.

Wesoff, Eric, "First Solar Acquires GE's PrimeStar Solar IP," *Greentech Media*, August 6, 2013.

Wesoff, Eric, "Jigar Shah's Letter to Gordon Brinser of SolarWorld," *Greentech Media*, December 20, 2011.

Wheat, Frank, *California Desert Miracle: The Fight for Desert Parks and Wilderness* (El Cajon, CA: Sunbelt, 1999).

Wilderness Society, "Solar Facility Will Improve Toxic Lands while Providing Much Needed Jobs and Energy to Local Economy" (press release), May 13, 2010.

Wilken, Jason A., Gail Sondermeyer, Dennis Shusterman, Jennifer McNary, Duc J. Vugia, Ann McDowell, Penny Borenstein, et al., "Coccidioidomycosis among Workers Constructing Solar Power Farms, California, USA, 2011–2014," *Emerging Infectious Diseases* 21, no. 11 (2015): 1997.

Wilson, E. O., and Thomas Lovejoy, "A Mojave Solar Project in the Bighorns' Way," *New York Times*, September 11, 2015.

Winner, Langdon, "Upon Opening the Black Box and Finding It Empty: Social Constructivism and the Philosophy of Technology," *Science, Technology, & Human Values* 18, no. 3 (1993): 362–78.

Winner, Langdon, *The Whale and the Reactor: A Search for Limits in an Age of High Technology* (Chicago, IL: University of Chicago Press, 1986).

Wiser, Ryan, D. Millstein, T. Mai, Jordan Macknick, and A. Carpenter, "The Environmental and Public Health Benefits of Achieving High Penetrations of Solar Energy in the United States," *Energy* 113 (2016): 472–86.

Wolsink, Maarten, "Entanglement of Interests and Motives: Assumptions behind the NIMBY-Theory on Facility Siting," *Urban Studies* 31, no. 6 (1994): 851–66.

Wolsink, Maarten, "Invalid Theory Impedes Our Understanding: A Critique on the Persistence of the Language of NIMBY," *Transactions of the Institute of British Geographers* 31, no. 1 (2006): 85–91.

Wolsink, Maarten, "Planning of Renewables Schemes: Deliberative and Fair Decision-Making on Landscape Issues Instead of Reproachful Accusations of Non-Cooperation," *Energy Policy* 35, no. 5 (2007): 2692–2704.

Woodhouse, Edward, and Steve Breyman, "Green Chemistry as Social Movement?" *Science, Technology, & Human Values* 30, no. 2 (2005): 199–222.

Woody, Todd, "Desert Vistas vs. Solar Power," *New York Times*, December 21, 2009.

Woody, Todd, "It's Green against Green in Mojave Desert Solar Battle," *Yale Environment 360*, February 1, 2010.

Woody, Todd, "Solar Millennium Files for Bankruptcy as Solar Shakeout Continues," *Forbes*, December 21, 2011.

World Intellectual Property Organization, "Photovoltaic Thin Film Cells," FRINNOV, Paris, 2009.

World Trade Organization, "Anti-dumping, Subsidies, Safeguards," Article VI of the General Agreement on Tariffs and Trade, Geneva, 1994.

World Trade Organization, "Certain Measures Affecting the Renewable Energy Generation Sector," Dispute DS412, 2014.

World Trade Organization, "China: Measures Related to the Exportation of Rare Earths, Tungsten and Molybdenum," Dispute Settlement DS431, 2015.

Worster, Donald, *Under Western Skies: Nature and History in the American West* (Oxford University Press, 1992).

Wyden, Ron, "China's Grab for Green Jobs," October 14, 2011, https://www.wyden.senate.gov/download/staff-report-chinas-grab-for-green-jobs_-examination-of-the-surge-of-solar-goods-exports-from-china-.

Wyden, Ron, "China's Subsidy of Solar Panels," *New York Times*, October 28, 2012.

Yang, Hong, Xiannjin Huang, and Julian R. Thompson, "Correspondence: Tackle Pollution from Solar Panels," *Nature* 509 (2014): 563.

Yang, Yuangen, Feili Li, Xiangyang Bi, Li Sun, Taoze Liu, Zhisheng Jin, and Congqiang Liu, "Lead, Zinc, and Cadmium in Vegetable Crops in a Zinc Smelting Region and Its Potential Human Toxicity," *Bulletin of Environmental Contamination and Toxicology* 87 (2011): 586–90.

Zanatta, Francesco, "Automatic Washing Device for Continuous Surfaces, in Particular Solar Thermal Collectors, Photovoltaic Panels, Continuous Glazed Building Walls and Similar Surfaces," U.S. Patent 9,192,966, issued November 24, 2015.

Zeng, Chao, Adriana Ramos-Ruiz, Jim A. Field, and Reyes Sierra-Alvarez, "Cadmium Telluride (CdTe) and Cadmium Selenide (CdSe) Leaching Behavior and Surface Chemistry in Response to pH and O_2," *Journal of Environmental Management* 154 (2015): 78–85.

Zhang, Sarah, "A Huge Solar Plant Caught on Fire, and That's the Least of Its Problems," *Wired*, May 23, 2016.

Zweibel, Ken, "The Impact of Tellurium Supply on Cadmium Telluride Photovoltaics," *Science* 328, no. 5979 (2010): 699–701.

Zweibel, Ken, "Thin Film Photovoltaics," Solar Energy Research Institute [now the National Renewable Energy Laboratory], Golden, CO, 1989.

Zweibel, Ken, and Vasilis Fthenakis, "Cadmium Facts and Handy Comparisons," National Renewable Energy Laboratory, Golden, CO, 2006.

Index